Lecture Notes in Computer Science 2189

Edited by G. Goos, J. Hartmanis, and J. van Leeuwen

Springer
Berlin
Heidelberg
New York
Barcelona
Hong Kong
London
Milan
Paris
Tokyo

Frank Hoffmann   David J. Hand   Niall Adams
Douglas Fisher   Gabriela Guimaraes (Eds.)

Advances in Intelligent Data Analysis

4th International Conference, IDA 2001
Cascais, Portugal, September 13-15, 2001
Proceedings

Springer

Volume Editors

Frank Hoffmann
Royal Institute of Technology, Centre for Autonomous Systems
10044 Stockholm, Sweden
E-mail: hoffmann@nada.kth.se

David J. Hand
Niall Adams
Imperial College, Huxley Building
180 Queen's Gate, London SW7 2BZ, UK
E-mail: {d.j.hand,n.adams}@ic.ac.uk

Douglas Fisher
Vanderbilt University, Department of Computer Science
Box 1679, Station B, Nashville, TN 37235, USA
E-mail: dfisher@vuse.vanderbilt.edu

Gabriela Guimaraes
New University of Lisbon, Department of Computer Science
2825-114 Caparica, Portugal
E-mail: gg@di.fct.unl.pt

Cataloging-in-Publication Data applied for

Die Deutsche Bibliothek - CIP-Einheitsaufnahme

Advances in intelligent data analysis : 4th international conference ;
proceedings / IDA 2001, Cascais, Portugal, September 13 - 15, 2001. Frank
Hoffmann ... (ed.). - Berlin ; Heidelberg ; New York ; Barcelona ; Hong Kong ;
London ; Milan ; Paris ; Tokyo : Springer, 2001
   (Lecture notes in computer science ; Vol. 2189)
   ISBN 3-540-42581-0

CR Subject Classification (1998): H.3, I.2, G.3, I.5.1, I.4.5, J.2, J.1, J.3

ISSN 0302-9743
ISBN 3-540-42581-0 Springer-Verlag Berlin Heidelberg New York

Springer-Verlag Berlin Heidelberg New York
a member of BertelsmannSpringer Science+Business Media GmbH

http://www.springer.de

© Springer-Verlag Berlin Heidelberg 2001
Printed in Germany

Typesetting: Camera-ready by author, data conversion by Boller Mediendesign
Printed on acid-free paper   SPIN: 10840583   06/3142   5 4 3 2 1 0

Preface

These are the proceedings of the fourth biennial conference in the *Intelligent Data Analysis* series. The conference took place in Cascais, Portugal, 13–15 September 2001. The theme of this conference series is the use of computers in intelligent ways in data analysis, including the exploration of intelligent programs for data analysis. Data analytic tools continue to develop, driven by the computer revolution. Methods which would have required unimaginable amounts of computing power, and which would have taken years to reach a conclusion, can now be applied with ease and virtually instantly. Such methods are being developed by a variety of intellectual communities, including statistics, artificial intelligence, neural networks, machine learning, data mining, and interactive dynamic data visualization. This conference series seeks to bring together researchers studying the use of intelligent data analysis in these various disciplines, to stimulate interaction so that each discipline may learn from the others. So as to encourage such interaction, we deliberately kept the conference to a single track meeting. This meant that, of the almost 150 submissions we received, we were able to select only 23 for oral presentation and 16 for poster presentation. In addition to these contributed papers, there was a keynote address from Daryl Pregibon, invited presentations from Katharina Morik, Rolf Backhofen, and Sunil Rao, and a special 'data challenge' session, where researchers described their attempts to analyse a challenging data set provided by Paul Cohen. This acceptance rate enabled us to ensure a high quality conference, while also permitting us to provide good coverage of the various topics subsumed within the general heading of intelligent data analysis.

We would like to express our thanks and appreciation to everyone involved in the organization of the meeting and the selection of the papers. It is the behind-the-scenes efforts which ensure the smooth running and success of any conference. We would also like to express our gratitude to the sponsors: Fundação para a Ciência e a Tecnologia, Ministério da Ciência e da Tecnologia, Faculdade de Ciências e Tecnologia, Universidade Nova de Lisboa, Fundação Calouste Gulbenkian and IPE Investimentos e Participações Empresariais, S.A.

September 2001

Frank Hoffmann
David J. Hand
Niall Adams
Gabriela Guimaraes
Doug Fisher

Organization

IDA 2001 was organized by the department of Computer Science, New University of Lisbon.

Conference Committee

General Chair:	Douglas Fisher (Vanderbilt University, USA)
Program Chairs:	David J. Hand (Imperial College, UK)
	Niall Adams (Imperial College, UK)
Conference Chair:	Gabriela Guimaraes (New University of Lisbon, Portugal)
Publicity Chair:	Frank Höppner (Univ. of Appl. Sciences Emden, Germany)
Publication Chair:	Frank Hoffmann (Royal Institute of Technology, Sweden)
Local Chair:	Fernando Moura-Pires (University of Evora, Portugal)
Area Chairs:	Roberta Siciliano (University of Naples, Italy)
	Arno Siebes (CWI, The Netherlands)
	Pavel Brazdil (University of Porto, Portugal)

Program Committee

Niall Adams (Imperial College, UK)
Pieter Adriaans (Syllogic, The Netherlands)
Russell Almond (Educational Testing Service, USA)
Thomas Bäck (Informatik Centrum Dortmund, Germany)
Riccardo Bellazzi (University of Pavia, Italy)
Michael Berthold (Tripos, USA)
Liu Bing (National University of Singapore)
Paul Cohen (University of Massachusetts, USA)
Paul Darius (Leuven University, Belgium)
Fazel Famili (National Research Council, Canada)
Douglas Fisher (Vanderbilt University, USA)
Karl Froeschl (University of Vienna, Austria)
Alex Gammerman (Royal Holloway, UK)
Adolf Grauel (University of Paderborn, Germany)
Gabriela Guimaraes (New University of Lisbon, Portugal)
Lawrence O. Hall (University of South Florida, USA)
Frank Hoffmann (Royal Institute of Technology, Sweden)
Adele Howe (Colorado State University, USA)
Klaus-Peter Huber (SAS Institute, Germany)
David Jensen (University of Massachusetts, USA)
Joost Kok (Leiden University, The Netherlands)
Rudolf Kruse (University of Magdeburg, Germany)
Frank Klawonn (University of Applied Sciences Emden, Germany)

Hans Lenz (Free University of Berlin, Germany)
David Madigan (Soliloquy, USA)
Rainer Malaka (European Media Laboratory, Germany)
Heikki Mannila (Nokia, Finland)
Fernando Moura Pires (University of Evora, Portugal)
Susana Nascimento (University of Lisbon, Portugal)
Wayne Oldford (University of Waterloo, Canada)
Albert Prat (Technical University of Catalunya, Spain)
Peter Protzel (Technical University Chemnitz, Germany)
Giacomo della Riccia (University of Udine, Italy)
Rosanna Schiavo (University of Venice, Italy)
Kaisa Sere (Abo Akademi University, Finland)
Roberta Siciliano (University of Naples, Italy)
Rosaria Silipo (Nuance, USA)
Floor Verdenius (ATO-DLO, The Netherlands)
Stefan Wrobel (University of Magdeburg, Germany)
Hui XiaoLiu (Brunel University, UK)
Nevin Zhang (Hong Kong University of Science and Technology, Hong Kong)

Sponsoring Institutions

Fundação para a Ciência e a Tecnologia, Ministério da Ciência e da Tecnologia
Faculdade de Ciências e Tecnologia, Universidade Nova de Lisboa
Fundação Calouste Gulbenkian
IPE Investimentos e Participações Empresariais, S.A.

Table of Contents

The Fourth International Symposium on Intelligent Data Analysis

The IDA'01 Robot Data Challenge

Feature Characterization in Scientific Datasets

Elizabeth Bradley[1], Nancy Collins[1*], and W. Philip Kegelmeyer[2]

[1] University of Colorado, Department of Computer Science, Boulder, CO 80309-0430
lizb,collinn@cs.colorado.edu,
[2] Sandia National Laboratories, P.O. Box 969, MS 9951, Livermore, CA, 94551-0969
wpk@ca.sandia.gov

Abstract. We describe a preliminary implementation of a data analysis tool that can characterize features in large scientific datasets. There are two primary challenges in making such a tool both general and practical: first, the definition of an interesting feature changes from domain to domain; second, scientific data varies greatly in format and structure. Our solution uses a hierarchical feature ontology that contains a base layer of objects that violate basic continuity and smoothness assumptions, and layers of higher-order objects that violate the physical laws of specific domains. Our implementation exploits the metadata facilities of the SAF data access libraries in order to combine basic mathematics subroutines smoothly and handle data format translation problems automatically. We demonstrate the results on real-world data from deployed simulators.

1 Introduction

Currently, the rate at which simulation data can be generated far outstrips the rate at which scientists can inspect and analyze it. 3D visualization techniques provide a partial solution to this problem, allowing an expert to scan large data sets, identifying and classifying important features and zeroing in on areas that require a closer look. Proficiency in this type of analysis, however, requires significant training in a variety of disciplines. An analyst must be familiar with domain science, numerical simulation, visualization methods, data formats, and the details of how to move data across heterogeneous computation and memory networks, among other things. At the same time, the sheer volume of these data sets makes this analysis task not only arduous, but also highly repetitive. A logical next step is to automate the feature recognition and characterization process so scientists can spend their time analyzing the science behind promising or unusual regions in their data, rather than wading through the mechanistic details of the data analysis. This paper is a preliminary report on a tool that does so.

General definitions of features are remarkably hard to phrase; most of those in the literature fall back upon ill-defined words like "unusual" or "interesting" or

---

* Supported by the DOE ASCI program through a Level 3 grant from Sandia National Laboratories, and a Packard Fellowship in Science and Engineering.

F. Hoffmann et al. (Eds.): IDA 2001, LNCS 2189, pp. 1–12, 2001.

"coherent." Features are often far easier to *recognize* than to *describe*, and they are also highly domain-dependent. The structures on which an expert analyst chooses to focus — as well as the manner in which he or she reasons about them — necessarily depend upon the physics that is involved, as well as upon the nature of the investigation. Meteorologists and oceanographers are interested in storms and gyres, while astrophysicists search for galaxies and pulsars, and molecular biologists classify parts of molecules as alpha-helices and beta-sheets. Data types vary — pressure, temperature, velocity, vorticity, etc. — and a critical part of the analyst's expert knowledge is knowing which features appear in what data fields.

In this paper, we describe a general-purpose feature characterization tool and validate it with several specific instances of problems in one particular field: finite element analysis data from computer simulations of solid mechanics problems. One of our goals is to produce a practical, useful tool, so we work with data from deployed simulators, in a real-world format: ASCI's SAF, a *lingua franca* used by several of the US national labs to read and write data files for large simulation projects. This choice raised some interoperability issues that are interesting from an IDA standpoint, as discussed in section 2 below. The SAF interface provides access to a geometric description of a computational mesh, including the spatial positions of the mesh points (generally xy or xyz) and the type of connectivity, such as triangles or quads, plus information about the physics variables, such as temperature or velocity. Given such a snapshot, our goal is to characterize the features therein and generate a meaningful report. We began by working closely with domain scientists to identify a simple ontology[1] of distinctive coherent structures that help them understand and evaluate the dynamics of the problem at hand. In finite-element applications, as in many others, there are two kinds of features that are of particular interest to us:

- those that violate the continuity and smoothness assumptions that are inherent in both the laws of physics and of numerical simulation: spikes, cracks, tears, wrinkles, etc. — either in the mesh geometry or in the physics variables.
- those that violate higher-level physical laws, such as the requirement for normal forces to be equal and opposite when two surfaces meet (such violations are referred to as "contact problems").

Note that we are assuming that expert users *can* describe these features mathematically; many of the alternate approaches to automated feature detection that are described in section 5 do not make this assumption. The knowledge engineering process is described in section 3.1 and the algorithms that we use to encapsulate the resulting characterizations, which rely on fairly basic mathematics, are described in section 3.2. We have tested these algorithms on roughly a half-dozen data sets; the results are summarized in section 4.

---

[1] Formally, an ontology seeks to distill the most basic concepts of a system down into a set of well defined nouns and verbs (objects and operators) that support effective reasoning about the system.

2 Data Formats and Issues

DMF[15] is a joint interoperability project involving several US national labs. Its goal is to coordinate the many heterogeneous data handling libraries and analysis tools that are used by these organizations, and to produce standards and libraries that will allow others to exploit the results. This project is motivated by the need to perform simulations at the system level, which requires formerly independent programs from various disciplines to exchange data smoothly. The attendant interoperability problems are exacerbated by the growing sophistication and complexity of these tools, which make it more difficult to adapt them to new data formats, particularly if the new format is richer than the old. The specific DMF data interface that we use, called SAF[11], exploits *metadata* — that is, data about the data — to solve these problems. Used properly, metadata can make a dataset *self-describing*. SAF, for example, captures not only the data values, but also the geometry and topology of the computational grid, the interpolation method used inside each computational element, and the relationships between various subsets of the data, among other things. Its interface routines can translate between different data formats automatically, which confers tremendous leverage upon tools that use it. They need only handle one type of data and specify it in their metadata; SAF will perform any necessary translation. In our project, this is important in both input and output. Not only must we handle different kinds of data, but we must also structure and format the results in appropriate ways. As discussed at length in the scientific visualization literature, different users need and want different data types and formats, so reporting facilities must be flexible. Moreover, the consumer of the data might not be a person, but rather another computer tool in a longer processing pipeline. For example, output generated by the characterization routines developed in this paper might be turned into a simple ascii report or formatted into an html page for viewing with a browser by a human expert, and simultaneously fed to a visualization tool for automatic dataset selection and viewpoint positioning. For all of these reasons, it is critical that data be stored in a format that supports the generation and use of metadata, and SAF is designed for exactly this purpose.

Metadata is a much broader research area, and SAF was not the first data model to incorporate and use it. Previous efforts included PDBlib, FITS, HDF, netCDF, VisAD, and DX, among others[4,16,5,8,9] — data formats that enabled analysis tools to reason about metadata in order to handle the regular data in an appropriate manner. While metadata facilities are of obvious utility to the IDA process, they are also somewhat of a Pandora's Box; as simulation tools increase in complexity, effective analysis of their results will require a corresponding increase in the structure, amount, and complexity of the metadata. This raises a host of hard and interesting ontology problems, as well as the predictable memory and speed issues, which are beyond the scope of the current paper.

The SAF libraries are currently in alpha-test release[2]. Because of this, few existing simulation, analysis, and visualization tools understand SAF's native interface. Our early development prototypes, for instance, used the SAF library directly for data access, but had to convert to the OpenDX file format for visualization: the very kind of translation that SAF is intended to obviate. Because visualization is so critical to data analysis, there has been some recent progress in adapting existing visualization tools to parse SAF input. In the first stages of our project, however, such tools did not exist, so we used OpenDX for visualization. We recently began converting to a SAF-aware visualization tool called EnSight[1], but this has not been without problems. Data interface libraries are subject to various chicken-and-egg growing pains. The tools need not understand a format until an interesting corpus of data exists in that format; scientists are understandably unwilling to produce data in a format for which no analysis tools exist. Intelligent data analysis tools that take care of low-level interoperability details can remove many barriers from this process.

3   Intelligent Analysis of Simulation Data

3.1   Knowledge Engineering

In order to automate the feature characterization process, we first needed to understand how human experts perform the analysis. We spent several days with various project analysts at Sandia National Laboratories, observing as they used existing tools on different kinds of data. We focused in on what they found important, how they identified and described those features, how they reasoned about which data fields to examine for a given stage of the process, and how the entire process changed if they were trying to prove or disprove a particular hypothesis. Most of the features of interest to these experts, we found, are clued from local geometry of the simulation mesh; inverted elements with non-positive volume, spikes, wrinkles, dimples, and so on. A smaller set of features of interest are extrema in the physics variables: hot spots and the like. We used this information to specify a simple ontology: that is, a set of canonical features (spikes, tears, cracks, etc.), together with mathematical descriptions of each — the statistical, geometric, and topological properties that define them. We also studied how the experts wrote up, reported, and used their results.

The Sandia analysts view the mechanical modeling process in two stages. The first is model debugging, wherein they ensure that the initial grid is sound, that the coupling is specified correctly between various parts of the model, and that the modeling code itself is operating correctly. The second is the actual simulation, where they examine the data for interesting physical effects: vibrational modes, areas that exceed the accepted stress tolerances, etc. We found that features play important roles in both phases, and that the sets of features actually

---

[2] We are a designated alpha-test group, and a secondary goal of this project is to provide feedback to the DMF developers, based on our experiences in designing an intelligent data analysis tool around this format.

overlapped. A spike in the results, for instance, can indicate either a numerical failure or a real (and interesting) physical effect. In some cases, reasoning about features let analysts identify model errors that were undetectable by traditional numerical tests like overflow, divide-by-zero, etc. One scientist described a simulation of an automobile engine compartment, including the front bumper. Due to a numerically innocuous error, one of the grid points moved to a location well beyond the back end of the entire car. This obviously non-physical situation — which was immediately visible to the analyst as a feature — flagged the model as faulty, even though no numerical fault occurred.

Note that features can involve the mesh coordinates, the physics variables, and sometimes both. Vertical relief, for instance, is a property of surface geometry, not the value of the physics variables upon that surface. Conversely, calculation of the highest temperature on a surface depends solely on the physics variables. Often, analysts are interested in features that involve both: say, the temperature or wind speed at the highest point on the landscape, or the position of the hottest point. Often, too, their underlying assumptions about geometry and about physics are similar, which can lead to some terminology confusion. A spike in temperature and a spike on the surface are similar in that both violate smoothness assumptions, but the mathematics of their characterization is quite different. This is actually a symptom of a deeper and more interesting property of features: like data analysis itself, they are hierarchical. All surfaces, whether numerical or physical, are generally continuous and smooth, so tears and spikes are likely to be considered to be features in *any* domain. If one knows more about the physics of the problem, other features become interesting as well. In contact problems, for instance — where two surfaces touch one another — the normal forces at the intersection of the two surfaces should be equal and opposite and surfaces should certainly not interpenetrate. Violations of these physical realities are interesting features. To capture these layers of meaning, our feature ontology is hierarchical. It contains a baseline set of features that rest on assumptions that are true of *all* physical systems, together with layers of higher-order features that are specific to individual domains (and sub-domains and so on). Currently, we have finished implementing two such layers: the baseline one mentioned above (spikes *et al.*) and a contact-problem one, which defines deviation from equal-and-opposite as a feature. Both are demonstrated in section 4.

3.2 Algorithms

Given the feature ontology described in the previous section, our next task was to develop algorithms that could find instances of those features in DMF data snapshots and generate meaningful reports about their characteristics. In order to make our work easily extensible, we structured the overall design so as to provide a general-purpose framework into which characterization routines specific to the features of a given domain can be easily installed. In particular, we provide several basic building-block tools that compute important statistical, geometrical, and topological information — about the mesh itself and about the values of the physics variables that are associated with each point in the mesh.

Their results are stored using the SAF library format, complete with metadata that allow them to be combined in different ways to assess a wide variety of features in a range of domains. Often, there is more than one way to find a single feature; a surface spike, for instance, can be characterized using statistics (a point that is several σ away from the mean) or geometry (a point where the slope changes rapidly).

Our current set of basic building blocks is fairly straightforward:

- `normals()`, which takes a DMF dataset and computes the unit-length normal vector to each mesh element.
- `topological-neighbors()`, which takes a DMF dataset and an individual mesh element m and returns a list of mesh elements that share an edge or a vertex with m.
- `geometric-neighbors()`, which takes a DMF dataset, an individual mesh element m and a radius r, and returns a list of mesh elements whose vertices fall entirely within r of the centroid of m.
- `statistics()`, which takes a DMF dataset and a specification of one variable (one of the mesh coordinates or physics variables) and computes the maximum, minimum, mean, and standard deviation of its values.
- `displacements()`, which takes a DMF dataset, finds all neighboring[3] pairs of vertices, measures the xyz distance between them, and reports the maximum, minimum, mean, and standard deviation of those distances

In addition, we provide various vector calculus facilities (e.g., dot products) and distance metric routines.

As an example of how these tools work, consider Fig. 1. The vectors computed by `normals()` are shown emanating from the center of each mesh face. In a regular mesh, finding topological neighbors could be trivial. SAF, how-

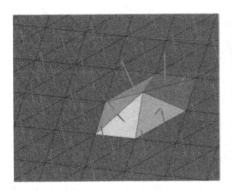 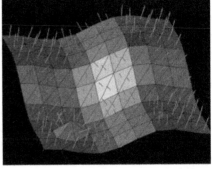

Fig. 1. 3D surface mesh examples, showing the vectors computed by the `normals()` function.

---

[3] *Topologically* neighboring

ever, is designed to be able to represent irregular and adaptive meshes as well, so the current version of SAF only provides neighbor information implicitly. For this reason, we preprocess the DMF data at the beginning of the characterization run and place it in a data structure that makes the topological information explicit. Our current design maintains a single list of vertices, including xyz position and the values of any associated physics variables. Three other lists point into this vertex list — a face list, an edge list, and a normal list — making it easy to look for shared edges or vertices and deduce neighbor relationships. In the examples in Fig. 1, each triangle has three "face neighbors" and at least three other "vertex neighbors," all of which are returned by `topological-neighbors`. The `geometrical-neighbors` function is a bit more complicated; it calls `topological-neighbors`, measures the Euclidean distances between the vertices of the resulting triangles and the centroid of the original element, discards any element whose vertices do not all fall within the specified distance, and iteratively expands on the others. The `statistics()` and `displacements()` routines use simple traditional methods. The left-hand surface in Fig. 1, for instance, is completely flat, with the exception of the bump in the foreground, and the `statistics()` results reflect the appropriate mean height of the surface and a very small standard deviation. The right-hand surface fluctuates somewhat, so the standard deviation is larger. In both cases, the `displacements()` results would likely be uninformative because the edge lengths of the elements are fairly uniform.

There are a variety of ways, both obvious and subtle, to improve on the toolset described above. We are currently focusing on methods from computational geometry[12] (e.g., Delaunay triangulation) and computational topology, such as the α-shape[7], and we have developed the theoretical framework and some preliminary implementations of these ideas[13,14]. Since features are often easier to *recognize* than to *describe*, we are also exploring the use of machine learning techniques to discover good values for the heuristic parameters that are embedded in these computational geometry and topology algorithms.

4   Results and Evaluation

We have done preliminary evaluations of the algorithms described in the previous section using half a dozen datasets. For space reasons, only two of those datasets are discussed here; please see our website[4] for further results, as well as color versions of all images in this paper. The first dataset, termed `irregular-with-spike`, is shown in Fig. 2. It consists simply of an irregular surface mesh; no physics variables are involved. Such a dataset might, for instance, represent the surface of a mechanical part. As rendered, this surface contains an obvious feature — a vertical spike — to which the eye is immediately drawn. Such a feature may be meaningful for many domain-dependent and -independent reasons: as an indicator of numerical problems or anomalies in the physics models, or perhaps

---

[4] http://www.cs.colorado.edu/~lizb/features.html

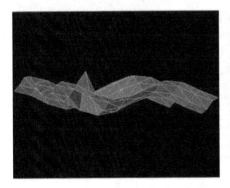 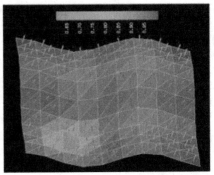

Fig. 2. A 3D surface mesh dataset that contains a spike. By dotting the normals of neighboring faces and comparing the result to the local average of the surface normals, we can detect anomalies in the slope of the surface. Results of this algorithm are used to shade the mesh elements in the right-hand image. Lighter elements are members of a *surface spike* feature.

a real (and surprising) physical effect. All of these reasons essentially boil down to an assumption of continuity and smoothness in the surface.

The task of our spike detection algorithm is to find places where that smoothness assumption is violated. To accomplish this, we begin by using the `normals()` and `topological-neighbors()` functions to find the normal vector to each mesh face and dot it with its neighbors' normals. While this does allow us to detect sudden variations in slope, it is inadequate for evaluating the results because anomalies are always *relative to a whole*. A 1 cm bump in a surface whose mean bumpiness is 5cm, for instance, may not be interesting; a 1 cm bump in the Hubble Space Telecope mirror, however, is most definitely an issue. Moreover, it is impossible to characterize a feature — e.g., to report the *size* of a bump — without having a baseline for the measurement. For these reasons, one must incorporate scale effects. We do so using the `geometric-neighbors()` function. In particular, we compute the *average* difference between neighboring face normals over some[5] region and compare the *individual* differences to that average. (This is essentially a spatial equivalent to the kinds of moving average or low-pass filtering algorithms that are used routinely in time-series analysis.) The right-hand image in Fig. 2 depicts the results. The normal vectors to each mesh element are shown as before, and each mesh element is grey-shaded according to how much the difference between its normal and those of its nearest neighbors differs from the local average of the surface normals. The cluster of lighter mesh elements in the bottom left corner of the right-hand image are part of a feature — a "surface spike" — whose distinguishing characteristic is lack of smoothness in slope.

---

[5] The size and shape of this region are an obvious research issue; our current solution chooses an arbitrary square, and we are investigating how best to improve this.

Fig. 3 shows the `chatter` dataset, a simulation of a hard cylindrical pin pushing into a deformable block. Each point in this dataset gives the xyz position

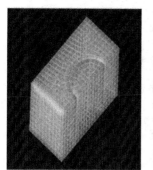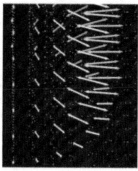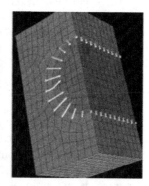

Fig. 3. A simulation of a hard cylindrical pin pushing into a deformable block. The left-hand figure shows the geometry; at the center is a 3D closeup view of the forces at a collection of grid points in and near the arch. In this rendering, the forces are difficult to see, let alone interpret; if the pin and block are shown in a view that suppresses information that is unrelated to the feature, however — as in the right-hand image — it is easy to identify places where the normal forces do not balance.

of a vertex and the force at that vertex. The issues that arise in this example are more complicated. Because there are physics variables involved, we are not only interested in features that violate mesh smoothness and continuity. In this particular case, we are also looking for *contact problems*: places on the surface between the two objects where the normal forces are not equal and opposite. Finding and describing contact problems with the tools described in the previous section is somewhat more involved than in the previous example, but still quite straightforward. We first find all mesh faces that lie on the contact surface between the two objects and determine which faces in the pin and the block touch one another. (Each object is a separate mesh, so this amounts to traversing the boundary of each, checking for xyz proximity of vertices and opposing faces, and building the appropriate association table.) We then compute the normals $\hat{n}_i$ to these faces, project the force vector $\boldsymbol{f}_i$ at each face along the corresponding $\hat{n}_i$ in order to eliminate its tangential[6] component, and finally compare the normal force vectors of adjacent faces to see if they are indeed of equal magnitude and in the opposite direction. The right-hand image shows the results, including a contact problem at one vertex, indicated by the bent vector in the middle of the arch.

---

[6] Tangential forces also play roles in different kinds of contact problems; see our website for more details.

Unlike the previous example, this dataset is complex and very hard to visualize: parts of the object obscure other parts, and it can be difficult or impossible to make sense of the geometry, as is clear from the center image in Fig. 3. In situations like this, automated feature characterization is critical, as it can find and highlight geometry that is effectively invisible to a human user — and even choose display parameters based on that investigation, in order to best present its results to that user (e.g., focusing in on the area around a feature and choosing a view that brings out its characteristic geometry, as in the right-hand image of the pin/block system).

These methods are not only very effective, but also quite extensible; one can detect a variety of other features using different combinations of these same basic tools. Tears and folds, for instance, can be flagged when geometric neighbors are not topological neighbors. Because we use the SAF format for the inputs and outputs of our toolkit routines, it is very easy to generate, modify, and use new combinations of those tools. The detection method for a specific feature, of course, is not uniquely defined; the `displacements()` and `statistics()` routines can also be useful in finding spikes, but geometric methods that rely on normal vectors are more precise. They are also more expensive, however, and when we begin working with the truly immense datasets that are the target applications of this project, we will likely use the former as a first pass on larger datasets, in order to identify areas for deeper study with the latter.

5   Related Work

This work draws upon ideas and techniques from a wide variety of disciplines, ranging from mathematics to artificial intelligence. Space requirements preclude a thorough discussion here; rather, we will just summarize a few of the methods and ideas that most closely relate to this paper. Many tools in the intelligent data analysis literature[2] focus on assessing different kinds of analysis tools and using that knowledge to build toolkits that adapt the analysis to the problem at hand. Our work is similar in spirit to many of these; indeed, our SAF-based framework solves the pernicious interoperability problems that motivate many of these toolkits. A handful of groups in the IDA, pattern recognition, and machine learning communities specifically target reasoning about features in scientific data. Notable instances are the spatial aggregation framework of Yip and Zhao[17], Mitchell's GENIE program[6], and the AVATAR pattern recognition tool[3,10], which invisibly watches as a user investigates a given 4D physics simulation dataset and deduces what s/he finds "interesting" in that data. Like these algorithms, our tool is designed to be both powerful and general, rather than domain-specific. Unlike GENIE and AVATAR, however, we assume that features can be described in closed form, and we are very interested both in those descriptions and in the process of discovering them. (Indeed, section 3.1 is essentially a chronicle of that knowledge engineering procedure.)

6   Conclusion

The goal of the intelligent data analysis tool described here is to distill a succinct description of the interesting and important features out of a massive simulation dataset. An automated tool like this, which produces a compact, useful description of the dynamics, couched in the language of the domain, frees human experts to devote their time and creative thought to other demanding tasks. It can not only classify the features in a data set, but also signal places where the expert analyst should take a closer look, and even aid in the presentation of the view — a critical prerequisite to effective visualization of a spatially complex datasets, where anything but a selective, narrowed focus will overwhelm the user with information. Equally important, it allows scientists to interact with their data at the *problem* level, encapsulating the details of and changes to the underlying infrastructure tools. Of course, this automated feature characterization tool will never *replace* the human expert. It does not do anything that cannot already be done by hand; it simply automates many of the more onerous, repetitive, and/or detailed parts of the analysis process.

While the feature ontology and the characterization algorithms described in this paper are specific to finite-element simulation data, the general ideas, the notion of a layered ontology, the mathematics routines that we use to implement the characterization process, and the compositional structure of the framework within which they are combined are far more broadly applicable. The results described here are only preliminary, of course; full assessment of the strengths and weaknesses of this approach will only be possible with much more experience and testing. To this end, we have begun working on turbulent convection problems, both numerical and experimental, where experts begin by reasoning about three basic structures in the 3D vorticity data: *tubes*, *sheets*, and *cells*, which play well-defined roles in fluid transport, and which can easily[7] be described using the same simple tools that we describe in section 3.2. The metadata facilities of the SAF libraries are an important part of what makes this work easily generalizable; to apply our tool to a new kind of data, we simply need to write the appropriate transliteration routine and pass it to SAF. This interoperability also makes the *results* of our tool's analysis more broadly applicable: our output contains not only raw data, but also metadata that describes the structure, format, and content of that data. This allows consumers of these results, whether human experts or other computer tools, to understand and use them. Incorporating the data analysis tool described here into an existing scientific computing environment would further streamline this process, and we plan to investigate this when the next Ensight release — which will allow such modifications — becomes available.

Acknowledgments: Andrey Smirnov and Stephanie Boyles contributed code and ideas to this paper as well.

---

[7] to the great dismay of the project analyst

References

1. http://www.ensight.com/.
2. M. Berthold and D. Hand, editors. *Intelligent Data Analysis: An Introduction.* Springer-Verlag, 2000.
3. K. Bowyer, L. Hall, N. Chawla, T. Moore, and W. Kegelmeyer. A parallel decision tree builder for mining very large visualization datasets. In *Proceedings of the IEEE 2000 Conference on Systems, Man, and Cybernetics*, October 2000.
4. S. Brown and D. Braddy. PDBLib user's manual. Technical Report M270, Rev. 2, Lawrence Livermore National Laboratory, January 1993.
5. S. A. Brown, M. Folk, G. Goucher, and R. Rew. Software for portable scientific data management. *Computers in Physics*, 7(3):304–308, 1993.
6. S. P. Brumby, J. Theiler, S. Perkins, N. Harvey, J. J. Szymanski, J. J. Bloch, , and M. Mitchell. Investigation of image feature extraction by a genetic algorithm. In *Proceedings of SPIE*, 1999.
7. H. Edelsbrunner and E. Muecke. Three-dimensional alpha shapes. *ACM Transactions on Graphics*, 13(1):43–72, 1994.
8. W. H. *et al.* A lattice model for data display. In *Proceedings IEEE Visualization*, pages 310–317, 1994.
9. R. Haber, B. Lucas, and N. Collins. A data model for scientific visualization with provisions for regular and irregular grids. In *Proceedings IEEE Visualization*, pages 298–305, 1991.
10. L. Hall, K. Bowyer, N. Chawla, T. Moore, and W. P. Kegelmeyer. AVATAR — adaptive visualization aid for touring and recovery. Sandia Report SAND2000-8203, Sandia National Laboratories, January 2000.
11. M. C. Miller, J. F. Reus, R. P. Matzke, W. J. Arrighi, L. A. Schoof, R. T. Hitt, and P. K. Espen. Enabling interoperation of high performance, scientific computing applications: Modeling scientific data with the sets & fields (SAF) modeling system. In *International Conference on Computational Science (ICCS-2001)*, 2001.
12. F. P. Preparata and M. I. Shamos. *Computational Geometry: An Introduction.* Springer-Verlag, New York, 1985.
13. V. Robins, J. Meiss, and E. Bradley. Computing connectedness: An exercise in computational topology. *Nonlinearity*, 11:913–922, 1998.
14. V. Robins, J. Meiss, and E. Bradley. Computing connectedness: Disconnectedness and discreteness. *Physica D*, 139:276–300, 2000.
15. L. Schoof. The ASCI data models and formats (DMF) effort: A comprehensive approach to interoperable scientific data management and analysis. In *4th Symposium on Multidisciplinary Applications and Interoperable Computing*, Dayton, OH, August 2000.
16. D. Wells, E. Greisen, and R. Harten. FITS: A flexible image transport system. *Astronomy and Astrophysics Supplement Series*, 44:363–370, 1981.
17. K. Yip and F. Zhao. Spatial aggregation: Theory and applications. *Journal of Artificial Intelligence Research*, 5:1–26, 1996.

Relevance Feedback in the Bayesian Network Retrieval Model: An Approach Based on Term Instantiation

Luis M. de Campos[1], Juan M. Fernández-Luna[2], and Juan F. Huete[1]

[1] Dpto. de Ciencias de la Computación e Inteligencia Artificial, E.T.S.I. Informática.
Universidad de Granada, 18071 – Granada, Spain.
{lci,jhg}@decsai.ugr.es
[2] Departamento de Informática, Escuela Politécnica Superior.
Universidad de Jaén, 23071 – Jaén, Spain.
jmfluna@ujaen.es

Abstract. Relevance feedback has been proven to be a very effective query modification technique that the user, by providing her/his relevance judgments to the Information Retrieval System, can use to retrieve more relevant documents. In this paper we are going to introduce a relevance feedback method for the *Bayesian Network Retrieval Model*, founded on propagating partial evidences in the underlying Bayesian network. We explain the theoretical frame in which our method is based on and report the results of a detailed set of experiments over the standard test collections Adi, CACM, CISI, Cranfield and Medlars.

1   Introduction

When a user formulates a query to an *Information Retrieval System* (IRS), this software attempts to find those documents that closely match the representation of the user's information need. The posterior examination of the ranking offered by the IRS may show that a few amount of documents are relevant according to the user's judgments. This fact could be due, among other factors, to the user's incapacity of expressing his information need by means of an exact query. In that moment, the user could apply several of the *query modification* techniques that have been developed in the field of *Information Retrieval* (IR) to improve the query, and therefore, obtain those documents that fulfill the user's relevance criteria.

Among the query modification techniques, we are going to highlight the *Relevance Feedback* [12], that generates a new query starting from the *relevance judgments* (i.e., which retrieved documents are relevant and which are not relevant to the initial query) provided by the user, when she/he inspects the ranking of documents returned by the IRS for the original query. The IRS updates that original query modifying the importance of the terms it contains (*term reweighting*), and adding new terms which are considered useful to retrieve more relevant documents (*query expansion*). This method has been successfully applied in a

F. Hoffmann et al. (Eds.): IDA 2001, LNCS 2189, pp. 13–23, 2001.

great variety of IR models [13], among them, the Croft and Turtle's *Inference Network* model [14], based on *Bayesian Networks*.

Founded on probabilistic methods, Bayesian networks [9] have been proven to be a good model to manage uncertainty, even in the IR environment, in which they have been successfully applied as an extension of the probabilistic IR model, because they offer important advantages to deal with the intrinsic uncertainty with which IR is pervaded [5,14]. Also based on these probabilistic tools, the Bayesian Network Retrieval model (BNR) was introduced in [2] as an alternative to the existing methods based on Bayesian networks [7,10,14].

In this paper we want to explain a proposal for relevance feedback within the BNR model, showing the theoretical frame in which it is based on and the experimental results obtained when it has been applied to several well known standard test collections.

To carry out this task, this paper is organized as follows: Section 2 introduces the BNR model, showing its most important features. Section 3 deals with a theoretical overview of our proposals of relevance feedback for the BNR model. Later, in Section 4, we will explain how we have put in practice some of the ideas developed in the previous section. Section 5 shows the experimental results obtained with CACM and four additional collections. Finally, Section 6 discusses about the conclusions of this work.

2   Description of the Bayesian Network Retrieval Model

A Bayesian network is a *Directed Acyclic Graph* (DAG), where the nodes represent the variables from the problem we want to solve. In that kind of graph, the knowledge is represented in two ways [9]: *(a)* Qualitatively, showing the (in)dependencies between the variables, and *(b)* Quantitatively, by means of conditional probability distributions which shape the relationships. In IR problems we can distinguish between two different sets of variables (nodes in the graph): The set of terms in the glossary from a given collection and the set of documents, grouped in the *term* and *document* subnetworks, respectively. Each term, T_i, consists on a binary random variable taking values in the set $\{\bar{t}_i, t_i\}$, where $\bar{t}_i$ stands for 'the term T_i is not relevant', and t_i represents 'the term T_i is relevant'. Similarly, a variable referring to a document D_j has its domain in the set $\{\bar{d}_j, d_j\}$, where in this case, $\bar{d}_j$ and d_j respectively mean 'the document D_j is not relevant for a given query', and 'the document D_j is relevant for a given query'.

Focusing on the structure of the network, the following guidelines have been considered to determine the topology of the graph [2]:

• For each term that has been used to index a document, there is a link between the node representing that keyword and the node associated with the document it belongs to.

• The relationships between documents only occur through the terms included in these documents.

- Documents are conditionally independent given the terms that they contain. Thus, if we know the relevance (or irrelevance) values for all the terms indexing document D_i then our belief about the relevance of D_i is not affected by knowing that another document D_j is relevant or irrelevant.

These assumptions have some effects on the structure of the network: On one hand, links joining terms and documents must be directed from term nodes to document nodes in the graph and, on the other hand, there are not links between document nodes.

With regard to the relationships among term nodes, the model allows terms being dependent among each other. In order to capture the relationships between terms in the collection, a Bayesian network learning algorithm is used to construct the term subnetwork. Taking into account considerations of efficiency in the learning and inference stages, some restrictions on the structure of the learned graph have been imposed. Particularly, the relationships between terms are represented by means of a polytree (a DAG in which there is no more than one undirected path connecting each pair of nodes). The used polytree learning algorithm is described in [3].

Once we know the structure of the graph, the final step to completely specify a Bayesian network is to estimate the probability distributions stored in each node. Three different cases have to be considered:

- Term nodes having no parent: In this case we store marginal distributions, estimated as follows: $p(t_i) = \frac{1}{M}$ and $p(\bar{t}_i) = \frac{M-1}{M}$, M being the number of terms in a given collection.

- Term nodes with parents: For each node, we need to store a set of conditional probability distributions, one for each possible configuration of the parent set. These distributions are simply estimated using frequencies of cooccurrence of terms in the documents of the collection.

- Document nodes: In this case, the estimation of the conditional probabilities is more problematic because of the huge number of parents that a document node has. For example, if a document has been indexed with 30 terms, we need to estimate and store 2^{30} (approx. 1.07×10^9) probabilities. Therefore, instead of explicitly computing these probabilities, the BNR model uses a *probability function*, which returns a conditional probability value when it is called during the propagation stage, each time that a conditional probability is required. In this paper, we are going to use the following probability function, based on the cosine measure [12]: for a given configuration $\pi(D_j)$ of the set of terms in document D_j, the probability of relevance of document D_j is

$$p(d_j \mid \pi(D_j)) = \alpha_j \sum_{T_i \in D_j, T_i = t_i} \text{tf}_{ji} \text{idf}_i^2 \qquad (1)$$

where tf_{ji} is the frequency of the term T_i in the document D_j, idf_i is its inverse document frequency and α_j is a normalizing constant computed as $1/\sqrt{\sum_{T_i \in D_j} \text{tf}_{ji}^2 \text{idf}_i^2}$.

Once the entire network is completely built, and given a query submitted to the system, the retrieval process starts by placing the evidences, i.e., the terms belonging to the query, in the term subnetwork by setting their states to '*the term is relevant'*. The propagation process is run, obtaining for each document its probability of relevance given that the terms in the query are also relevant. Then, the documents are sorted by their posterior probability to carry out the performance evaluation process.

Taking into account the size of the Bayesian networks used by the BNR model, even for small document collections, general purpose inference algorithms can not be applied due to efficiency considerations. To solve this problem, the propagation process in the whole network is substituted by an inference process in two steps, but ensuring that the results are the same that the ones obtained using exact propagation in the entire network [6]: First, we propagate the query just only in the term subnetwork by means of the Pearl's Exact Propagation Algorithm in Polytrees [9], and compute, for each term, its probability of being relevant given the query submitted to the IRS, $p(t_i \mid Q)$. In the second step, we compute the probability that each document is relevant given the query by means of the evaluation of the following formula:

$$p(d_j \mid Q) = \alpha_j \sum_{T_i \in D_j} \text{tf}_{ji} \text{idf}_i^2 \, p(t_i \mid Q) \qquad (2)$$

For more details about the BNR model and its comparison with other retrieval engines also based on Bayesian networks, please refer to [2].

3   Description of the Methodology for Relevance Feedback

In the BNR model, the proposed methodology for relevance feedback is based on the following idea, centered basically in the way in which new evidential information is entered in the Bayesian network: By evaluating a set of retrieved documents obtained as a consequence of running a query, the user obtains new pieces of evidence for some of the variables in the network, that may help to discriminate which documents are relevant to our information need.

For example, whenever a term T_i indexes a document judged relevant by the user, perhaps we should increase the belief supporting the assert "term T_i is relevant"; we speak about *partial evidence* because, by using only document's judgments of relevance we can not be completely certain about the truth of the previous assertion. Nevertheless, this information must be combined with the previous evidence supported by the original query, in order to formulate a new query to be processed by the I.R. system.

Suppose that, after analyzing the retrieved documents, we have a strong belief supporting the assert "term T_i is relevant", which is quantified, e.g., as being ten times more than the belief supporting the assert "term T_i is not relevant". This information can be considered as a ratio of likelihood values, i.e., $P(Obs|\bar{t}_i)$: $P(Obs|t_i) = 1 : 10$, where Obs represents the observation. In order to include this information in the IRS we can use the concept of *partial evidence* [9]: If for a node X (term or document node) we find a new piece of evidence supporting its relevance or irrelevance, the node will receive a message[1] denoted by $\lambda(X)$,

---

[1] sent by a dummy node, an imaginary child for node X.

encoding a pair of likelihood values, i.e., $\lambda(X) = (P(Obs|\bar{x}), P(Obs|x))$. This message, in a normalized form, will be combined with the whole information that node X obtains from its parents and children in the network.

Remark: The use of λ vectors can be extended for the rest of the nodes in the network, so that we can assume that every node will receive a λ vector: for instantiated nodes (we know for sure their relevance values) we use $\lambda(X) = (0, 1)$ to indicate that X is relevant and $\lambda(X) = (1, 0)$ indicating that X is not relevant; for uninstantiated nodes that do not receive any evidence, we use the vector $\lambda(X) = (1, 1)$.

The particular implementation of these ideas gives rise to two different mechanisms for relevance feedback in the BNR model, which basically differ in the way that new evidences are included in the model: Document-based and term-based relevance feedback. In this paper we are going to focus in term-based relevance, although we briefly describe the document-based approach too.

Document-based relevance feedback

This method intends to capture the idea that, by evaluating the retrieved documents, the set of evidences obtained by the user are focused on document nodes and not on the particular terms indexing these documents. In order to include these new evidences in the IRS, it would be sufficient to instantiate each judged document as relevant or non-relevant. Therefore, the new query would be $Q_1 = (Q, d_1, d_2, \ldots, d_k, \bar{d}_{k+1}, \ldots, \bar{d}_{|R|})$, where $|R|$ is the number of documents that have been judged by the user.

In the propagation process, and as a consequence of document instantiation, it could be considered that each evaluated document sends a message to its parents (the set of terms indexing the document) encoding that it has been judged as relevant or non-relevant by the user. However, due to the huge number of parents that documents have, exact computation is problematic and we use approximate methods (see [4,6] for details).

Term-based relevance feedback

In this case we are considering that, by evaluating the retrieved documents, the user can obtain new evidences for term nodes exclusively, i.e., the relevance judgments about the retrieved documents translates into partial evidences about the relevance of the terms indexing these documents.

Term-based relevance feedback needs to classify the terms indexing the observed documents in three groups: those terms that only occur in relevant documents (*positive terms*), those that only occur in non-relevant documents (*negative terms*), and those that occur in both types of documents (*neutral terms*). In the next section we discuss term-based relevance feedback in detail, and distinguish between query term reweighting and query expansion. i.e., whether the terms belong to the initial query or not.

4   Term-Based Relevance Feedback in Detail

We are going to introduce some notation that will be used in this section: n_r and $n_{\bar{r}}$ represent the number of retrieved relevant and non-relevant documents, respectively; n_t and $n_{\bar{t}}$ denote the number of retrieved documents where the term T has been observed and not observed, respectively; n_{rt} is the number of retrieved relevant documents in which the term T has been seen; $n_{\bar{r}t}$ is the number of retrieved non-relevant documents including term T; $n_{r\bar{t}}$ is the number of retrieved relevant documents that do not include term T; $n_{\bar{r}\bar{t}}$ is the number of retrieved non-relevant documents that do not include term T.

4.1   Query Term Reweighting

Terms, T_q, belonging to the initial query were instantiated as relevant. Therefore, after judging the retrieved documents we can still consider them as relevant or, otherwise, we need to decrease the belief supporting the relevance of the terms, i.e., penalize them. Some preliminary experiments have shown that a much better performance is obtained if we only penalize those terms that exclusively occur in non-relevant documents (*negative query terms* $(-qt)$). Nevertheless, since they belong to the original query, we still will assign some belief supporting their relevance. Therefore, for those terms considered completely relevant, i.e., they appear, at least, in one relevant document (*positive query terms* $(+qt)$, and *neutral query terms* $(= qt)$) we use the λ message $\lambda(T_q) = (0, 1)$ and for penalized terms we use a λ message such as $\lambda(T_q) = (\gamma_t, 1)$, with $0 < \gamma_t < 1$. We have tried two different approaches for determining the value γ_t: (i) Use a fixed γ value for all the negative query terms $(\lambda_\gamma(T_q) = (\gamma, 1))$, and (ii) consider a γ_t value sensible to the number of non-relevant documents in which the term occurs: the higher this number is, the higher γ_t will be. The following expression satisfies this requirement, assessing γ_t values verifying $0.5 \leq \gamma_t < 1$:

$$\lambda_{tr}(T_q) = (1 - \frac{1}{n_{\bar{r}t} + 1}, 1) \tag{3}$$

4.2   Query Expansion

Now, we want to determine the impact of the addition of new terms, T_e, to the original query. We call these terms *expansion terms*, which may be positive $(+et)$, negative $(-et)$ or neutral $(= et)$.

- The negative expansion terms will be directly instantiated to non-relevant, i.e., $\lambda(T_e) = (1, 0)$, because we consider that they are not useful at all to retrieve relevant documents.

- With regard to neutral expansion terms, we propose to use the λ vector $\lambda(T_e) = (1, 1)$ (which is equivalent to not considering them). We have also experimented with other alternatives, taking into account the number of documents relevant and not relevant where each term appears, but we obtained worse results.

- For positive expansion terms, the first approach would be to instantiate all of them by using $\lambda(T_e) = (0, 1)$, as normal query terms. Experimental results demonstrate that this is not a good choice, because they are being treated as query terms, thus changing the original sense of the query. The second approach is based on the use of a contingency table (relating relevance of the documents with appearance of the term) in order to measure the evidence supporting the relevance or non relevance for a given term. We can see the probability of obtaining a relevant document when we consider the term T_e being relevant, i.e., $p(r|t_e)$ (where r stands for relevant document), as an evidence favouring the relevance of T_e. Analogously, the evidence supporting the non-relevance of T_e could be measured by the probability of obtaining a relevant document given that the term is non relevant, i.e., $p(r|\bar{t}_e)$. These two values constitute the λ vector for each positive expansion term:

$$\lambda(T_e) = (p(r|\bar{t}_e), p(r|t_e)) \text{ or equivalently } \lambda(T_e) = (\frac{p(r|\bar{t}_e)}{p(r|t_e)}, 1) \qquad (4)$$

Now, different ways to estimate these probabilities give rise to different expansion term weighting methods:

- $qe1$: Using a maximum likelihood estimator

$$p(r|t) = \frac{n_{rt}}{n_t} \quad \text{and} \quad p(r|\bar{t}) = \frac{n_{r\bar{t}}}{n_{\bar{t}}} \qquad (5)$$

- Using a Bayesian estimator

$$p(r|t) = \frac{n_{rt} + s_t \frac{n_r}{|R|}}{n_t + s_t} \quad \text{and} \quad p(r|\bar{t}) = \frac{n_{r\bar{t}} + s_{\bar{t}} \frac{n_r}{|R|}}{n_{\bar{t}} + s_{\bar{t}}} \qquad (6)$$

where the parameters s_t and $s_{\bar{t}}$ represent the equivalent sample size. The reason to propose this method is that the quality of the maximum likelihood estimation may be low, because we are dealing with few data (the $|R|$ retrieved documents).

Now, we need some alternatives to fix these parameters s_t and $s_{\bar{t}}$. The first one is obtained by setting them to a given fixed value, configuring the $qe2$ expansion term weighting method. Another option is to relate s_t and $s_{\bar{t}}$ with the amount of data available to estimate $p(r|t)$ and $p(r|\bar{t})$, respectively (less data implies less reliability and therefore a greater equivalent sample size). So, the $qe3$ method uses

$$s_t = |R| - n_t = n_{\bar{t}} \text{ and } s_{\bar{t}} = |R| - n_{\bar{t}} = n_t \qquad (7)$$

Finally, another alternative for computing s_t and $s_{\bar{t}}$ can be obtained by taking into account the total number of documents in which a term occurs and does not occur in the whole set of documents in the collection (as opposed to the set of retrieved documents). For instance, if a term is in three documents, and these three documents have been retrieved, s_t should be 0, because we are using all the possible data to estimate the probability. The opposite case is when the term indexes a great amount of documents, e.g. one hundred documents, and only one is retrieved: s_t should be high. This behaviour can be obtained by computing the ratio between the frequency of the term in the collection and n_t, in the following way (N is the number of documents in the collection):

$$s_t = \frac{\log(\text{tf}_T + 1)}{\log(n_t + 1)} + 1 \text{ and } s_{\bar{t}} = \frac{\log(N - \text{tf}_T + 1)}{\log(n_{\bar{t}} + 1)} + 1 \qquad (8)$$

We use logarithms because in certain cases the values computed for the parameters could be extremely high and they have to be smoothed. We denote this last approach as $qe4$.

5   Experimental Results

To test our feedback methods, we carried out several, more detailed, experiments with the CACM collection and some other experiments with four additional collections: Adi, Cranfield, CISI and Medlars, with the aim of observing the behaviour of our feedback methods in different environments.

Several remarks have to be done before starting: The number of documents that the IRS gives back to the user is fifteen ($|R| = 15$). The performance measure employed is the percentage of change of the average *precision* for the three intermediate points of *recall* (0.2, 0.5 and 0.8) with respect to the results obtained after submitting to the system the original queries. This value is presented in the last column of all the tables below, where the intermediate columns display the method used to compute the λ values for each type of query or expansion term. Moreover, we evaluate the feedback performance by using the *Residual Collection* method [1], which removes from the collection all the documents that the user has seen in the first relevance judgment step, and we take into account those queries in which no relevant document has been retrieved.

Table 1 shows the results of term reweighting using CACM. The first three experiments try to show how a low, a medium and a high value for the γ parameter in the vector λ_γ for negative query terms, could affect the feedback performance. As we can observe in this table, the higher γ is, the better the results are, although the increase is very low. We run a fourth experiment in which we use the vector λ_{tr}, showing that its performance is very similar to those obtained in the previous experiments, with the main advantage that we do not have to tune the γ parameter.

Exp	+qt	-qt	=qt	%C
1	$(0,1)$	$(0.2,1)$	$(0,1)$	28.23
2	$(0,1)$	$(0.5,1)$	$(0,1)$	29.83
3	$(0,1)$	$(0.8,1)$	$(0,1)$	30.05
4	$(0,1)$	λ_{tr}	$(0,1)$	29.90

Table 1. Term reweighting results with CACM.

Table 2 shows the results of the experiments about query expansion with CACM. In order to focus on the impact of added terms, the original query will remain unchanged, instantiating all its terms to relevant. The first two experiments show that expanding the original query with all the positive expansion terms as being original terms is not a good idea, because these new query terms impair the original query. The third experiment applies the $qe1$ weighting approach, and the result has been considerably improved because each positive expansion term is added with a λ vector different from $(0,1)$. The $qe1$ weighting approach is still very poor, but its modifications by means of the Bayesian estimation (the fourth and fifth experiments, using $qe2$) give a good qualitative jump. The next two experiments show the performance of the $qe3$ and $qe4$

Exp	qt	-et	=et	+et	%C
1	$(0,1)$	$(1,1)$	$(1,1)$	$(0,1)$	-69.42
2	$(0,1)$	$(1,0)$	$(1,1)$	$(0,1)$	-59.73
3	$(0,1)$	$(1,0)$	$(1,1)$	$qe1$	3.0
4	$(0,1)$	$(1,0)$	$(1,1)$	$qe2 : s_t, s_{\bar{t}} = 5$	46.36
5	$(0,1)$	$(1,0)$	$(1,1)$	$qe2 : s_t, s_{\bar{t}} = n_r$	48.40
6	$(0,1)$	$(1,0)$	$(1,1)$	$qe3$	38.76
7	$(0,1)$	$(1,0)$	$(1,1)$	$qe4$	59.45
8	$(0,1)$	$(1,1)$	$(1,1)$	$qe4$	12.19

Table 2. Query expansion results with CACM.

weighting approaches. Observe that $qe4$ exhibits the best percentage of change of all the experiments. The last experiment tries to reinforce the idea of how the negative expansion plays an important role in the query expansion of our feedback model: We instantiate the negative expansion terms to $(1,1)$, i.e. these terms are not added to the query, adding only the positive expansion terms using the $qe4$ approach. In this case the performance decreases considerably.

Table 3 displays the results obtained (with CACM) by combining term reweighting and query expansion. In these experiments, we have used vector λ_{tr} for negative query term, positive and neutral query terms use vector $(0,1)$, negative expansion terms use vector $(1,0)$, and we compare the different weighting techniques for positive expansion terms. It is clear that the combination of both techniques is very convenient to improve the performance of our relevance feedback method. The best weighting technique for positive expansion terms is $qe4$, whose performance is clearly superior to the rest.

Exp	+et	%C
1	$qe2 : s_t, s_{\bar{t}} = 5$	65.25
2	$qe2 : s_t, s_{\bar{t}} = n_r$	67.11
3	$qe3$	58.30
4	$qe4$	72.08

Table 3. Term reweighting and query expansion experiments with CACM.

The results of the experiments obtained with Adi, Cranfield, CISI and Medlars are displayed in Table 4. The main characteristics of these collections (number of documents, terms and queries), as well as the average precision at the three intermediate points of recall obtained by the BNR model for all the queries are also shown in this table. All the experiments have been carried out using term reweighting and query expansion, and the fixed parameters are the same used in Table 3. Adi behaves more or less as CACM does (obtaining the best performance with $qe4$). Medlars offers the worst percentages of change of the five collections. The reason could be that the retrieval rate in this collection is good. Again, $qe4$ is the best weighting method. Cranfield shows an excellent behavior,

Col.	Doc.	Terms	Quer.	3p Avg.	$qe1$	$qe2: s_t, s_{\bar{t}} = 5$	$qe2: s_t, s_{\bar{t}} = n_r$	$qe3$	$qe4$
Adi	82	828	35	0.36	104.2	71.7	95.0	68.6	105.0
CACM	3204	7562	52	0.34	6.33	65.3	67.1	58.3	72.1
CISI	1460	4985	76	0.17	8.2	41.0	42.9	42.7	39.0
Cran.	1398	3857	225	0.42	99.1	98.1	99.9	94.6	107.7
Med	1033	7170	30	0.63	−32.6	10.1	7.9	8.4	12.4

Table 4. Main characteristics of the used collections. Term reweighting and query expansion experiments.

even using $qe1$, with the same pattern as CACM. Finally, CISI exhibits a quite uniform behaviour across the different weighting methods.

6   Concluding Remarks

In this paper we have introduced a relevance feedback method for the Bayesian Network Retrieval model based on partial evidences. We have presented the theoretical frame over it is based on, and empirically shown how this method has a robust behaviour with four of the five standard test collections that we have used. The future works will be centered in the development of new relevance feedback methods for the BNR model based on the underlying concept of partial evidences, trying to improve the performance obtained with the methods introduced here.

Acknowledgments: This work has been supported by the Spanish Comisión Interministerial de Ciencia y Tecnología (CICYT) under Project TIC2000-1351.

References

1. Y. K. Chang, C. Cirillo, and J. Razon. *Evaluation of feedback retrieval using modified freezing, residual collection and test and control groups*, pages 355–370. Prentice Hall, Inc., Englewood Cliffs, NJ, 1971.
2. L. M. de Campos, J. M. Fernández, and J. F. Huete. Building bayesian network-based information retrieval systems. In 11th *International Workshop on Database and Expert Systems Applications: 2nd Workshop on Logical and Uncertainty Models for Information Systems (LUMIS)*, 543–552, 2000.
3. L. M. de Campos, J. M. Fernández, and J. F. Huete. Query expansion in information retrieval systems using a bayesian network-based thesaurus. In *Proceeding of the 14th Uncertainty in Artificial Intelligence Conference*, 53–60, 1998.
4. L. M. de Campos, J. M. Fernández, and J. F. Huete. Document instantiation for relevance feedback in the Bayesian network retrieval model. Submitted to the ACM-SIGIR'01 Workshop on Mathematical and Formal Methods in IR.
5. R. Fung and B. D. Favero. Applying bayesian networks to information retrieval. *Communications of the ACM*, 38(2):42–57, 1995.
6. J. M. Fernández. Modelos de Recuperación de Información Basados en Redes de Creencia (in Spanish). Ph.D. Thesis, Universidad de Granada, 2001.

7. D. Ghazfan, M. Indrawan, and B. Srinivasan. Towards meaningful bayesian networks for information retrieval systems. In *Proceedings of the IPMU'96 Conference*, 841–846, 1996.

8. D. Harman. Relevance feedback revisited. In *Proceedings of the 16th ACM–SIGIR Conference*, 1–10, 1992.

9. J. Pearl. *Probabilistic Reasoning in Intelligent Systems: Networks of Plausible Inference*. Morgan and Kaufmann, San Mateo, 1988.

10. B. A. Ribeiro-Neto and R. R. Muntz. A belief network model for IR. In H. Frei, D. Harman, P. Schäble, and R. Wilkinson, editors, *Proceedings of the 19th ACM–SIGIR Conference*, 253–260, 1996.

11. G. Salton and C. Buckley. Improving retrieval performance by relevance feedback. *Journal of the American Society for Information Science*, 41:288–297, 1990.

12. G. Salton and M. J. McGill. *Introduction to modern Information Retrieval*. McGraw-Hill, Inc., 1983.

13. A. Spink and R. M. Losee. Feedback in information retrieval. *Annual Review of Information Science and Technology*, 31:33–78, 1996.

14. H. R. Turtle and W. B. Croft. Inference networks for document retrieval. In J.-L. Vidick, editor, *Proceedings of the 13th ACM–SIGIR Conference*, 1–24, 1990.

Generating Fuzzy Summaries from Fuzzy Multidimensional Databases

Anne Laurent

Université Pierre et Marie Curie
LIP6
8, rue du capitaine Scott
F-75015 Paris, France
Anne.Laurent@lip6.fr

Abstract. Fuzzy Multidimensional Databases and OLAP (On Line Analytical Processing) are very efficient for data mining ([3,8]). In [9], we have defined an extension of these databases in order to handle imperfect information and flexible multidimensional queries. An architecture is proposed for knowledge discovery. It integrates fuzzy multidimensional databases and learning systems. We investigate here the problem of generating relevant fuzzy summaries from fuzzy multidimensional databases.

1 Introduction

Multidimensional databases (hereafter MDB) are well suited for data mining tasks. They provide a framework for hierarchies handling and they also enhance the calculus of aggregates. Moreover, we have extended this model in our recent work in order to deal with imperfect data and flexible queries.

On the other hand, fuzzy summaries represent a good tool to extract relevant knowledge from databases. A fuzzy summary is of the following type: "*Q objects are S: τ*", where Q is a *quantifier* represented by a fuzzy set[1] (for instance *Most*), S is the *summarizer* represented by another fuzzy set, and τ is the degree of truth of the summary. The use of fuzzy labels in order to describe data is very interesting since it allows data to be described in a natural language instead of a numerical way. For instance, saying that *"Sales were bad"* is more understandable than *"The average sale is 82720 units"*.

We propose thus to generate fuzzy summaries from fuzzy multidimensional databases. We show here that the model defined in [9] for fuzzy multidimensional databases provides the operators needed for the generation of such fuzzy summaries. This solution has been implemented and tested using `Oracle Express`.

In the sequel, we first describe our model of fuzzy multidimensional databases, and the existing work on fuzzy summaries. Then we detail the problems related to the generation of fuzzy summaries from fuzzy multidimensional databases, and we propose some solutions to enhance the process of summary discovery. Finally, we conclude and give some related perspectives.

---

[1] A fuzzy set A of a universe U is defined by its membership function f_A which associates each $u \in U$ with a degree of membership $f_A(u) \in [0, 1]$.

F. Hoffmann et al. (Eds.): IDA 2001, LNCS 2189, pp. 24–33, 2001.

2   Fuzzy Multidimensional Databases

There is no unique model for multidimensional databases. Roughly speaking, a multidimensional database is a set of *hypercubes* (hereafter *cubes*), defined on a set of dimensions, which may be organized hierarchically. One dimension of particular interest is chosen to be the *measure* and its values are stored in the cube cells. Operations are defined to visualize and manipulate cubes (*e.g.* rotation, selection by slice and dice, roll-up, drill-down).

In a previous paper ([9]), we have extended this model in order to handle imperfect data and flexible operations. This model is briefly described below.

Entities. Data in the cubes may be imprecise and/or uncertain. For this reason, we introduce the notion of *element*. An *element* is a pair (v, d) where v is a label standing for a fuzzy set (or a precise value) and d ($d \in [0, 1]$) is the degree of confidence in this value (estimated correctness). As in classical database models, we call *domain* of a dimension a finite set of such elements.

In a fuzzy cube, each cell is described by its position on each dimension. $\overrightarrow{x}$ stands for such a cell, and we denote by $(v(\overrightarrow{x}), d(\overrightarrow{x}))$ the cell element. For dimensions, we denote by d_i an entity and $(v(d_i), d(d_i))$ is the pair representing the corresponding element, where $d(d_i)$ indicates at which extent the corresponding slice belongs to the cube.

Each cell $\overrightarrow{x}$ of a fuzzy cube is associated with a degree $\mu(\overrightarrow{x}) \in [0, 1]$ corresponding to the degree to which it belongs to the cube.

For instance, a fuzzy cube C is defined as a relation $C : PRODUCT \times MONTH \times DISTRICT \rightarrow SALES \times [0, 1]$.   Fig. 1 shows an example of such a fuzzy cube. In this cube, sales of *canoes* in *January* in *Boston* have been *bad*. This information is known with confidence 1. The slice corresponding to *january* belongs totally to the cube, whereas the slice corresponding to *canoes* belongs only with degree 0.7 to the cube.

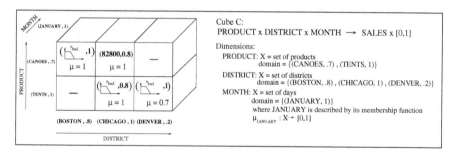

Fig. 1. A fuzzy cube

Hierarchies. Hierarchies may be defined on dimensions, in order to represent data at different levels of granularity (see Fig. 2). These hierarchies may be defined using either fuzzy relations or fuzzy partitions ordered by a relation $\prec$.

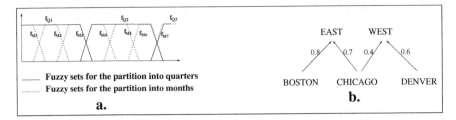

Fig. 2. Fuzzy Hierarchies

Operations. Operations are defined on fuzzy cubes. Some of these operations are briefly presented here, but they are not detailed. See [9] for further information.

Selection on cell values (dice) - This operation is defined in order to select cells from the cube matching a given criterion. In our model, the criterion may be imprecise. For instance, *medium* sales are selected from the previous example cube. This operation results in a new cube where the degrees $\mu(\overrightarrow{x})$ have been modified in order to take into account the extent to which the cell $\overrightarrow{x}$ matches the selection criterion.

Selection on dimension values (slice) - Elements on dimensions belong gradually to the cube, depending on the $d(d_i)$ values. This operation modifies these degrees depending on how the element on the dimension matches the selection criterion. For instance, all the slices corresponding to *eastern* districts are selected from the cube C.

Aggregation - We call *aggregation* the operation consisting in summarizing a cube C by a value, fuzzy or not. For instance, we consider two kinds of counting. The first one sums all the μ degrees from the cells, while the second one counts all the cells having a μ degree greater than a given threshold t. For both counting methods, a cell is taken into account only if all the corresponding slices belong to the cube with a degree greater than a given value χ:

$$\sum_{\overrightarrow{x}} \mu_C(\overrightarrow{x}) \prod_{i=1}^{k} \delta_\chi(d(d_i)) \text{ or } \sum_{\overrightarrow{x}} \delta_t(\mu_C(\overrightarrow{x})) \prod_{i=1}^{k} \delta_\chi(d(d_i)) \text{ where } \delta_\chi(x) = \begin{cases} 1 \text{ if } x \geq \chi \\ 0 \text{ otherwise} \end{cases}$$

Roll-Up - Rolling up a cube consists in considering it at a higher level of granularity. For instance, the *sales* cube can be rolled up on the geographical dimension from the district level to the region one.

3   Fuzzy Summaries

Fuzzy summaries have been studied since the 1980s. Basically, they have the form "$Q\ y_i\ are\ S\ :\ \tau$" or "$Q\ B\ y_i\ are\ S\ :\ \tau$" where Q is a quantifier, S is a summarizer, y_i $(1 \leq i \leq n)$ are the objects to summarize, τ is the degree of truth, and B is a pre-selection condition on the objects. For instance, we consider the summaries: *"Most sales are high"* or *"A few important experts are convinced"*.

Quantifiers and summarizers are described as fuzzy sets. We note μ_Q and μ_S the corresponding membership functions. τ can be computed as $\tau = \mu_Q(\frac{1}{n}\sum_{i=1}^{n} \mu_S(y_i))$ or $\tau = \mu_Q(\frac{1}{n}\sum_{i=1}^{n} \top(\mu_S(y_i), \mu_B(y_i)))$ where $\top$ is a t-norm[2].

Many papers have dealt with the generation of fuzzy summaries. [4] considers relational databases and studies the discovery of fuzzy functional dependencies as a way of generating fuzzy summaries. These dependencies are defined as an extension of functional dependencies by considering a measure of similarity between objects instead of their equality. In this framework, $X \rightarrow Y$ holds if and only if for each pair of tuples in the relation, if values on X are similar at a degree α then values on Y are similar at a degree β.

In other approaches ([5], [10]), the authors consider each tuple as a summary. The attributes are rewritten at different levels of granularity by considering fuzzy labels. [10] generates fuzzy summaries having the following shape: *"Q tuples are a_1 and ... and a_m"* where a_i are fuzzy labels.

Finally, [2] proposes the generation of gradual rules, like *"the younger the employees, the smaller their bonus"*.

4   Fuzzy Summaries from Fuzzy MDB

In [9], we obtained some results for generating of fuzzy summaries. In this first approach, we generate fuzzy summaries like *"Most sales are medium: τ"* as follows:

> **For each** *summarizer S*
> $\quad C_1 \leftarrow Count(all\ cells)$
> $\quad$ *Select cells matching the summarizer*
> $\quad C_2 \leftarrow Count(selected\ cells)$
> $\quad r \leftarrow C_1/C_2$
> $\quad$ **For each** *quantifier Q:* $\tau_Q = \mu_Q(r)$

Those summaries are like *"Q cells are $S : \tau_Q$"*. The user can either display all the summaries, or choose to display the one maximizing τ_Q. We call these summaries *simple* because the summarizer is a single item and because no pre-selection is required. The next section presents the extension of this approach to the generation of *complex* summaries.

4.1   Support Calculus and Complex Summaries

The *covering* (also called *support*) is not directly handled by this first approach. This notion is strongly related to association rules ([1]), like *confidence*. For a summary like *"Q objects which are A are B"*, the support is the proportion of objects being both A and B with respect to the total count of the database, and the confidence is the proportion of objects which fulfill the criterion B among

---

[2] A t-norm is a function $\top : [0,1] \times [0,1] \rightarrow [0,1]$ having the following properties: commutativity, associativity, and monotonicity, with 1 as a neutral element.

objects which fulfill the criterion A. A is called the *left part* of the summary and B the *right part*. The criterion A may be complex, defined as a conjunction of criteria on different dimensions of the cube. The *simple* case occurs when there is no criterion A. The user is then able to specify the minimal required support, either by giving a crisp value, or by specifying a fuzzy quantifier (for instance *more than a quarter*). The corresponding summaries are generated as expressed in a sequence of flexible queries:

$C_0 \leftarrow$ *Count(Non empty cells)*
If $A = \emptyset$ **Then** $C_1 \leftarrow C_0$ *and go to step 5*
Select cells that match criterion A
$C_1 \leftarrow$ *Count(selected cells)*
Select candidate cells matching criterion B
$C_2 \leftarrow$ *Count(selected cells)*
Support $\leftarrow C_2/C_0$
Confidence $\leftarrow C_2/C_1$
$\tau \leftarrow \mu_Q(Confidence)$

4.2   Limitation

Generating all possible summaries is impossible, because of the number of available dimensions, quantifiers and summarizers. Indeed, given q quantifiers and a cube C having k dimensions (including the measure), with c_k potential summarizers per dimension, the generation of all the summaries like *"Q y are S"* produces qc_k possible summaries for each dimension. Therefore there exist $\sum_{j=1}^{k} qc_j$ possible summaries.

We consider now *complex* summaries like *"Q y which are B are S"* where criterion B is defined with respect to the $k-1$ remaining dimensions. The pre-selection is constituted by a set of i criteria ($i \in [0, k-1]$). The number of possible combinations to express B depends on the number of available criteria for each of the i chosen dimensions. Given i ranging from 0 to $k-1$, we have thus $\sum_{i=0}^{k-1} \prod_{l=0}^{i} c_l$ ways to pre-qualify the data, once the dimension on the right part is chosen. The number of possible summaries is then: $q \sum_{j=1}^{k} \left(c_j \sum_{i<j} \left(\prod_{l=0}^{i} c_l \right) \right)$.

4.3   Fuzzy Summaries and Hierarchies

The multidimensional model is well adapted to handle data at different levels of granularity. Hierarchies allow the navigation through these levels. In the literature, some studies (among which [10]) generate a hierarchy of fuzzy concepts. In our approach, a method of discretization of numerical attributes has been used in order to construct fuzzy decision trees and may be used to build hierarchies. However, in this paper we consider that hierarchies are given. These hierarchies allow gradual memberships to several categories.

Multi-level Summaries. When considering the possibility of generating summaries at all the available levels of granularity, the number of such summaries is very high. The idea here is to roll up the cube before generating the summaries.

The user is then able to indicate which summaries and which dimensions will be explored further. However, as highlighted in [6], rolling up the cube at a high level of granularity may produce irrelevant summaries. [7] also points out this problem for the discovery of association rules, since high supports are likely to appear at high levels of granularity, whereas interesting rules are likely to be at lower levels of granularity, with low supports, and will thus not be discovered.

Our model allows the discovery of summaries at different levels of granularity, since rolling up by counting is provided. In this approach, a generalized cube is computed by counting. The resulting cube may be pre-computed in order to reduce the response times. Different summaries are generated at this level. The user may then specify the summaries and dimensions to explore further.

Intra-dimension Summaries. Summaries like "Most middle sales in *eastern cities* are from *Chicago*" are called *Intra-dimension Summaries*. The model proposed here is well adapted to the generation of such summaries since it allows both browsing between granularity levels and flexible queries.

As seen previously, the number of possible summaries that may be generated is very high, thus some means to reduce the generation cost are required. We consider summaries whose left part is constituted by a combination of terms relative to the same dimension at different levels of granularity. The right part is characterized by a criterion describing the value of the measure. The elements constituting the left part must be compatible. In the case of a hierarchy defined by a relation R, the transitive closure R_T is computed. An element a is said compatible with another element b if $f_{R_T}(a, b) > 0$. The following algorithm describes the process for generating summaries in the naive way:

> **For each** *dimension*
> > **For each** *element e_s of the highest level*
> > > **For each** *element e_i from lowest level compatible with e_s*
> > > > **For each** *value c describing the measure*
> > > > > *Find the best quantifier Q and generate the summary*
> > > > > *"Q c objects concerning e_s are e_i"*

We consider q quantifiers and c possible summarizers to describe the measure. For a given dimension D_i, we denote by s_k the set of elements from high levels and $i(s)$ $(s \in s_k)$ the set of elements from low levels compatible with the value s. We denote by $|i(s)|$ the number of these elements. The number of possible summaries is then: $qc\prod_{s \in s_k} |i(s)|$

As seen previously, generating all the possible summaries is computationally impossible. We introduce algorithms to reduce the computation cost.

4.4   Performance Enhancement

The generation cost is due to the multiple possibilities of combining summarizers on the dimensions. In the case of the generation of all the possible summaries, the number of selection queries and counting queries is too high to obtain good performances.

User expectations can be taken into account to reduce the number of quantifiers and selection criteria to be considered. The reduction of the number of quantifiers is less important than the reduction of the number of selection criteria since the selection operation is computationally much more expensive.

The user may also reduce the number of combinations to be considered by selecting the dimensions to take into account in the two parts of the summaries. For instance, the user who wants only *simple* summaries or who chooses only l dimensions among the k available ones ($l \ll k$) reduces the cost of the research. For example, the user may choose to consider only the summaries where the summarizer S corresponds to a criterion on the measure. When considering multidimensional databases, this choice is very efficient since cubes have been constructed for a particular goal which guides the choice of this measure as a dimension of particular interest.

The interaction with the user for the generation of summaries or association rules is used in many systems ([6],[5],[7]). In our system, the user may choose: the dimensions to take into account, the interesting summarizers, the covering, the quantifiers, the level of granularity, the summaries and dimensions to be refined.

Frequent Items Generation. The discovery of association rules ([1]) has been well studied in the literature and several algorithms have been developed in order to reduce the cost of the generation of the rules. The application of these methods to multidimensional databases and hypercubes has been studied in [11]. Moreover, works concerning the generation of fuzzy summaries have also planned to use such methods ([6]). The usual operators used to select and count cube cells are t-norms, which are monotonically decreasing functions. Variants of APriori methods can thus be applied in the framework of fuzzy summaries to enhance the performances by pruning candidates. Fuzzy summaries are discovered by following the classical algorithm processing in two steps, generation of frequent item sets having a sufficient support and extraction of rules having a sufficient confidence:

> $\mathcal{L}_1 \leftarrow$ *frequent item sets of size 1*
> **For each** k *from* $k = 2$ *and while* $\mathcal{L}_{k-1} \neq \emptyset$
> $\quad \mathcal{C}_k \leftarrow$ *candidate sets of size* k *{using* $\mathcal{L}_{k-1}{}^3$*}*
> $\quad$ **For each** *candidate* C:
> $\quad\quad$ *Selection of corresponding cells*
> $\quad\quad$ *Counting*
> $\quad\quad$ **If** *counting* $>$ *min_sup* **Then** $\mathcal{L}_k \leftarrow \mathcal{L}_k \bigcup C$
> $\quad$ *Generation of the summaries*

The computations of the degrees of truth are issued from previous steps. All the counting values are stored, and thus no other computation is required except the computation of the corresponding ratios. This algorithm provides thus a method to reduce the cost of the generation of fuzzy summaries. The

---

[3] Two frequent item sets generate a candidate if they share $k-2$ terms. The candidate is then obtained by taking the $k-2$ shared terms and the two distinct ones from each of the two sets.

minimal support is specified by the user (in this approach, *covering* and *support* refer to the same concept). The generation of candidate summaries from the frequent item sets is processed by the following steps:

> $C\_Resumes \leftarrow \emptyset$ {*Set of candidate summaries*}
> {*Generation of the set of candidates*}
> **For each** *frequent item set*
> $\quad$ $C\_Part \leftarrow$ *Set of pairs from partitions of size 2 of the frequent item set*
> $\quad$ $C\_Resumes \leftarrow C\_Resumes \bigcup C\_Part$
> **For each** *candidate* (A, B)
> $\quad$ *Search for counts of selections on* A, *on* B *and on* $A \wedge B$
> $\quad$ {*all the counts have already been computed*}
> $\quad$ **For each** *quantifier* Q
> $\quad\quad$ $\tau_Q \leftarrow \mu_Q(Count(A \wedge B)/Count(A))$
> {*the summary is* "Q objects which are A are B : τ_Q" }

Properties of Fuzzy Summaries. Fuzzy summaries are characterized by their quantifier and by the summarizers themselves. We mainly use here two methods to reduce the cost of their generation.

Quantifiers -   We consider two fuzzy subsets representing two disjoint fuzzy quantifiers. If the degree of truth for one of them equals 1, then it will be 0 for the second one. We may also extend this idea by considering that, as soon as a *high* degree of truth is discovered for a quantifier, the calculus for the other quantifiers having an *almost null intersection* is useless.

However, the most time-consuming computation is not the computation of the degree of truth, which is only a matter of membership computation, but the selection, which scans all the cube cells. Thus we rather use properties of the summarizers.

Summarizers -   Even if the discovery of exceptions is very important, most users are particularly interested in discovering general trends. Algorithms for the discovery of association rules are devoted to such tasks and are processed for high confidence degrees (for instance 0.8). Meanwhile, the discovery of a summary with a high degree of truth for a given criterion and a quantifier which considers high proportions (more than half of the objects) allows to break the discovery process for all the other criteria. We can then reduce the set of all the candidate summaries:

Given a quantifier Q verifying $\forall x \in [0,1]$, $\mu_Q(x) > 0 \Rightarrow x > 0.5$, *and a set of fuzzy summarizers* $\{c_i\}_{i=1}^n$ *represented by their membership functions* μ_{c_i} $(1 \leq i \leq n)$ *on a universe U such that* $\forall u \in U, \forall i\ (i = 1, \ldots, n)$ $\mu_{c_i}(u) > 0.5 \Rightarrow \nexists c_j \neq c_i$ *such that* $\mu_{c_j}(u) > 0.5$, *we have:*
If there exists a summary "Q objects which are A are c_i: τ" *such that* $\tau > 0$, *then there does not exist* $c_j \neq c_i$ *such that* "Q objects which are A are c_j: τ" *is true with* $\tau > 0$.

Indeed, if one summarizer covers more than half of the examples, then no other one will have the same covering. Let us consider a user willing to generate summaries like *"Most objects verifying A are B"*, where *Most* is a quantifier verifying $\forall x, \mu_{Most}(x) > 0 \Rightarrow x > 0.5$. The two classical steps of the generation of summaries are mixed in order to skip useless time-consuming selection and counting operations as soon as possible. Thus for each set of summarizers reaching minimal support, the algorithm computes immediately the candidate summaries. A summary is produced if the confidence is greater than min_conf (which is greater than 0.5).

The following algorithm shows how we could enhance the previous one for the generation of frequent item sets and summaries of size 1:

> $totalcount \leftarrow Count$ *non empty cube cells,* $R \leftarrow \emptyset$ *{Set of summaries}*
> *{Generation of frequent item set of size 1}*
> **For each** *dimension*
> > **For** *each element e in the dimension*
> > > *Selection on the value of the dimension*
> > > $Count \leftarrow$ *(selected cells)*
> > > **If** *(prop* $\leftarrow count/totalcount)$ > min_sup *and* min_conf
> > > **Then** $R \leftarrow R \bigcup \{(\emptyset, e), prop\}$
> > > **Else** *EndFor {Go to next dimension}*

Properties of Fuzzy Multidimensional Databases. The multidimensional model presents some advantages for the generation of fuzzy summaries. Indeed, it provides hierarchies that allow gradual membership to different concepts and computes aggregates efficiently. The advantages of this model for the reduction of the generation cost are mainly due to the technical performances for the computation of aggregates. This computation is enhanced when using a $MOLAP$ model (Multidimensional OLAP), where data are physically stored as multidimensional data, unlike $ROLAP$ (Relational OLAP) approaches that require to access to the relational database at each query.

Moreover, these aggregates may be pre-computed. The choice of these precomputations is then the key point to have an optimal balance between response times to queries and physical data storage space. In this framework, contingency cubes are pre-computed.

Finally, cubes are built for a specific aim, and the choice of the measure in the cells is very important. Therefore, in summaries like *"Q objects which are A are B"*, the measure will constitute the term B. The analyst will also guide the choice of A by manipulating the cube with rotations, selections, and projections.

Implementation The fuzzy multidimensional database model we have defined has been implemented by enhancing the multidimensional database management system `Oracle Express Server`. The discovery of relevant summaries is currently being implemented and enhanced using the algorithms presented in this paper. This process has been implemented using `Oracle Express Objects`, `Java` and `C++`.

Tests have been performed on a cube describing *sales* results. We present here some examples obtained for simple summaries:

> *Few sales are* Medium: *0*
> *About half sales are* Medium: *0.84*
> *Most sales are* Medium: *0.16*
> *Most(1) sales are* Medium: *0.48*
> *Most(2) sales are* Medium: *0*

where Most, Most(1) and Most(2) are different ways of defining the fuzzy quantifier *Most*.

Acknowledgments The author would like to acknowledge B. Bouchon-Meunier for her help and her useful suggestions.

5 Conclusion

In this paper, we present an approach based on fuzzy multidimensional databases to automatically extract fuzzy summaries from large databases. Multidimensional databases are a good means to browse through granularity levels. Moreover, they have been extended to the management of imperfect data and flexible queries in our recent work. We also study the link between association rules and fuzzy summaries, especially in the case of multidimensional databases.

The fuzzy multidimensional database management system is already implemented, and the results of this paper are currently processed. They complete the global architecture we have previously proposed for knowledge discovery from multidimensional databases ([8]), which has already been tested on large data repositories. The perspectives associated with this work concern the implementation and optimization of performance by integrating all the results we proposed here.

References

1. R. Agrawal, T. Imielinski and A. Swami, Mining association rules between sets of items in large databases, in *Proc. of ACM SIGMOD Conf.*, pp 207-216, 1993.
2. P. Bosc, O. Pivert and L. Ughetto, On Data Summaries Based on Gradual Rules, in *LNCS*, vol. 1625, Springer-Verlag, 1999.
3. M.S. Chen, J. Han and P.S. Yu, Data Mining: An Overview from a Database Perspective, in *IEEE TKDE*, 8(6):866-883, 1996.
4. J.C. Cubero, J.M. Medina, O. Pons and M.A. Vila, Data Summarization with Linguistic Labels: A Loss Less Decomposition Approach, in *Proc. of IFSA*, 1997.
5. D. Dubois and H. Prade, Fuzzy sets in data summaries - Outline of a new approach, in *Proc. of IPMU*, Granada (Spain), pp 1035-1040, 2000.
6. J. Kacprzyk, R.R. Yager and S. Zadrozny, A fuzzy logic based approach to linguistic summaries of databases, in *J. of Applied Mathematics and Computer Science*, 2000.
7. M. Kamber, J. Han and J. Y. Chiang, Metarule-Guided Mining of Multi-Dimensional Association Rules, in *KDD*, 1997.
8. A. Laurent, B. Bouchon-Meunier, A. Doucet, S. Gançarski and C. Marsala, Fuzzy Data Mining from Multidimensional Databases, in *Studies in Fuzziness and Soft Computing*, 54:278-283, Springer-Verlag, 2000.
9. A. Laurent, De l'OLAP Mining au F-OLAP Mining, in *Proc. of Extraction et Gestion des Connaissances*, vol. 1, pp 189-200, Hermes, 2001.
10. G. Raschia and N. Mouaddib, Evaluation de la qualité des partitions de concepts dans un processus de résumés de bases de données, in *LFA'00*, Cépaduès Editions, pp 297-305, 2000.
11. H. Zhu, On-Line Analytical Mining of Association Rules, *Master's Degree Thesis*, Simon Fraser University, 1998.

A Mixture-of-Experts Framework for Learning from Imbalanced Data Sets*

Andrew Estabrooks[1] and Nathalie Japkowicz[2]

[1] IBM Toronto Lab, Office 1B28B,
1150 Eglinton Avenue East,
North York, Ontario, Canada, M3C 1H7
aestabro@ca.ibm.com
[2] SITE, University of Ottawa,
150 Louis Pasteur, P.O. Box 450 Stn. A,
Ottawa, Ontario, Canada K1N 6N5
nat@site.uottawa.ca

Abstract. Re-Sampling methods are some of the different types of approaches proposed to deal with the class-imbalance problem. Although such approaches are very simple, tuning them most effectively is not an easy task. In particular, it is unclear whether oversampling is more effective than undersampling and which oversampling or undersampling rate should be used. This paper presents an experimental study of these questions and concludes that combining different expressions of the re-sampling approach in a mixture of experts framework is an effective solution to the tuning problem. The proposed combination scheme is evaluated on a subset of the REUTERS-21578 text collection (the 10 top categories) and is shown to be very effective when the data is drastically imbalanced.

1 Introduction

In a concept-learning problem, the data set is said to present a class imbalance if it contains many more examples of one class than the other. Such a situation poses challenges for typical classifiers such as Decision Tree Induction Systems or Multi-Layer Perceptrons that are designed to optimize overall accuracy without taking into account the relative distribution of each class [4,2]. As a result, these classifiers tend to ignore small classes while concentrating on classifying the large ones accurately. Unfortunately, this problem is quite pervasive as many domains are cursed with a class imbalance. This is the case, for example, with text classification tasks whose training sets typically contain much fewer documents of interest to the reader than on irrelevant topics. Other domains suffering from class imbalances include target detection, fault detection, or fraud detection problems, which, again, typically contain much fewer instances of the event of interest than of irrelevant events.

---

* We would like to thank Rob Holte and Chris Drummond for their useful comments. This research was funded, in part, by an NSERC grant. The work conducted in this paper was conducted at Dalhousie University.

F. Hoffmann et al. (Eds.): IDA 2001, LNCS 2189, pp. 34–43, 2001.

The purpose of this study is to propose a technique for dealing with the class imbalance problem, and, in the process of doing so, that of finding a way to take advantage of the extra, albeit, unlabeled data that are often left unused in classification studies.

Several approaches have previously been proposed to deal with the class imbalance problem including a simple and yet quite effective method: re-sampling e.g. [6,5,1]. This paper deals with the two general types of re-sampling approaches: methods that *oversample* the small class in order to make it reach a size close to that of the larger class and methods that *undersample* the large class in order to make it reach a size close to that of the smaller class. Because it is unclear whether oversampling is more effective than undersampling and which oversampling or undersampling rate should be used, we propose a method for combining a number of classifiers that oversample and undersample the data at different rates in a mixture of experts framework. The mixture-of-experts is constructed in the context of a decision tree induction system: C5.0, and all re-sampling is done randomly. This proposed combination scheme is, subsequently, evaluated on a a subset of the REUTERS-21578 text collection and is shown to be very effective in this case.

The remainder of this paper is divided into four sections. Section 2 describes an experimental study on a series of artificial data sets to explore the effect of oversampling and undersampling and oversampling or undersampling at different rates. This study suggests a mixture-of-experts scheme which is described in Section 3. Section 4 discusses the experiment conducted with that mixture-of-experts scheme on a series of text-classification tasks and discusses their results. Section 5 is the conclusion.

2   Experimental Study

We begin this work by studying the effects of oversampling versus undersampling and oversampling or undersampling at different rates. All the experiments in this part of the paper are conducted over artificial data sets defined over the domain of 4 x 7 DNF expressions, where the first number represents the number of literals present in each disjunct and the second number represents the number of disjuncts in each concept. We used an alphabet of size 50. For each concept, we created a training set containing 240 positive and 6000 negative examples. In other words, we created an imbalance ratio of 1:25 in favor of the negative class.

2.1   Re-sampling versus Downsizing

In this part of our study, three sets of experiments were conducted. First, we trained and tested C5.0 on the 4x7 DNF 1:25 imbalanced data sets just mentioned. Second, we randomly oversampled the positive class, until its size reached the size of the negative class, i.e., 6000 examples. The added examples were straight copies of the data in the original positive class, with no noise added. Third, we undersampled the negative class by randomly eliminating data points from the negative class until it reached the size of the positive class or, 240 data points. Here again, we used a straightforward random approach for selecting the points to be eliminated. Each experiment was repeated 50 times on different 4x7 DNF concepts and using different oversampled or removed examples. After each training session, C5.0 was tested on separate testing sets containing 1,200 positive and 1,200 negative examples. The average accuracy results are reported in Figure 1(a) (in terms of error rates) while Figure 1(b) reports the results obtained with the F_1-, F_2- and $F_{0.5}$-measure.[1]

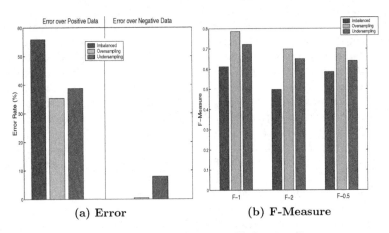

(a) Error   (b) F-Measure

Fig. 1. Oversampling versus Undersampling

The left side of Figure 1 (a) shows the results obtained on the positive testing set while its right side shows the results obtained on the negative testing set. As can be expected, the results show that the number of false negatives (results over the positive class) is a lot higher than the number of false positives (results over the negative class). Similarly, Figure 1 (b) shows that re-sampling is more helpful for recall than it is for precision, though it does help both. Altogether, the results suggest that both naive oversampling and undersampling are helpful

---

[1] The F_B-measure is defined as: $F_B = \frac{(B^2+1) \times P \times R}{B^2 \times P + R}$ where P represents precision, and R, recall, which are respectively defined as follows: $P = \frac{TruePositives}{TruePositives+FalsePositives}$; $R = \frac{TruePositives}{TruePositives+FalseNegatives}$. Thus, the higher the F_B-measure, the better. In other words, precision corresponds to the proportion of examples classified as positive that are truly positive; recall corresponds to the proportion of truly positive examples that are classified as positive; the F_B-measure combines the precision and recall by a ratio specified by B. If $B = 1$, then precision and recall are considered as being of equal importance. If $B = 2$, then recall is considered to be twice as important as precision. If $B = 0.5$, then precision is considered to be twice as important as recall. We report the F-measure, here because text classification tasks—the application we are ultimately interested in in this paper—are typically evaluated using that measure rather than accuracy.

for reducing the error caused by the class imbalance on this problem although oversampling appears more accurate than undersampling.[2]

2.2. Re-sampling and Down-Sizing at Various Rates

In order to find out what happens when different sampling rates are used, we continued using the imbalanced data sets of the previous section, but rather than simply oversampling and undersampling them by equalizing the size of the positive and the negative set, we oversampled and undersampled them at different rates. In particular, we divided the difference between the size of the positive and negative training sets by 10 and used this value as an increment in our oversampling and undersampling experiments. We chose to make the 100% oversampling rate correspond to the fully oversampled data sets of the previous section but to make the 90% undersampled rate correspond to the fully undersampled data sets of the previous section.[3] For example, data sets with a 10% oversampling rate contain $240 + (6,000 - 240)/10 = 816$ positive examples and 6,000 negative examples. Conversely, data sets with a 0% undersampling rate contain 240 positive examples and 6,000 negative ones while data sets with a 10% undersampling rate contain 240 positive examples and $6,000 - (6,000 - 240)/10 = 5424$ negative examples. A 0% oversampling rate and a 90% undersampling rate correspond to the fully imbalanced data sets designed in the previous section while a 100% undersampling rate corresponds to the case where no negative examples are present in the training set.

Once again, and for each oversampling and undersampling rate, the rules learned by C5.0 on the training sets were tested on testing sets containing 1,200 positive and 1,200 negative examples. The results of our experiments are displayed in Figure 2 for the case of oversampling and undersampling respectively. They represent the averages of 50 trials. The results are reported in terms of error rates over the positive and the negative testing sets.[4]

These results suggest that different sampling rates have different effects on the accuracy of C5.0 on imbalanced data sets for both the oversampling and the undersampling method. In particular, the following observation can be made:

> Oversampling or undersampling until a cardinal balance of the two classes is reached is not necessarily the best strategy: best accuracies are reached before the two sets are cardinally balanced.

In more detail, this observation comes from the fact that in both the oversampling and undersampling curves of figure 2 the optimal accuracy is not obtained when the positive and the negative classes have the same size. In the oversampling curves, where class equality is reached at the 100% oversampling rate,

---

[2] Note that the usefulness of oversampling versus undersampling is problem dependent. [1], for example, finds that oversampling is, sometimes, more effective than undersampling, although in many cases, the opposite can be observed.

[3] This was done so that no classifier was duplicated in our combination scheme.

[4] This time results in terms of the F-measure are not reported for the sake of clarity of our figure.

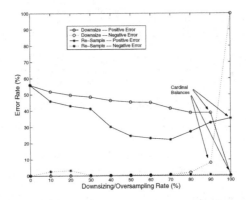

Fig. 2. Oversampling and Downsizing at Different Rates

the average error rate obtained on the data sets over the positive class at that point is 35.3% (it is of 0.45% over the negative class) whereas the optimal error rate is obtained at a sampling rate of 70% (with an error rate of 22.23% over the positive class and of 0.56% over the negative class). Similarly, although less significantly, in the undersampling curves, where class equality is reached at the 90% undersampling rate[5], the average error rate obtained at that point is worse than the one obtained at a sampling rate of 80% since although the error rate is the same over the positive class (at 38.72%) it went from 1.84% at 90% oversampling over the negative class to 7.93%.

In general, it is quite likely that the optimal sampling rates can vary in a way that might not be predictable for various approaches and problems.

3   The Mixture-of-Experts Scheme

The results obtained in the previous section suggest that it might be useful to combine oversampling and undersampling versions of C5.0 sampled at different rates. On the one hand, the combination of the oversampling and undersampling strategies may be useful given the fact that the two approaches are both useful in the presence of imbalanced data sets (cf. results of Section 2.1) and may learn a same concept in different ways.[6] On the other hand, the combination of classifiers using different oversampling and undersampling rates may be useful since we may not be able to predict, in advance, which rate is optimal (cf. results of Section 2.2).

---

[5] The increase in error rate taking place at the 100% undersampling point is caused by the fact that at this point, no negative examples are present in the training set.

[6] In fact, further results comparing C5.0's rule sizes in each case suggest that the two methods, indeed, do tackle the problem differently (see, [2]).

We will now describe the combination scheme we designed to deal with the class imbalance problem. This combination scheme will be tested on a subset of the REUTERS-21578 text classification domain.

3.1   Architecture

A combination scheme for inductive learning consists of two parts. On the one hand, we must decide *which* classifiers will be combined and on the other hand, we must decide *how* these classifiers will be combined. We begin our discussion with a description of the architecture of our mixture of experts scheme. This discussion explains which classifiers are combined and gives a general idea of how they are combined. The specifics of our combination scheme are motivated and explained in the subsequent section.

In order for a combination method to be effective, it is necessary for the various classifiers that constitute the combination to make different decisions [3]. The experiments in Section 2 of this paper suggest that undersampling and oversampling at different rates will produce classifiers able to make different decisions, including some corresponding to the "optimal" undersampling or oversampling rates that could not have been predicted in advance. This suggests a 3-level hierarchical combination approach consisting of the *output level*, which combines the results of the oversampling and undersampling experts located at the *expert level*, which themselves each combine the results of 10 classifiers located at the *classifier level* and trained on data sets sampled at different rates. In particular, the 10 oversampling classifiers oversample the data at rates 10%, 20%, ... 100% (the positive class is oversampled until the two classes are of the same size) and the 10 undersampling classifiers undersample the negative class at rate 0%, 10%, ..., 90% (the negative class is undersampled until the two classes are of the same size). Figure 3 illustrates the architecture of this combination scheme that was motivated by Shimshoni's Integrated Classification Machine [7].[7]

3.2   Detailed Combination Scheme

Our combination scheme is based on two different facts:

Fact #1: Within a single testing set, different testing points could be best classified by different single classifiers. (This is a general fact that can be true for any problem and any set of classifiers).

Fact #2: In class imbalanced domains for which the positive training set is small and the negative training set is large, classifiers tend to make many false-negative errors. (This is a well-known fact often reported in the literature on the class-imbalance problem and which was illustrated in Figure 1, above).

---

[7] However, [7] is a general architecture. It was not tuned to the imbalance problem, nor did it take into consideration the use of oversampling and undersampling to inject principled variance into the different classifiers.

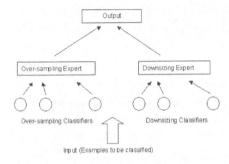

Fig. 3. Re-Sampling versus Downsizing

In order to deal with the first fact, we decided not to average the outcome of different classifiers by letting them vote on a given testing point, but rather to let a single "good enough" classifier make a decision on that point. The classifier selected for a single data point needs not be the same as the one selected for a different data point. In general, letting a single, rather than several classifiers decide on a data point is based on the assumption that the instance space may be divided into non-overlapping areas, each best classified by a different expert. In such a case, averaging the result of different classifiers may not yield the best solution. We, thus, created a combination scheme that allowed single but different classifiers to make a decision for each point.

Of course, such an approach is dangerous given that if the single classifier chosen to make a decision on a data point is not reliable, the result for this data point has a good chance of being unreliable as well. In order to prevent such a problem, we designed an elimination procedure geared at preventing any unfit classifier present at our architecture's classification level from participating in the decision-making process. This elimination program relies on our second fact in that it invalidates any classifier labeling too many examples as positive. Since the classifiers of the combination scheme have a tendency of being naturally biased towards classifying the examples as negative, we assume that a classifier making too many positive decision is probably doing so unreliably.

In more detail, our combination scheme consists of

- a combination scheme applied to each expert at the expert level
- a combination scheme applied at the output level
- an elimination scheme applied to the classifier level

The expert and output level combination schemes use the same very simple heuristic: if one of the non-eliminated classifiers decides that an example is positive, so does the expert to which this classifier belongs. Similarly, if one of the two experts decides (based on its classifiers' decision) that an example is positive, so does the output level, and thus, the example is classified as positive by the overall system.

The elimination scheme used at the classifier level uses the following heuristic: the first (most imbalanced) and the last (most balanced) classifiers of each expert are tested on an unlabeled data set. The number of positive classifications each classifier makes on the unlabeled data set is recorded and averaged and this average is taken as the threshold that none of the expert's classifiers must cross. In other words, any classifier that classifies more unlabeled data points as positive than the threshold established for the expert to which this classifier belongs needs to be discarded.[8]

It is important to note that, at the expert and output level, our combination scheme is heavily biased towards the positive under-represented class. This was done as a way to compensate for the natural bias against the positive class embodied by the individual classifiers trained on the class imbalanced domain. This heavy positive bias, however, is mitigated by our elimination scheme which strenuously eliminates any classifier believed to be too biased towards the positive class.

4   Experiments on a Text Classification Task

Our combination scheme was tested on a subset of the 10 top categories of the REUTERS-21578 Data Set. The data was first divided according to the ModApte Split which consists of considering all labeled documents published before 04/07/87 as training data and all labeled documents published after that date as testing data.[9] All the unlabeled documents available from the REUTERS data set were used in our elimination scheme. The training data was, further, divided into 10 problems each containing a positive class composed of 100 documents randomly selected from one of the 10 top REUTERS categories and a negative class composed of all the documents belonging to the other 9 top REUTERS categories.[10] Thus, the class imbalances in each of these problems ranged from an imbalance ratio of 1:60 to one of 1:100 in favour of the negative class. The testing set consisted of all the texts published after 04/07/87 belonging to the 10 top categories.

---

[8] Because no labels are present, this technique constitutes an educated guess of what an appropriate threshold should be. This heuristic was tested in [2] on the text classification task discussed below and was shown to improve the system (over the combination scheme not using this heuristic).

[9] This split is commonly used in the literature on text classification.

[10] Although the REUTERS Data set contains more than 100 examples in each of its top 10 categories, we believe it more realistic to use a restricted number of positive examples. Indeed, very often in practical situations, we only have access to a small number of articles labeled "of interest" whereas huge number of documents "of no interest" are available.

The results obtained by our scheme on these data were pitted against those of C5.0 ran with the Ada-boost option.[11] The results of these experiments are reported in Figure 4 as a function of the averaged (over the 10 different classification problems) F_1, F_2 and $F_{0.5}$ measures.

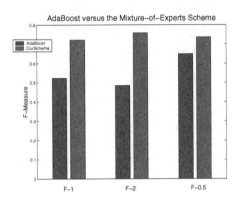

Fig. 4. Average results obtained by Ada-Boost and the Mixture-of-Experts scheme on 10 text classification problems

The results in Figure 4 show that our combination scheme is much more effective than Ada-boost on *both* recall and precision.[12] Indeed, Ada-boost gets an F_1 measure of 52.3% on the data set while our combination scheme gets an F_1 measure of 72.25%. If recall is considered as twice more important than precision, the results are even better. Indeed, the mixture-of-experts scheme gets an F_2-measure of 75.9% while Ada-boost obtains an F_2-measure of 48.5%. On the other hand, if precision is considered as twice more important than recall, then the combination scheme is still effective, but not as effective with respect to Ada-boost since it brings the $F_{0.5}$-measure on the reduced data set to only 73.61%, whereas Ada-Boost's performance amounts to 64.9%.

The generally better performance displayed by our proposed system when evaluated using the F_2-measure and its generally worse performance when evaluated using the $F_{0.5}$-measure are not surprising, since we biased our system so that it classifies more data points as positive. In other words, it is expected that our system will correctly discover new positive examples that were not discovered by Ada-Boost, but will incorrectly label as positive examples that are not positive. Overall, however, the results of our approach are quite positive with

---

[11] Our scheme was compared to C5.0 ran with the Ada-boost option combining 20 classifiers. This was done in order to present a fair comparison to our approach which also uses 20 classifiers. It turns out, however, that the Ada-boost option provided only a marginal improvement over using a single version of C5.0 (which itself compares favorably to state-of-the-art approaches for this problem) [2].

[12] [2] shows that it is also more effective than uncombined re-sampling.

respect to both precision and recall. Furthermore, it is important to note that this method is not particularly computationally intensive. In particular, its computation costs are comparable to those of commonly used combination methods, such as AdaBoost.

5   Conclusion and Future Work

This paper presented an approach for dealing with the class-imbalance problem that consisted of combining different expressions of re-sampling based classifiers in an informed fashion. In particular, our combination system was built so as to bias the classifiers towards the positive set so as counteract the negative bias typically developed by classifiers facing a higher proportion of negative than positive examples. The positive bias we included was carefully regulated by an elimination strategy designed to prevent unreliable classifiers to participate in the process. The technique was shown to be very effective on a drastically imbalanced version of a subset of the REUTERS text classification task.

There are different ways in which this study could be expanded in the future. First, our technique was used in the context of a very naive oversampling and undersampling scheme. It would be useful to apply our scheme to more sophisticated re-sampling approaches such as those of [6] and [5]. Second, it would be interesting to find out whether our combination approach could also improve on cost-sensitive techniques previously designed. Finally, we would like to test our technique on other domains presenting a large class imbalance.

References

1. Domingos, Pedro (1999): Metacost: A general method for making classifiers cost sensitive, *Proceedings of the Fifth International Conference on Knowledge Discovery and Data Mining*, 155–164.
2. Estabrooks, Andrew (2000): *A Combination Scheme for Inductive Learning from Imbalanced Data Sets*, MCS Thesis, Faculty of Computer Science, Dalhousie University.
3. Hansen, L. K. and Salamon, P. (1990): Neural Network Ensembles, *IEEE Transactions on Pattern Analysis and Machine Intelligence*, 12(10), 993–1001.
4. Japkowicz, Nathalie (2000): The Class Imbalance Problem: Significance and Strategies, *Proceedings of the 2000 International Conference on Artificial Intelligence (IC-AI'2000)*, 111–117.
5. Kubat, Miroslav and Matwin, Stan (1997): Addressing the Curse of Imbalanced Data Sets: One-Sided Sampling, *Proceedings of the Fourteenth International Conference on Machine Learning*, 179–186.
6. Lewis, D. and Gale, W. (1994): Training Text Classifiers by Uncertainty Sampling, *Proceedings of the Seventh Annual International ACM SIGIR Conference on Research and Development in Information Retrieval*.
7. Shimshoni, Y. and Intrator, N. (1998): Classifying Seismic Signals by Integrating Ensembles of Neural Networks, *IEEE Transactions On Signal Processing, Special issue on NN*.

Predicting Time-Varying Functions with Local Models

Achim Lewandowski and Peter Protzel

Chemnitz University of Technology
Dept. of Electrical Engineering and Information Technology
Institute of Automation
09107 Chemnitz, Germany
achim.lewandowski@alewand.de
peter.protzel@e-technik.tu-chemnitz.de

Abstract. Data analysis applications which have to cope with changing environments require adaptive models. In these cases, it is not sufficient to train e.g. a neural network off-line with no further learning during the actual operation. Therefore, we are concerned with developing algorithms for approximating time-varying functions from data. We assume that the data arrives sequentially and we require an immediate update of the approximating function. The algorithm presented in this paper uses local linear regression models with adaptive kernel functions describing the validity region of a local model. As we would like to *anticipate* changes instead of just following the time-varying function, we use the time explicitly as an input. An example is given to demonstrate the learning capabilities of the algorithm.

1   Introduction

Many data analysis applications we are working on are characterized by highly nonlinear systems which also vary with time. Some systems change slowly e.g. due to aging, other systems could show *sudden* changes in the underlying processes. The prediction of these systems requires accurate models, which leads to the problem of approximating nonlinear and time-variant functions. In contrast to time-series prediction, we need to predict not just the next output value at time t+1 but the complete functional relationship between outputs and inputs.

Global models (e.g. multilayer perceptrons with sigmoid activation functions) generally suffer from the Stability-Plasticity-dilemma, i.e. a model update with new training data in a certain region of the input space can change the model also in other regions of the input space which is undesirable. In contrast, local models are only influenced by data points which fall into their validity regions. In the recent years, several algorithms were developed which work with local regression models. The restriction of a model to be only a local learner can be implemented in an easy way by using weighted regression whereby the weights are given by a kernel function. Schaal and Atkeson [1] introduced an algorithm which can adjust the number and the shape of the kernel functions to yield

F. Hoffmann et al. (Eds.): IDA 2001, LNCS 2189, pp. 44–52, 2001.

a better approximation in the case of a time-invariant function. Recently, another algorithm (Vijayakumar and Schaal [2]) was presented, which can deal with some special cases of high-dimensional input data, again for time-invariant functions. While these algorithms are capable to learn online (i.e. pattern-by-pattern update), optimal results will be achieved only if the sample is repeatedly presented.

The algorithm presented in this paper is a modification and extension of our previous approach (Lewandowski et al. [3]) which is based on competition between the kernel functions. This approach was capable of *following* slow changes, but suffered from the fact, that it never *anticipated* changes. In the following, we describe the idea of using the time explicitly as an input for each local model in order to make an actual model prediction.

It should be noticed that our and the other mentioned algorithms are not so-called "lazy learners" (Atkeson, Moore and Schaal [4]). For every request to forecast the output of a new input, the already constructed local models are used. In contrast, a lazy learner working with kernel functions would construct the necessary model(s) just at the moment, when a new input pattern arrives.

2   The Algorithm

2.1   Statistical Assumptions

We assume a standard regression model

$$y = f(x, t) + \epsilon \,, \tag{1}$$

with x denoting the n-dimensional input vector (including the constant input 1) and y the output variable. The errors are assumed to be independently distributed with variance σ^2 and expectation zero. The unknown function $f(x, t)$ is allowed to depend on the time t. Each local model i $(i = 1, \ldots, m)$ is given by a linear model

$$y = \beta_i^T z \,, \tag{2}$$

whereby with $z = (x^T, t)^T$ the time t is explicitly included as an additional input. The parameters for a given sample are estimated by weighted regression, whereby the weighting kernel for model i,

$$w_i(z) = w_i(x, t) := \exp\left(-\frac{1}{2}(x - c_i)^T D_i (x - c_i)\right) \,. \tag{3}$$

depends only on the original inputs. The weighting kernel does *not* take the time t into account. The matrix D_i describes the shape of the kernel function and c_i the location.

Suppose that all input vectors z are summarized row-by-row in a matrix Z, the outputs y in a vector Y and the weights for model i in a diagonal matrix W_i, then the parameter estimators are given by

$$\hat{\beta}_i = (Z^T W_i Z)^{-1} Z^T W_i Y \,. \tag{4}$$

With $P_i = (Z^T W_i Z)^{-1}$ a convenient recursion formula exists. After the arrival of a new data point (z, y), according to Ljung and Söderström [5] the following incremental update formula exists:

$$\hat{\beta}_i^{new} = \hat{\beta}_i + e_i w_i(z) P_i^{new} z \,, \tag{5}$$

with

$$P_i^{new} = \frac{1}{\lambda} \left(P_i - \frac{P_i z z^T P_i}{\frac{\lambda}{w_i(z)} + z^T P_i z} \right) \tag{6}$$

and

$$e_i = y - \hat{\beta}_i^T z \,. \tag{7}$$

Please notice, that a forgetting factor λ ($0 < \lambda \leq 1$) is included. This forgetting factor is essential for our algorithm to allow faster changes during the kernel update, as the influence of older observations must vanish, if the approximating function is expected to follow the changes sufficiently fast.

2.2   Update of Receptive Fields

Assume, that a new query point $z = (x_1, \ldots, x_n, t)$ arrives. Each local model gives a prediction $\hat{y}_i$. These individual predictions are combined into a weighted mean, according to the activations of the belonging kernels:

$$\hat{y} = \frac{\sum w_i(z) \hat{y}_i}{\sum w_i(z)} \,. \tag{8}$$

Models with a low activation near zero have no influence on the prediction.

We will now describe the way how new kernels are integrated and old kernel functions are updated. The basic idea is that models, which give better predictions for a certain area than their neighbors, should get more space while the worse models should retire. This is done by changing the width and shape of the belonging kernel functions.

A new kernel function is now inserted, when no existing kernel function has an activation $w_i(z)$ exceeding a user-defined threshold $\eta > 0$. As a side effect, the user could be warned that the prediction $\hat{y}$ of the existing models must be judged carefully. The center c_i of the new kernel function is set to x and the matrix D_i is set to a predefined matrix D, usually a multiple of the identity matrix ($D = kI$). High values of k produce narrow kernels.

If no new kernel function is needed, the neighboring kernel functions are updated. We restrict the update to kernel functions which exceed a second threshold $\delta = \eta^2$. After the true output y is known, the absolute errors $|e_i|$ of the predictions of all activated models are computed. If the model with the highest activation belongs to the lowest error, nothing happens. If for example only one kernel function is activated sufficiently, the kernel shapes remain unchanged.

Otherwise, if $\bar{e}$ denotes the mean of these absolute errors, the matrix D_i of a model i, which is included in the shape update is slightly changed. Define

$$g_i = \min(\frac{|e_i| - \bar{e}}{\bar{e}}, 1) , \tag{9}$$

then the shape matrix is updated by the following formula:

$$D_i^{new} = D_i + \frac{\rho^{g_i} - 1}{(x - c_i)^T D_i (x - c_i)} D_i (x - c_i)(x - c_i)^T D_i . \tag{10}$$

The parameter $\rho \geq 1$ controls the amount of changes. The essential part of the activation, $a_i = (x - c_i)^T D_i (x - c_i)$, is given after the update by

$$a_i^{new} = \rho^{g_i} a_i . \tag{11}$$

The activation of kernel functions belonging to models which perform better than the average will therefore rise for this particular x. The update of D_i is constructed in that way, that activations for other input values v, which fulfil

$$(v - c_i)^T D_i (x - c_i) = 0 , \tag{12}$$

are not changed. Inputs v, for which the vector $v - c_i$ is in a certain sense perpendicular to $x - c_i$, receive the same activation as before.

After the kernel update *all* linear models are updated as described in the last section. For a model with very low activation ($w_i(z) \approx 0$) the parameter estimator will not change, but according to equation (6) P_i will grow by the factor $1/\lambda$, so that the belonging model gets more and more sensitive if it has not seen data for a long time. We could further use a function $\lambda = \lambda(x)$, expressing our prior believe of variability.

3   Empirical Results

Vijayakumar and Schaal [2] and Schaal and Atkeson [1] used the following time-invariant function to demonstrate the capabilities of their algorithm:

$$s = \max \left(e^{-10x^2}, e^{-50y^2}, 1.25e^{-5(x^2+y^2)} \right) + N(0, 0.01) . \tag{13}$$

In their example, a sample of 500 points, uniformly drawn from the unit square was used to fit a model. The whole sample was repeatedly presented pattern-by-pattern whereby from epoch to epoch the sample was randomly shuffled. As a test set to measure the generalization capabilities, a 41×41 grid over the unit square was used, with the true outputs without noise.

For our purposes we will modify this approach. Each data point will be presented just once. The first 3000 data points will be generated according to (13). From $t = 3001$ to $t = 7000$ the true function will change gradually according to

$$s = \max \left(\frac{7000 - t}{4000} e^{-10x^2}, e^{-50y^2}, 1.25e^{-5(x^2+y^2)} \right) + N(0, 0.01) . \tag{14}$$

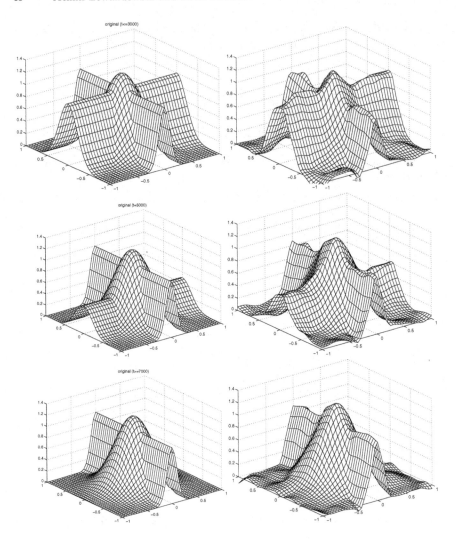

Fig. 1. True function (left) and approximation (right) for $t = 3000$ (top), $t = 5000$ (middle) and $t = 10000$ (bottom)

From $t = 7001$ to $t = 10000$ the function will be time-invariant again:

$$s = \max\left(e^{-50y^2}, 1.25 e^{-5(x^2 + y^2)}\right) + N(0, 0.01) \,. \tag{15}$$

We set $D = 120I$, $\lambda = 0.999$, $\eta = 0.10$ and $\rho = 1.05$. Figure 1 shows the true function (left) and our approximation (right) for $t = 3000$ (top), $t = 5000$ (middle) and $t = 10000$ (bottom). The approximation quality is good, even for the case $t = 5000$ where the function is in a changing state.

We started three other runs to judge the progress we achieved by our approach to include the time as an input and performing kernel shape updates simultaneously. For the first run we excluded the time. For the second run we used the time but fixed ρ at $\rho = 1$, so that no kernel shape update was performed. For the last run we didn't use the time input and performed no kernel shape update. The same 10000 data points were always used. Figure 2 shows the development of the mean squared errors of all four runs.

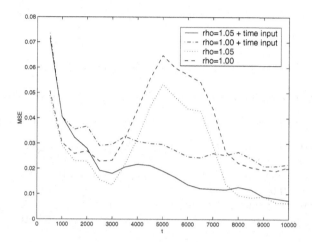

Fig. 2. Mean squared error with and without time input, and with ($\rho = 1.05$) and without ($\rho = 1.00$) kernel shape updates

As our function is time-independent for the first 3000 values, the error starts on a lower level if the time t is not explicitly included in the models. But during the phase where the function is changing, our new approach (solid line) achieves significantly better results. For t between 3000 and 4000 the error increases slightly but the two models which don't use the time, are much more affected. For the interval from $t = 4000$ to $t = 7000$ the error decreases although the function is still changing. The figure shows additionally, that the combination *with* kernel shape update and *with* time input yields the best overall performance.

Figure 3 shows a contour plot $(w_i(z) \equiv 0.7)$ of the activation functions of our local models, after all data points have been presented ($\rho = 1.05$, using the time input). It is clearly visible, that the kernel functions have adjusted their shape according to the appearance of our example function.

Additionally we investigated the influence of the user-defined values of $D = kI$ and η. For every combination of $k \in \{20, 40, \ldots, 220\}$ and $\eta \in \{0.05, 0.10, 0.15\}$ the overall performance was measured as the mean of the mean squared errors for $t = 500, 1000, 1500, \ldots, 10000$. We used as before $\rho = 1.05$ and $\lambda = 0.999$. Figure 4 shows that only for $k \leq 60$ the results are not satisfactory. In these

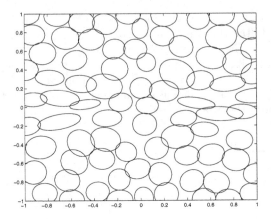

Fig. 3. Contour plot for $t = 10000$ $(w_i(z) \equiv 0.7, \rho = 1.05)$

cases the kernel functions are quite wide, and as a consequence, too few local models for a good approximation are inserted. Unfortunately, the level of non-linearity is not known in advance. The number of generated models varies from 13 ($k = 20$, $\eta = 0.05$) to 158 ($k = 220$, $\eta = 0.15$), but the overall performance is nearly the same for values of k between $k = 100$ and $k = 220$. At least for this example, the progressive insertion of models seems to be the better strategy.

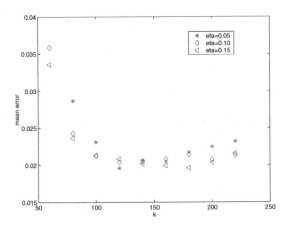

Fig. 4. Overall performance for different values of η and $D = kI$

Finally we investigated the influence of the parameter ρ, which controls the sensitivity of the kernel shape updates. We used again $D = 120I$, $\eta = 0.10$ and $\lambda = 0.999$. Figure 5 shows that the range of satisfactory results is quite broad.

Setting ρ is like setting a learning rate. Values, which are too large, will produce chaotic behavior. If the forgetting factor λ is too small, the matrices P_i may explode. This is a well-known problem of the Recursive Least Squares algorithm using a forgetting factor [6]. An adaptive choice of λ may help to avoid this dilemma.

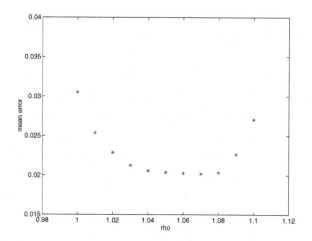

Fig. 5. Overall performance for different values of ρ

4   Conclusions

The increasing need in industrial applications to deal with nonlinear and time-variant systems is the motivation for developing algorithms that approximate nonlinear functions which vary with time. We presented an algorithm that works with local regression models and adaptive kernel functions based on competition. There is no need to store any data points, which we consider as an advantage. The results visualized for an example show a quick adaptation and good convergence of the model after presenting each data point from the time-varying function sequentially and only once. Including the time as an additional input improved the approximation significantly.

For future work, it will be necessary to develop self-adjusting parameters. In our algorithm, the choice of the initialization matrix D in combination with the threshold value η and an appropriate choice of ρ and λ are crucial for the performance and generalization capabilities. We are currently working on an approach to optimize these parameters online.

References

1. Schaal, S., Atkeson, C.: Constructive incremental learning from only local information. Neural Computation, 10(8) (1998) 2047–2084
2. Vijayakumar, S., Schaal,S.: Locally weighted projection regression. Technical Report (2000).
3. Lewandowski, A., Tagscherer, M., Kindermann, L., Protzel, P.: Improving the fit of locally weighted regression models. Proceedings of the 6th International Conference on Neural Information Processing, Vol. 1 (1999) 371–374
4. Atkeson, C., Moore, A., Schaal, S.: Locally weighted learning. Artificial Intelligence Review, 11(4) (1997) 76–113
5. Ljung, L., Söderström, T.: Theory and practice of recursive identification. MIT Press, Cambridge (1986)
6. Nelles, O.: Nonlinear System Identification. Springer, Berlin Heidelberg (2001)

Building Models of Ecological Dynamics Using HMM Based Temporal Data Clustering – A Preliminary Study

Cen Li[1], Gautam Biswas[2], Mike Dale[3], and Pat Dale[3]

[1] Dept of Computer Science, Middle Tennessee State University
Box 48, Murfreesboro, TN 37132.
cli@mtsu.edu
[2] Dept of Electrical and Computer Engineering, Vanderbilt University
Box 1679 Station B, Nashville, TN 37235.
biswas@vuse.vanderbilt.edu
[3] Environmental Sciences, Griffith University, Qld 4111, Australia.
m.dale@mailbox.gu.edu.au, p.dale@mailbox.gu.edu.au

Abstract. This paper discusses a temporal data clustering system that is based on the Hidden Markov Model(HMM) methodology. The proposed methodology improves upon existing HMM clustering methods in two ways. First, an explicit HMM model size selection procedure is incorporated into the clustering process, i.e., the sizes of the individual HMMs are dynamically determined for each cluster. This improves the interpretability of cluster models, and the quality of the final clustering partition results. Secondly, a partition selection method is developed to ensure an objective, data-driven selection of the number of clusters in the partition. The result is a heuristic sequential search control algorithm that is computationally feasible. Experiments with artificially generated data and real world ecology data show that: (i) the HMM model size selection algorithm is effective in re-discovering the structure of the generating HMMs, (ii) the HMM clustering with model size selection significantly outperforms HMM clustering using uniform HMM model sizes for re-discovering clustering partition structures, (iii) it is able to produce interpretable and "interesting" models for real world data.

Keywords: Temporal data clustering, Hidden Markov Model, Bayesian model selection

1 Introduction

In many real world applications, the dynamic characteristics, i.e., how a system interacts with the environment and evolves over time, are of interest. Dynamic characteristics of these systems are best described by temporal features whose values change significantly during the observation period. In this paper, we discuss the development of a temporal data clustering system, and the application

F. Hoffmann et al. (Eds.): IDA 2001, LNCS 2189, pp. 53–62, 2001.

of this system in the study of the ecological effects of mosquito control through the modification of drainage patterns by runneling in a region. More specifically, scientists have collected data from 30 sites on a salt marsh area south of Brisbane, Australia. Some of these sites are located around the area where runneling has been applied, and others are associated with pools which were not treated. At each site, four measurements of species performance, including two measurements for the properties of the grass *Sporobolus* and two measurements to the succulent plant *Sarcocornia*, and 11 environmental measurements, e.g., the water depth and water salinity, were collected. The measurements were collected every three months since mid-1985. The objective of the study is to derive models of ecological dynamics, with respect to the usage of the mosquito control technique. Given the location of the sites, it is believed that more than one model may be present in data.

We assume that the temporal data we are dealing with satisfy the Markov property. When state definitions directly correspond to feature values, a Markov chain model representation of the data may be appropriate [8]. When the state definitions are not directly observable, or it is not feasible to define states by exhaustive enumeration of feature values, i.e., data is described with multiple continuous valued temporal features, the hidden Markov model (HMM) is appropriate. The hidden states of a HMM represent the set of valid states in the dynamic process. While the complete set of states and the exact sequence of states a system goes through may not be observable, it can be estimated by studying the observed behaviors of the system. Our work focuses on applying the HMM methodology for unsupervised learning of temporal data.

Our ultimate goal is to develop through the extracted HMM models, an accurate and explainable representation of the system dynamics. It is important for our clustering system to determine the best number of clusters to partition the data, and the best model structure, i.e., the number of states in a model, to characterize the dynamics of the homogeneous data within each cluster. We approach these tasks by (i) developing an explicit HMM model size selection procedure that dynamically modifies the size of the HMMs during the clustering process, and (ii) solving the HMM model size selection and partition selection problems in terms of a Bayesian model selection problem.

2   The HMM Clustering Problem

A HMM is a non-deterministic stochastic Finite State Automata. We use first order HMMs, where the state of a system at a particular time t, S_t, is only dependent on the state of the system at the immediate previous time point, i.e., $P(S_t|S_{t-1}, S_{t-2}, ..., S_1) = P(S_t|S_{t-1})$. In addition, we assume all the temporal feature values are continuous, therefore, we use the continuous density HMM (CDHMM) representation where all temporal features have continuous values. A CDHMM with n states for data having m temporal features can be characterized in terms of three sets of probabilities [7]: (i) the initial state probabilities, (ii) the transition probability, and (iii) the emission probabilities. The initial state probability vector, $\boldsymbol{\pi}$ of size n, defines the probability of any of the states being the initial state of a given sequence. The transition probability matrix, A of size $n \times n$, defines the probability of transition from state i at time t, to state j at the next time step. The emission probability matrix, B of size $n \times m$, defines the probability of feature value at any given state. For CDHMM, the emission probability density function for each state is defined by a multivariate Gaussian distribution.

3   The HMM Clustering Problem

Unlike the earlier work on HMM clustering ([4], [9]), our work does not make the assumption that the model size (i.e., the number of states in the HMM model) for individual clusters are available. One of our key objectives is to derive this information directly from data during the clustering process. We adopt the Bayesian clustering approach to automatically derive the best number of clusters for the given data. Our HMM clustering algorithm can be described in terms of four nested search steps:

Step 1: Determine the number of clusters in a partition;

Step 2: Distribute the objects to clusters in a given partition size;

Step 3: Compute the HMM model sizes for individual clusters in the partition;

Step 4: Estimate the HMM parameter configuration for the cluster models.

For a HMM of a chosen size, step four is invoked to estimate model parameters that optimize a chosen criterion. The step implements the well known Maximum Likelihood (ML) parameter estimation method, the *Baum-Welch* procedure [7], which is a variation of the more general EM algorithm [1].

4   Bayesian Clustering Methodology

In model-based clustering, it is assumed that data is generated by a mixture of underlying probability distributions. The mixture model, M, is represented by K component models and a hidden, independent discrete variable C, where each value i of C represents a component cluster, modeled by λ_i. Given observations $X = (x_1, \cdots, x_N)$, let $f_k(x_i|\theta_k, \lambda_k)$ be the density of an observation x_i from the kth component model, λ_k, where θ_k is the corresponding parameters of the model. The likelihood of the mixture model given data is expressed as: $P(X|\theta_1, \cdots, \theta_K, P_1, \cdots, P_K) = \prod_{i=1}^{N} \sum_{k=1}^{K} P_k \cdot f_k(x_i|\theta_k, \lambda_k)$, where P_k is the probability that an observation belongs to the kth component($P_k \geq 0, \sum_{k=1}^{K} P_k = 1$). Bayesian clustering casts the model-based clustering problem into the Bayesian model selection problem. Given partitions with different component clusters, the goal is to select the best overall model, M, that has the highest *posterior probability*, $P(M|X)$.

From Bayes theorem, the posterior probability of the model, $P(M|X)$, is given by: $P(M|X) = \frac{P(M)P(X|M)}{P(X)}$, where $P(X)$ and $P(M)$ are prior probabilities of the data and the model respectively, and $P(X|M)$ is the marginal likelihood of the data. For the purpose of comparing alternate models, we have

$P(M|X) \propto P(M)P(X|M)$. Assuming none of the models considered is favored a priori, $P(M|X) \propto P(X|M)$. That is, the posterior probability of a model is directly proportional to the marginal likelihood. Therefore, the goal is to select the mixture model that gives the highest marginal likelihood.

In our work, marginal likelihood computation is approximated using the Bayesian information criterion (BIC) ([3], [2]). BIC is defined as

$$logP(M|X) \approx logP(X|M,\hat{\theta}) - \frac{d}{2}logN,$$

where d is the number of parameters in the model, N is the number of data objects, and $\hat{\theta}$ is the ML parameter configuration of model M. $logP(X|M,\hat{\theta})$, the data likelihood, tends to promote larger and more detailed models of data, whereas the penalty term, $-\frac{d}{2}logN$ favors smaller models with less parameters. BIC selects the best model for the data by balancing these two terms.

5   Bayesian HMM Clustering

In our nested four step Bayesian HMM clustering approach, steps one and three have been cast into model selection problem. Step three finds the optimal HMM for a group of data, and step one finds the optimal clustering partition model for the entire set of data. BIC has been used as the model selection criterion in both steps. Next, we describe how the general Bayesian model selection criterion is adapted for the HMM model size selection and the cluster partition selection problems. In addition, we describe how the characteristics of these criterion functions are used to design our heuristic clustering search control structure.

5.1   Criterion for HMM Size Selection

The HMM model size selection process picks the HMM with the number of states that best describe the data. From our discussion earlier, for a HMM λ, trained on data X, the best model size is the one, when coupled with its ML configuration, gives the highest posterior probability.

Applying the BIC approximation, marginal likelihood of the HMM, λ_k, for cluster k is computed as:

$$logP(X_k|\lambda_k) \approx \sum_{j=1}^{N_k} logP(X_{kj}|\lambda_k,\hat{\theta}_k) - \frac{d_k}{2}logN_k,$$

where N_k is the number of objects in cluster k, d_k is the number of parameters in λ_k and $\hat{\theta}_k$ is the ML parameters in λ_k.

Figure 1(a) illustrates the characteristics of the BIC measure in HMM model size selection. Data generated on a 5-state HMM is modeled using HMMs of sizes ranging from 2 to 10. The dashed line shows the likelihood of data for the different size HMMs. The dotted line shows the penalty for each model. And the solid line shows BIC as a combination of the above two terms. We observe that as the size of the model increases, the model likelihood also increases and the model penalty decreases monotonically. BIC has its highest value corresponding to the size of the original HMM for data.

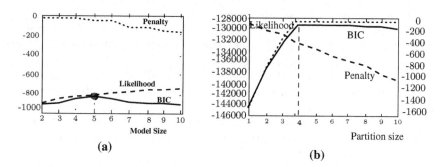

Fig. 1. BIC used for (a) HMM model size selection, and (b) cluster partition selection

5.2 Criterion for Partition Selection

In the Bayesian framework, the best clustering mixture model, M, has the highest *partition posterior probability* (PPP), $P(M|X)$. We approximate PPP with the marginal likelihood of the mixture model, $P(X|M)$.

For partition with K clusters, modeled as $\lambda_1, \cdots, \lambda_K$, the PPP computed using the BIC approximation is:

$$logP(X|M) \approx \sum_{i=1}^{N} log[\sum_{k=1}^{K} P_k \cdot P(X_i|\hat{\theta}_k, \lambda_k)] - \frac{K + \sum_{k=1}^{K} d_k}{2} logN,$$

where $\hat{\theta}_k$ and d_k are the ML model parameter configuration and the number of significant model parameters of cluster k, respectively. P_k is the likelihood of data given the model for cluster k. When computing the data likelihood, we assume that the data is complete, i.e., each object is assigned to one known cluster in the partition. Therefore, $P_k = 1$ if object X_i is in cluster k, and $P_k = 0$ otherwise. The best model is the one that balances the overall data likelihood and the complexity of the entire cluster partition.

Figure 1(b) illustrates the application of the BIC criterion to cluster partition selection. Given data from four randomly generated HMMs, the BIC scores are measured when data is partitioned into 1 to 10 clusters. When the number of clusters is small, the improvements in data likelihood dominate the BIC computation, therefore, the value increases monotonically as the size of the partition increases. The BIC measure peaks at partition size = 4, which is the true partition size. As the partition size is increased beyond four, improvements of data likelihood becomes less significant, and the penalty on model complexity and the model prior terms dominate the BIC measure. Its value decreases monotonically.

5.3 The Clustering Search Control Structure

The exponential complexity associated with the four nested search steps for our HMM clustering prompts us to introduce heuristics into the search process. We exploit the monotonic characteristics of the BIC function to develop a sequential search strategy for steps one and three of the clustering algorithm. We start with the simplest model, i.e., a one cluster partition for step one and a one state HMM for step three. Then, we gradually increase the size of the model, i.e., adding one cluster to the partition or adding one state to the HMM, and re-estimate the model. After each expansion, we evaluate the model using the BIC. If the score of the current model decreases from that of the previous model, we conclude that we have just passed the peak point, and accept the previous model as our "best" model. Otherwise, we continue with the model expansion process.

Partition expansion is achieved by adding a cluster of a few homogeneous objects that have the least likelihood of belonging to an existing cluster model in the current partition. Next, search step two distributes objects among the clusters such that the overall data likelihood given the partition is maximized. We assign object, x_i, to cluster, $(\hat{\theta}_k, \lambda_k)$, based on its object-to-HMM likelihood measure [7], $P(x_i|\hat{\theta}_k, \lambda_k)$. Individual objects are assigned to the cluster whose model provides the highest data likelihood. If after one round of object distribution, any object changes its cluster membership, models for all clusters are updated (model size and model parameters are re-estimated using steps three and four) to reflect the current data in the clusters. Then all objects are redistributed based on the set of new models. Otherwise, the distribution is accepted. We refer to this algorithm the Bayesian HMM Clustering (BHMMC) algorithm. A complete description of the algorithm can be found in [5]

6   Experimental Results

To demonstrate the effectiveness of our approach, we have conducted experiments on both artificially generated data and the ecological data. A number of performance indices have been defined and tested on this data.

6.1   Synthetic Data

Similarity between pairwise clusters was used as the defining factor for creating synthetic data. We define the similarity between pairwise HMMs using the Normalized Symmetrized Similarity (NSS) measure:

$$\bar{d}(\lambda_1, \lambda_2) = \frac{logP(S_2|\lambda_1) + logP(S_1|\lambda_2) - logP(S_1|\lambda_1) - logP(S_2|\lambda_2)}{2 \cdot |S_1| \cdot |S_2|},$$

where $|S_1|$ and $|S_2|$ are the number of data objects generated based on λ_1 and λ_2, respectively. Note that the NSS is a log measure, and since the likelihood values are $<= 1$, $d(\lambda_1, \lambda_2)$ is usually negative. The highest normalized symmetrized similarity, $\bar{d}(\lambda_1, \lambda_2) = 0$, is obtained when the two models are identical. We defined four NSS levels that correspond to four selected NSS value ranges: Level 1 corresponds to NSS value range (-100, 0), Level 2 corresponds to the range (-400, -300), Level 3 corresponds to the range (-700, -600), and Level 4 corresponds to the range (-1000, -900).

A set of 3-cluster partition models were constructed. The sizes of the three HMMs in each partition are four, six, and eight. The configurations of the three HMMs are as the following:

- all HMMs in set 1 are level one NSS apart,
- all HMMs in set 2 are level two NSS apart, and
- all HMMs in set 3 are level three NSS apart.

According to each of the three configurations, five different sets of HMMs satisfying the pairwise model similarity definitions were derived. For each of the 15 partition models, 40 data objects were derived from each of the three HMMs, and then combined to form the actual data set.

6.2   Performance Indices

In addition to the partition posterior probability, we evaluate the quality of the cluster partitions generated using **Partition Misclassification Count (PMC)** measure. PMC computes the sum of the misclassification counts in a derived partition when compared to the true generating partition. Given the true generating partition model, the smaller PMC score, the better the quality of the partition generated. For details on PMC, refer to [[5], [6]].

6.3   Experimental Results

Results from Synthetic Data To verify the advantage of dynamically determining the optimal HMM model sizes for individual cluster models during the clustering process, we compared BHMMC to HMM clustering with predetermined HMM model sizes, i.e., all clusters have the same, pre-determined HMM size. This latter approach is called the BHMMC-fixed algorithm. Three different HMM sizes, i.e., HMM size that is smaller than (three), equal to (six), and greater than (ten) the average HMM size (six) of the three generating HMMs, have been used in three separate experiments with the BHMMC-fixed algorithm.

Table 1 gives the PMC scores on results obtained using the BHMMC and BHMMC-fixed algorithms. It is observed that the BHMMC algorithm performs better than BHMMC-fixed. BHMMC-fixed with average model size produces a better partition structure than BHMMC-fixed with smaller or larger than average model size. The BHMMC algorithm results are better than BHMMC-fixed with average HMM size. On the average, partition structures obtained using BHMMC-fixed with a larger model size is slightly better than those obtained using BHMMC-fixed with a smaller model size.

Figure 2 compares the PPP scores on the clustering partitions obtained using these two algorithms. The overall quality of the partition models derived using the BHMMC algorithm is better than those derived using the BHMMC-fixed algorithms. The PPP values for partitions derived using BHMMC-fixed with 3-state models are significantly smaller than those obtained using BHMMC and BHMMC-fixed with average size models and 10-state models. The PPP values of

		Data Sets														
		Set 1					Set 2					Set 3				
		1	2	3	4	5	1	2	3	4	5	1	2	3	4	5
BHMMC		0	0	0	0	0	0	0	0	0	0	0	0	0	0	0
BHMMC -fixed	size 3	78	96	6	99	0	80	80	6	9	0	80	80	100	0	20
	size 6	0	0	0	0	18	0	0	80	10	80	0	0	0	0	018
	size 10	0	0	80	0	80	80	0	0	0	0	54	80	0	0	80

Table 1. Comparing the PMC scores on clustering results obtained using the BHMMC and the BHMMC-fixed algorithms

partitions generated in the latter three cases are quite close. The PPP criterion takes into account both the model fitness and the model complexity. Partition results obtained using BHMMC have a comparable, or slightly smaller, data likelihood than those obtained using BHMMC-fixed with the 10-state models. But their model complexity penalty is much smaller, which leads to higher PPP values.

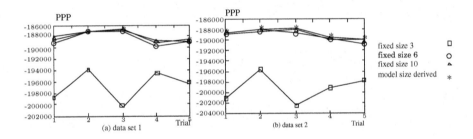

Fig. 2. PPP for partition models derived using BHMMC and BHMMC-fixed on data generated from 3-cluster generating partition models.

Results from the Ecology Data According to the domain experts, for the ecology data, only the plant variables, i.e., the first four temporal features, were used to construct the models. The effects of the runneling on the 11 environmental variables will be studied by superimposing them onto the models learned from the plant data. The four plant variables represent the density and the height of *Sporobolus* and *Sarcocornia*. All the plant variables are continuous valued.

In the first experiment, all four plant variables are used in the clustering process. The BHMMC algorithm produces a single cluster of four states. This corresponds to the case of a single process operating with four states. Careful examination of the derived HMM reveals that the model encodes a cycle: ($S_1 \rightarrow S_2 \rightarrow S_3 \rightarrow S_4$ and then back to S_1). In terms of the characteristics of the plants, the cycle of the states can be explained as the following: starting from state one, the *Sporobolus* are dense and tall, but the *Sarcocornia* are sparse. Then the situation changes where the *Sporobolus* become sparse and of moderate height, and *Sarcocornia* are short and sparse. Next, the *Sporobolus* have low density, and the *Sarcocornia* become sparse and tall. Finally, in state four, there is barely any *Sporobolus*, and the *Sarcocornia* become dense with moderate height. After this state, the system goes back to state one.

In the second experiment, only the two features describing *Sporobolus* were used in the clustering process. This time, the BHMMC algorithm produced two clusters both of size five. The process represented by cluster one is not very dynamic, i.e., the transition probability between states are low. There are two sub-processes appearing in this cluster. State one and two interact in that state two changes to state one. This corresponds to a change of *Sporobolus* from very sparse and short to even shorter and sparser. States three, four, and five interact among themselves. Both states three and five, both representing the state with very sparse *Sporobolus*, change to state four where *Sporobolus* have medium density and height.

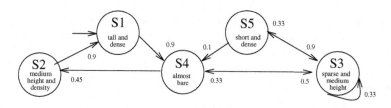

Fig. 3. The interpreted HMM model corresponding to cluster two of the partition derived from the ecology data

The second cluster is more dynamic in that the state transition probabilities are high. It contains two cycles that are almost disjoint. Figure 3 illustrates the interpreted HMM in this cluster. The process starts with state one where the *Sporobolus* are tall and dense. It goes to bare ground in state four, and recovers via state two. Another way of becoming bare ground is via states three, four, and five. When the system is in state three, it may remain in that state with roughly equal chance of shifting to states four and five.

The results represent a preliminary study of the ecology data using the temporal data clustering and modeling approach. The potential implications of these models need to be further validated by analyzing the effects of superimposing the environmental features onto the models.

7   Conclusions and Future Work

We have presented a Bayesian clustering methodology for temporal data using HMMs. The clustering process incorporates a HMM model size selection procedure which not only generates more accurate model structure for individual clusters, but also improves the quality of the partitions generated. We cast both HMM model size selection and cluster partition selection problems into the Bayesian model selection framework and have experimentally shown the effectiveness of the BIC measure when applied in both tasks. Experimental results using both artificially generated data and real world data show the effectiveness of the clustering methodology.

One problem with analyzing real world data is that the amount of data available for analysis may be limited. For the ecology data, the length of the temporal data for each feature is 56, which is considerably short when the data is used to derive HMM models and to estimate the often many parameters associated with the model. Data insufficiency can greatly affect the performance of a temporal data clustering and modeling system. This issue has been systematically studied in [5]. In addition, assumptions made about data may not hold. For example, the BHMMC algorithm assumes that the temporal features are independent. But for the ecology data, it is believed that the two properties (density and height) measured for each plant are correlated. Also, the type of HMM used in BHMMC is first order HMM. While this type of HMM is the most frequently used type of HMM in practice, HMMs of higher order may be considered since there is evidence that the ecology data has a long "memory", longer than the 3-month period between samples. These and other considerations stem from this study provide directions for our future work.

References

1. A. P. Dempster, N. M. Laird, and D. B. Rubin. Maximum likelihood from incomplete data via the em algorithm. *Journal of Royal Statistical Society Series B(methodological)*, 39:1–38, 1977.
2. D Heckerman, D. Geiger, and D. M. Chickering. A tutorial on learning with bayesian networks. *Machine Learning*, 20:197–243, 1995.
3. M. Jordan. *AAAI Tutorial on Graphical Models and Variational Approximation.* 1998.
4. T. Kosaka, S. Masunaga, and M. Kuraoka. Speaker-independent phone modeling based on speaker-dependent hmm's composition and clustering. In *Proceedings of the ICASSP' 95*, pages 441–444, 1995.
5. C. Li. *A Bayesian Approach to Temporal Data Clustering using the Hidden Markov Model Methodology.* PhD thesis, Vanderbilt University, December 2000.
6. C. Li and G. Biswas. Bayesian temporal data clustering using hidden markov model representation. In P. Langley, editor, *Proceedings of the 17th International conference on Machine Learning*, pages 543–550. Morgan Kaufmann Publishers, 2000.
7. L. R. Rabiner. A tutorial on hidden markov models and selected applications in speech recognition. *Proceedings of the IEEE*, 77(2):257–285, February 1989.
8. P. Sebastiani, M. Ramoni, P. Cohen, J. Warwick, and J. Davis. Discovering dynamics using bayesian clustering. In *Proceedings of the 3rd International Symposium on Intelligent Data Analysis*, August 1999.
9. P. Smyth. Clustering sequences with hidden markov models. *Advances in Neural Information Processing*, 1997.

Tagging with Small Training Corpora

Nuno C. Marques[1,2] and Gabriel Pereira Lopes[2]

[1] Universidade Aberta
nmm@univ-ab.pt
http://kholosso.di.fct.unl.pt/~nmm
[2] Centria, DI-FCT/UNL
nmm,gpl@di.fct.unl.pt

Abstract. The analysis of textual data may start by classifying words using a predefined tag set. However, it is still a problem for natural language text understanding the assignment of part-of-speech tags to words in unrestricted text (called POS-tagging). Most part of current taggers require huge amounts of hand tagged text for training (in the order of 10^5 pretagged words): it requires linguistically highly trained man power for a highly repetitive and boring job, and the results obtained have no optimal quality. Moreover, when one wants to change to another text genre the same kind of problem must be faced again. Our proposal goes in another direction. By carefully combining a large lexicon with an efficient neural network based generator of taggers we can generate POS-taggers using no more than 10^4 hand corrected tagged words for training. This training tagged text size can be feasibly hand corrected. Experimental results are presented and discussed for the SUSANNE Corpus . Results in three additional different Portuguese corpora are also discussed. 96% precision rates are obtained when unknown words occur in the test set. 98% precision rates are obtained when every word in the test set is known.

1 Introduction

Textual data is the most valuable source of information. It is currently available from the Internet, inside textual fields in databases and from large information retrieval systems. Unfortunately, traditional computational linguistics techniques are usually of limited relevance for extracting useful information from text. Text is written by humans and aims at transmitting human knowledge. But, even the simplest peace of text becomes a normally computationally intractable puzzle of ambiguous meanings.

One of the simplest text processing tasks is usually its tagging. For accomplishing it, generally a dictionary is used for associating a tag to each word appearing in the text. Generally, a fairly small set of tags (usually containing less than 100 entries) is used. The main problem with this process is word ambiguity. When a standard part-of-speech tag set is used, we have two to five possible tags for each word ([11]. The only way to overcome this ambiguity is

F. Hoffmann et al. (Eds.): IDA 2001, LNCS 2189, pp. 63–72, 2001.

to use word context. A classifier that resolves this ambiguity is usually called a POS Tagger.

The huge amount of work internationally developed on POS taggers (among others [8], [1], [14], just to cite a few authors) has always assumed either that there is a training corpus with more than 100, 000 words, manually annotated with suitable tags or human resources to built such an annotated corpus. For Portuguese there was neither such a corpus, nor the resources to build it. The effort needed to develop a rule based system such as the one presented in [14], also seemed to us too excessive[1].

The problem of the lack of training text is a universal one even for English: The huge amounts of pre-tagged texts available are only suitable for some domains. As a matter of fact taggers trained with the existing hand tagged corpora perform quite poorly in specific application areas. Moreover, the tag set to use is highly dependent on the kind of problem and text one wants to study. This is the reason why we think that the colossal efforts done either for hand tagging huge amounts of text or for hand building (or tuning) complex rule systems ([14]) are only of limited use. Since [8], it has also been clear that unsupervised training methods for tagging (such as the one proposed in [3] or [1]) derive their good performance from lexical information (word endings) and usually need background information, supplied to the tagger by a hand defined set of initial contextual tagging rules. In practice these systems are just hand built rule base taggers enriched with probability estimates collected from untagged text. Although these probability estimation improves slightly the results of pure rules, these taggers still need a huge amount of hand work on rule building.

In our first experiments ([9]), we've obtained very good results by using neural networks. In ([7]) we showed that neural network back propagation based training can be seen as a estimation device that does build rather good probability estimators from sparse data. The use of neural networks for tagging should be similar to other probability estimation methods also applied to tagging, such as the maximum entropy tagger ([12]), having the advantage of allowing a clear and richer modulation of context. The purpose of this paper is to show that a tagged corpus with 10, 000 hand tagged or hand corrected tagged words is what it is necessary for training a neural network and for achieving a precision greater or equal to 96%. Availability of additional pre-tagged corpora is only necessary for lexicon learning. So, we argue that our technique successfully overcomes the huge training corpus problem. Moreover this proposal is language and text genre independent.

2   The Statistics: Applying a Neural Network to Tagging

In [7] we have described a neural model for POS-tagging. Let us start with a set $T = \{t^j : 1 \leq j \leq N\}$ of tags, where N is the number of tags used. For any word

---

[1] Besides that, this approach would contradict our position that it shouldn't be the human user to provide a so specific, interdependent and text dependent set of rules. Rules should be descriptive and easily applicable, not specific and unreadable.

we assume that any of these tags is possible. If, for a given context (a context W is a sequence of words (w_i)), only one of the tags of T is valid, then we can use a tagging function to associate each word w_i, with its tag t^j valid in that context. If ST is the sequence of tags t_i given to the words w_i, in context W, we have that

$$ST = \Phi(W)$$

is the perfect tagging function.

But world is not perfect and so we must help ourselves using statistics. In fact if, for a given corpus Ω, we have a probability function, for a given sequence of tagged words:

$$p(W, ST|\Omega),$$

then this could be expressed by:

$$\Phi(W|\Omega) = \underset{ST}{\mathrm{argmax}}\, p(ST|W, \Omega) =$$

$$\underset{ST}{\mathrm{argmax}}\, p(W, ST|\Omega)$$

since:

$$p(ST|W, \Omega) = \frac{p(ST, W|\Omega)}{p(W)}$$

and $p(W)$ is constant for any ST.

Using these definitions it's usual to make a trigram approximation of the context, and apply Bayes rule to split lexical and contextual probabilities. Finally HK model [8] is applied to linguistics, giving rise to a Hidden Markov Model Viterbi tagger. But we are not going to use a HMM tagger. So we may use the much better linguistically motivated probability of the tag known the word. For a given word w_i, there is a certain probability that word will appear with a given tag t_i: $p(t_i|w_i)$. Considering those probabilities, for all tags in T, we may built the ambiguity vector $\overrightarrow{amb_i}$:

$$\overrightarrow{amb_i} = \begin{bmatrix} p(t_i^1|w_i) \\ ... \\ p(t_i^N|w_i) \end{bmatrix}.$$

The first approach we will use for our neuronal model is that this probability vector is an accurate representation of the word w_i. In order to determine these vectors we use an internal lexicon. We based ourselves on MLE estimators extracted from the training corpus to calculate these values. After counting the frequency of pairs (word, tag), we build the lexicon by estimating $\overrightarrow{amb_i}$ for each tag. The lexical vector is calculated by:

$$\overrightarrow{amb_i} = \begin{bmatrix} freq(t_i^1, w_i)/freq(w_i) \\ ... \\ freq(t_i^N, w_i)/freq(w_i) \end{bmatrix}.$$

We will call this lexicon the internal tagging lexicon.

The second approach is that the context w_{i-1}, w_i, w_{i+1} is sufficient for specifying the word tag t_i (this is a smaller approach than the one done by the HK model). Conjugating the first assumption, with the maximum likelihood decomposition proposed for POS-Tagging by Merialdo [8], we get:

$$\Phi_W^*(W_i = w_i|\Omega) = \underset{t_i}{\mathrm{argmax}}$$

$$\sum_{\forall (ST_{1,i-1},\ ST_{i+1,n})\ \in ST^{n-1}} p(t_i|\overrightarrow{amb_1}, ..., ST_{1,i-1}, ST_{i+1,n}, \Omega).$$

that is, we are just selecting the most probable tag, taking into account (i.e. summing with a positive function) all the other possible contexts. So, by the second assumption, the neuronal tagger for word w_i will be approached by:

$$\Phi^+(W_i = w_i) = p(t_i|\overrightarrow{amb_{i-1}}, \overrightarrow{amb_i}, \overrightarrow{amb_{i+1}}, \Omega)$$

From neural networks ([4]), we know that if the data acquired from the training corpus is identically distributed by the classes considered (i.e. the first assumption), then our second assumption allows us to say that a neural network trained with all known pairs $\overrightarrow{amb_{i-1}}, \overrightarrow{amb_i}, \overrightarrow{amb_{i+1}}$ as input, and having as output, a vector with the position for tag t_i with value 1 and the value 0 in other positions, will asymptotically approach the distribution Φ^+. That is, if we train a neural network tagger with all the possible trigrams from the corpus, we should acquire the best possible estimators for tagging that corpus.

3   The Neural Network Used

In earlier work ([9]) several neural network topologies were compared. One was based on Elman Networks (using a topology very similar to the one used by [13]), another one used a two layer feed-forward network and a third one used a single layer feed-forward neural network. According to the results obtained then, the simplest network (the one layer feed-forward neural network) achieved higher precision and was easier to train. According to these results we will simplify and, in this paper, will use and describe only the results taken by using a single layer network.

The topology used for solving the part-of-speech tagging problem, was a simple feed-forward neural net [4] having only input and output units. The input units were divided into three sets of context units. Each output unit represents one of the tags. A one-to-one relation is established between each value in the lexical probability vector (acquired from the internal lexicon) and each input unit in each set. The previous model was implemented using these three sets: vectors $\overrightarrow{amb_{i-1}}, \overrightarrow{amb_i}, \overrightarrow{amb_{i+1}}$ are assigned to the three sets of context units.

In this network each unit is associated to a part-of-speech tag t_i. The value acquired from the lexical probability vector for the word w_k $(\overrightarrow{amb_k})$, which was supplied by the internal tagging lexicon is assigned to the associated input unit.

Each word w_k can be represented in the first, second or third set of input units, depending if it is the word we have just classified, the word we wish to classify or the next word in the sentence.

For training purposes the output units are assigned the values 1 or 0, according to the part-of-speech category they were tagged in the corpus: 1 if the neuron represents the tag assigned to the word and 0 otherwise. The network was trained first using the standard backpropagation algorithm and then momentum backpropagation algorithm. The standard backpropagation was used with $\eta = 1.0$. The momentum backpropagation algorithm was used with $\eta = 1.0$, $\mu = 0.7$ and $c = 0.1$ [15].

Evaluation/tagging is similar: input values are acquired in the same way as during training. Then we propagate the input through the neural network and select the output unit with the larger activation value. We tag the word with the tag that is associated with the selected neuron.

We usually split our tagged corpus into a training corpus (used by the network training algorithm), an evaluation corpus (with 3 sentences, approximately 100 words[2]) and a test corpus (also with approximately 100 words). As usual in neural network training, to avoid over-fitting, we used the evaluation corpus to determine when we should stop the training process[3]. The test corpus is solely used to compute error rate.

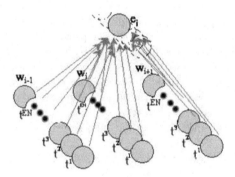

Fig. 1. Diagram of the neuron used to identify a given POS tag

4   The SUSANNE Corpus Experiment

Several authors have used the closed lexicon assumption in POS tagging ([1], [9]). With this approach tagging results are evaluated without unknown words:

---

[2] We have used more than the three sentences for test/evaluation, but results on the ten different runs showed that the use of just three sentences led to stable values

[3] That is, we minimized the learning error on another corpus different from the training corpus

the test corpus is used to build the lexical probabilities. That way it is possible to measure the influence of contextual information separated from the influence of lexical information, introduced namely by unknown words (unknown word identification is by itself a field of research).However, by extracting lexical information from the test corpus this kind of technique may be providing a somewhat optimistic measure for tagging precision. Indeed, we acquire outstanding results (98% precision) when we use a close lexicon approach. Unfortunately it is hard to measure how far these results are from reality. Aren't we simplifying the task for contextual probabilities? And how good must our lexicon be in order to achieve this probability?

In order to solve these questions we have prepared the following experiment using the SUSANNE[4] corpus to separately measure the influence of context and lexicon in our tagging results. Without loss of generality, we have remapped the 426 tags presented in the original SUSANNE tag set into a smaller tag set of 37 POS tags. The 37 POS set represents a fairly standard tag set in tagging literature[5]. This has also been done in order to increase the number of occurrences of each tag (some of the original tags occurred only once in corpus). For instance distinct unambiguous tags, such as tags MCn and MCr, denoting an Arabic numeral or a Roman numeral have been joined into the same POS class, numeral.

For evaluating the importance of the size of corpus into global precision we have selected several subsets of the corpus. Each subset had a different size (i.e. different percentages from the original corpus). At each of these subsets we randomly split the corpus in the 3 sets previously described: a test and evaluation set with 3 sentences each, and a training set with the rest of the words in that subset of the corpus. As usual we trained the tagger with training and evaluation sets and then used the test set to compute tagger precision. To compute confidence intervals, for the same subset, we repeat this process 10 times with different selections for test, evaluation and training sets (that is, we applied a 10-fold cross-validation method). These results give the line TrigNorm in figure 2.

We also evaluated the importance of the lexicon size in training results we have used the same subsets to compute the line TrigNorm from figure 2. The main difference now was that, for each subset, we have built our lexicon from the whole corpus (of course that in each run we excluded the selected test set, or else we would end with close lexicon results). As a result we have also measured how our tagger behaves if we use a the same context that was used for line TrigNorm and a very good lexicon (that is a lexicon extracted from approximately 130

---

[4] The SUSANNE Corpus is a freely available, English annotated subset of the Brown corpus (ftp://ota.ox.ac.uk/pub/ota/public/susanne). This corpus contains a total of 4200 sentences, or 142524 tagged words and is supplied by the University of Sussex.

[5] Since very few tagging efforts are made public, no "standard tag set" for comparison purposes has yet emerged. Probably the problem is that every human tagger – we included – has her/his particular view of what tags she/he should/ has to use in its work. Despite this we have tried to adapt SUSANNE tags to what is fairly common in literature.

000 words, systematically excluding the test set) for each run. These results are presented under line TrigFull in figure 2.

We also, measured unigram precision for both methods. This is a standard tagging baseline: it measures how far can we go without using any word context. For that we select, for each word w_k, the most probable tag in vector $\overrightarrow{amb_k}$. This information is presented in lines UnigFull and UnigNorm).

The analysis of graphic from figure 2 gives rise to an outstanding conclusion for most of the tagging research community. When using a sufficiently good statistical estimator, trigram contextual information can be extracted from as little as 5 700 words (92.6% ± 0.5 precision). Moreover top performance results (93.6% ± 0.5%) are acquired with only 22803 tagged words (that is a rather conservative engineered estimate since we achieve 93.2% ± 0.5 precision with only 11402 tagged words).

These results point also two other things: closed word taggers are indeed too optimistic (96% precision in a previous measure over the same corpus [10]). But, their improvement over unigram tagger (with 90% in the same closed lexicon experiment) is real and pointed us to the difference between the sizes of tagged corpora needed to estimate the lexical and the contextual probabilities.

These results seem interesting, but we haven't solved the huge hand tagged corpus problem yet: we still need good lexical probabilities. Fortunately good lexical probabilities don't need a huge tagged corpus, at least if we use a method similar to the one described in [11] and outlined in next section.

Finally let us end this section with a word about precision results: besides the obvious a priori tag set reduction, we didn't tune tagging precision at all (these were first run results). Further work on English language is needed to improve results: we are convinced that after some tuning of SUSANNE corpus we should achieve state of art results with a rather standard tag set. For example, by using the tags in a syntactical parser should help reducing incoherence, which the generic tagging goals of SUSANNE did hide. Also we should add code for better lemmatisation of specific text in SUSANNE (namely the treatment of formulas, proper names and titles)

5   Using an External Lexicon

In [11], we have also applied this tagger generator to Portuguese text. In the experience reported, we have described how to use an external lexicon to acquire ambiguity vectors for unknown words. This external lexicon is the POLARIS system lexicon ([6]). This system has a lexical database with more than 100, 000 base word forms, together with morphological inflection rules, as well as word derivation (suffixation and prefixation) rules and irregular word formation rules. By using the POLARIS morphological engine and morphological rules POLARIS is capable of recognizing more than 10^6 words. Unfortunately this lexicon is a dictionary-extracted lexicon. So, it only contains the list of possible parts-of-speech for each word. No tag probabilities are supplied.

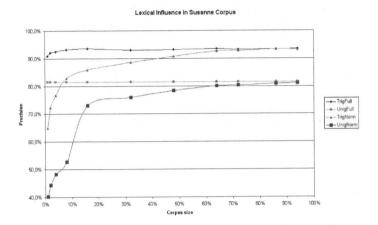

Fig. 2. Learning speed with a good dictionary and different sizes of corpora. Each dot represents a subset of the corpus. Unigram lines represent results without any context and trigram lines represent results with context. The -norm suffix indicates that the internal tagging lexicon was extracted only from that subset of corpus. -full suffix indicates that the full corpus was used for building the internal tagging lexicon.

In [11], we have presented a set of rules that can convert this ambiguity tag class into a lexical probability vector. In order to do so, we first check the word in the internal tagging lexicon. If the word doesn't have any related probability vector in the internal tagging lexicon then we use the POLARIS lexicon. Since Polaris only returns an ambiguity class, we have to use the training corpus to build the ambiguity class probability vector. Polaris lexicon is used to convert each word to its ambiguity class. Then we estimate a dictionary of ambiguity classes by the same process we used to build the internal tagging lexicon.

We have tested the use of this rules on two distinct Portuguese corpora: The Lusa corpus with internal news bulletins from the Portuguese news agency and the PGR corpus related to case-law texts. The Lusa corpus has 5 400 hand tagged words (with $2.748 \pm 2 \times 10^{-3}$ tags per word) and the PGR corpus has 18675 hand corrected tagged words (with 2.316 ± 10^{-3} tags per word). We counted how many times the external lexicon was used in the test corpus, when we use the training corpus to build the internal tagging dictionary. In the Lusa corpus 83.4% of words were present in the internal tagging dictionary and 12.7% only occur in the POLARIS lexicon, leaving 3.9% words were unknown. In PGR corpus these numbers were changed due to the higher number of tagged words: 92.6% of words were present in the internal tagging dictionary, 3.0% only occur in the POLARIS lexicon and 4.4% of words were unknown. This clearly shows that the importance of using an external dictionary rises with the decrease in the size of the training corpus.

We also evaluated global precision values for our two corpora accordingly with the type of lexicon used. For LUSA corpus, we had a precision of $95.5 \pm 0.2\%$ for internal tagging lexicon, $77 \pm 2\%$ for the words using the POLARIS lexicon and $61 \pm 3\%$ of precision for unknown words. The global average precision for LUSA was $91.9 \pm 0.3\%$. In the PGR corpus we had a precision of $97.30 \pm 0.09\%$ for internal tagging lexicon, $91.1 \pm 0.7\%$ for the words using the POLARIS lexicon and $56 \pm 2\%$ of precision for unknown words. The global average precision for PGR was $96.3 \pm 0.2\%$.

Once more, these results pointed the importance of using a large lexicon. The known word precision in corpus is good (95.5% for Lusa and 97.3% for PGR[6]) but those numbers are worst when the word is unknown in the training corpus. However, when we use contextual information, our external dictionary does indeed provide some help.

6   Conclusions

In this paper we have evaluated the main factors that determine the amount of corpora needed to learn tagging parameters. Since most of available tagging systems give different results and have distinct advantages each (that is probably why boosting techniques [5] give good results), we decided to concentrate ourselves on using just one tagging model. This allowed us to reach a rather generic conclusion: poor lexicons are the major cause for tagging errors. And as we show in [11] even when there is no dictionary, it's a lot easier to build a tagging dictionary than to repeatedly tag the same words in a corpus. The lexical dependency shown also points to the need of using morphological information. This is being addressed and will probably be included as additional input information to the neural network.

Finally let us stress the importance of being able to train a tagger with such a small size of tagged corpora. Even in a studied language such as English, when we start treating new types and genres of text, not only it is frequently required a new tagger, but also it is frequently needed a completely different tag set. Indeed, the tag sets used for tagging are useful for comparing works, and do have some interest for ulterior parsing: the LUSA tag set was used for parsing and for subcategorization extraction (see, for example [7]). Yet tag tuning was needed: in both these tasks we needed to refine the tags used in order to best fit our different goals. By using a 43 million automatically POS-tagged LUSA corpus, we could cross validate our tagging precision. Tagging incoherence was highlighted when we looked at the reasons why a given verb was classified as subcategorizing a given phrase type. And that could hardly be seen by human correctors. Even for a limited number of tasks, where we used a rather standard tag set, we needed to adjust the tags to our specific needs. For other problems completely different tag sets are surely needed. For instance, more semantically inspired tag sets are often more suited for some tasks. To conclude we will

---

[6] Since Lusa corpus is a noisiest one, as it should be expected, precision degradation in Lusa is higher than in PGR.

say that highly learning-efficient taggers as the one described, will enable the achievement of good tagging performances for textual data analysis tasks, for no matter the language, using less man power for repetitive and boring tasks.

References

1. Eric Brill. Unsupervised learning of disambiguation rules for part of speech tagging. In *Proceedings of the Very Large Corpora Workshop*, 1995.
2. H. Baayen and Richard Sproat. Estimating lexical priors for low-frequency morphologically ambiguous forms. *Computational Linguistics*, 22(2):155–166, 1996.
3. Doug Cutting, Julian Kupiec, Jan Pedersen, and Penelope Sibun. A practical part-of-speech tagger. In *Proceedings of the third ACL Conference on Applied Natural Language Processing*, pages 133 – 140, Trento, Italy, 1992.
4. Simon Haykin. *Neural Networks: A comprehensive Foundation*. Macmillan College Publishing Company, Inc., 1994.
5. V. Hoste and W. Daelemans. Comparing bagging and boosting for natural language processing tasks: a typically approach. In Bernard Lang, editor, *BENELEARN 2000: proceedings of the Tenth Belgian-Dutch Conference on Machine Learning*, pages 101–109, Tilburg University, 2000, 2000.
6. José Gabriel Lopes, Nuno Cavalheiro Marques, and Vitor Ramos Rocio. Polaris, a pOrtuguese Lexicon Acquisition and Retrieval Interactive System. In *Proceedings of the conference on Pratical Applications of PROLOG*, 1994.
7. Nuno Cavalheiro Marques. *Uma Metodologia Estatística para a Modelação da Sub-categorização Verbal*. PhD thesis, Faculdade de Ciências e Tecnologia da Universidade Nova de Lisboa, 2000.
8. Bernard Merialdo. Tagging english text with a probabilistic model. *Computacional Linguistics*, 20(2):155–171, 1994.
9. Nuno C. Marques and José Gabriel Lopes. Using neural networks for portuguese part-of-speech tagging. In *Proceedings of the Fifth International Conference on Cognitive Science and Natural Language Processing*, Dublin City University, Ireland, September 2-5 1996.
10. Nuno Cavalheiro Marques and José Gabriel Lopes. Neural networks, part-of-speech tagging and lexicon. Technical report, Departamento de Informática, Faculdade de Ciências e Tecnologia da Universidade Nova de Lisboa, Febuary 1997.
11. Nuno C. Marques and José Gabriel Lopes. A POS-Tagger Generator for Unknown Languages. In *Proceedings of the XVII Congreso de la SEPLN*, Jaén – Spain, to appear, September 2001.
12. Adwait Ratnaparkhi. *Maximum Entropy Models for Natural Language Ambiguity Resolution*. PhD thesis, University of Pennsylvania, 1998.
13. Helmut Schmid. Part-of-speech tagging with neural networks. In *Proceedings of the International Conference on Computational Linguistics*, Kyoto, Japan, 1994.
14. Christer Samuelsson and Atro Voutilainen. Tagging french - comparing a statistical and a constraint-based method. In *Proceedings of the European Chapter of the Annual Meeting of ACL*, 1997.
15. University of Stuttgart – Institute for Parallel and Distributed High Performance Systems (IPVR). *User Manual of the Stuttgart Neural Network Simulator*, 1994. Report No. 3//94.

A Search Engine
for Morphologically Complex Languages

Udo Hahn[1], Martin Honeck[2], and Stefan Schulz[2]

[1] Linguistische Informatik / Computerlinguistik, Universität Freiburg,
Werthmannplatz 1, D-79085 Freiburg
[2] Abteilung Medizinische Informatik, Universitätsklinikum Freiburg,
Stefan-Meier-Str. 26, D-79104 Freiburg

Abstract. Document retrieval on natural languages with a rich morphology — particularly in terms of derivation and (single-word) composition — suffers from serious performance degradation with the direct query-term-to-text-word matching paradigm that underlies the vast majority of current search engines. We propose an alternative approach in which morphologically complex word forms, which appear in the query as well as in the documents, are segmented into relevant subwords (such as stems, named entities, acronyms) and are subsequently submitted to the matching procedure. We evaluate our approach with the *AltaVista*[TM] *Search Engine* on a large medical document collection.

1 Introduction

Morphological variants of a search term occur in free-text documents as concatenations of the term's basic lexical form with additional substrings, so-called affixes. From an information retrieval (IR) perspective, such alterations of the search term's surface form have a negative impact on the recall performance of an IR system [2,6,7], since they prevent a direct match between the search term proper and those text tokens which are morphological variants of that search term. Hence, various attempts have been made to account for such phenomena in terms of morphological analysis.

Basically, three types of morphological processes have to be taken into account, *viz.* inflection, derivation and composition. *Inflection* is concerned with adding, e.g., number, gender and case information to nouns (e.g., *'document⊕s'*, *'document⊕'s'*),[1] or number, person, and tense information to verbs (e.g., *'search⊕es'*, *'search⊕ed'*, *'search⊕ing'*). These modifications are typically motivated by syntactic considerations. In particular, they do not change the part-of-speech and semantic characteristics of the basic word form. *Derivation* adds lexico-grammatical morphemes such that the part-of-speech characteristics of the basic word form are changed (e.g., *'search⊕er'* (verb to noun), *'search⊕able'* (verb to adjective), and minor changes of the semantic interpretation of the derived form relative to the basic one occur (e.g., *'X⊕er'* denotes 'someone who

---

[1] '⊕' denotes the string concatenation operator.

F. Hoffmann et al. (Eds.): IDA 2001, LNCS 2189, pp. 73–83, 2001.
© Springer-Verlag Berlin Heidelberg 2001

$X \oplus es$'). *Composition*, finally, is conceived as combining several basic lexico-semantic units to form a composite one. In English, e.g., nominal compounding surfaces in terms of a complex noun phrase (e.g., *'information theory seminar'*), while in German the same process constitutes a rather complex single-word form typically glued together by *'s'*, one of the dedicated grammatical infixes to form complex nominal compounds (*viz.* *'Information* $\oplus s \oplus theorie \oplus seminar$ *'*).

Morphological analysis is concerned with the reverse processing of inflection, derivation and composition in terms of deflection (or IR-specific stemming), dederivation or decomposition, respectively. The general goal is to map all occurring morphological variants to some canonical base form(s) — e.g., *'search'* in the examples from above. The efforts required for performing morphological analysis vary from language to language, and also depend on the breadth and generality of the chosen approach (i.e., whether all types of morphological variations are to be accounted for or not). The English language, e.g., is known for the limited number of inflection patterns, while those of, say, German, French or Russian are much more diverse.[2] Therefore, even simple forms of English general-purpose stemming algorithms available for IR applications [9,13] have no counterparts in these morphologically richer languages. When it comes to a broader scope of morphological analysis, including derivation and composition phenomena, even for the English language only restricted, domain-specific algorithms exist. This is particularly true for the medical domain which we consider in our experiments, too. From an IR view, a lot of specialized research has already been carried out for medical applications, with emphasis on the lexicosemantic aspects of dederivation and decomposition [11,10,17,16,3,1].

While one may argue that single-word compounds are quite rare in English (which is not the case in the medical domain either), this is certainly not true for the German language and related ones known for excessive single-word nominal compounding. This problem becomes even more pressing for technical sublanguages, as for medical German. The problem one faces from an IR point of view is that besides fairly standardized noun compounds, which already form a regular part of the sublanguage proper, a myriad of *ad hoc* compounds are formed on the fly which cannot be anticipated when formulating a retrieval query though they appear in relevant documents. Hence, morphological analysis is mandatory for optimal retrieval results. Furthermore, medical terminology on which we concentrate here is characterized by a typical mix of Latin and Greek roots with the corresponding host language (e.g., German), often referred to as *neo-classical compounding*. While this is not even a side issue for general-purpose morphological analyzers (it is simply irrelevant), dealing with such phenomena is crucial

---

[2] Evidence for this statement and the implications this has on text retrieval performance come form a large variety of highly inflectional and/or agglutinating languages such as Hebrew [2], Finnish [6], or Slovene [12]. Taking the perspective of *different* languages makes perfect sense here, since in medical practice routinely generated narratives (finding and admission reports, discharge summaries, etc.), no matter whether they appear in a clinical environment or at the practitioner's work place, are usually written in the medical expert's native language.

for any attempt to cope adequately with medical free-texts in an IR setting (cf. also [17]).

We here propose an approach to document retrieval which is based on the idea of segmenting query and document terms into basic subword units. Hence, this approach combines procedures for deflection, dederivation and decomposition. Subwords cannot be equated with linguistically significant morphemes, in general, since their granularity is coarser than that of morphemes (cf. the discussion in Section 2). We validate our claims in Section 4 on a substantial document collection from the medical domain (introduced in Section 3), using the *AltaVista*^{TM}*Search Engine* as a widely available routine IR testbed.

2   Morphological Analysis for Medical Information Retrieval

Morphological analysis for IR has requirements which differ from those for natural language processing (NLP) proper. Accordingly, the decomposition units vary, too. Within a canonical NLP framework, linguistically significant *morphemes* are chosen as nondecomposable entities and defined as the smallest content-bearing (*stem*) or grammatically relevant units (*affixes* such as prefixes, infixes and suffixes). As an IR alternative, we here propose *subwords* (and grammatical affixes) as the smallest units of morphological analysis. Subwords differ from morphemes only, if the meaning of a combination of linguistically significant morphemes is (almost) equal to that of another nondecomposable medical synonym. In this way, subwords preserve a sublanguage-specific composite meaning that would get lost if they were split up into their constituent morpheme parts. Hence, we trade linguistic atomicity against medical plausibility considerations and claim here that the latter are beneficial for boosting the system's retrieval performance. As an example, a medically justified minimal segmentation of *'diaphysis'* into *'diaphys⊕is'* will be preferred over a linguistically motivated one (*'dia⊕phys⊕is'*), because the first can be mapped to the quasi-synonym stem *'shaft'*. Such a mapping would not be possible with the overly unspecific morphemes *'dia'* and *'phys'*, which occur in numerous other contexts as well (e.g. *'dia⊕gnos⊕is'*, *'phys⊕io⊕logy'*). Hence, a decrease of the precision of the retrieval system would be highly likely due to over-stemming.

Accordingly, we distinguish the following decomposition classes and briefly mention the basic regularities underlying their formation in text types:

- *Subwords* like { *'gastr'*, *'hepat'*, *'nier'*, *'leuk'*, *'diaphys'*, ...} are the primary content carriers in a word. They can be prefixed, linked by infixes, and suffixed. As a particularity of the German medical language, proper names may appear as part of complex nouns (e.g., *'Parkinson⊕verdacht'* [*'suspicion of Parkinson's disease'*]) and are therefore included in this category.
- *Short words*, with four characters or less, like { *'ion'*, *'gene'*, *'ovum'*}, are classified separately applying stricter grammatical rules (e.g. they cannot be composed at all). Their stems (e.g. *'gen'* or *'ov'*) are *not* included in the

dictionary in order to prevent artificial ambiguities. The price one has to pay for this is the inclusion of derived and composed forms in the subword dictionary (e.g. *'anion'*, *'genet'*, *'ovul'*).

- *Acronyms* such as {*'AIDS'*, *'ECG'*, ...} and *abbreviations* (e.g., *'chron.'* [for *'chronical'*], *'diabet.'* [for *'diabetical'*]) are nondecomposable entities in morphological terms and do not undergo any further morphological variation, e.g., by suffixing.
- *Prefixes* like {*'a-'*, *'de-'*, *'in-'*, *'ent-'*, *'ver-'*, *'anti-'*, ...} precede a subword.
- *Infixes* (e.g., *'-o-'* in "*gastr⊕o⊕intestinal*", or *'-s-'* in *'Sektion⊕s⊕bericht'* [*'autopsy report'*]) are used as a (phonologically motivated) 'glue' between morphemes, typically as a link between subwords.
- *Derivational suffixes* such as {*'-io-'*, *'-ion-'*, *'-ie-'*, *'-ung-'*, *'-itis-'*, *'-tomie-'*, ...} usually follow a subword.
- *Inflectional suffixes* like {*'-e'*, *'-en'*, *'-s'*, *'-idis'*, *'-ae'*, *'-oris'*, ...} appear at the very end of a composite word form following the subwords or derivational suffixes.

The morphological segmentation procedure for German currently (in June 2001) incorporates a *subword dictionary* which is composed of 4,630 (short) subwords, 344 proper names and an *affix list* composed of 139 prefixes, 8 infixes and 154 (derivational as well as inflectional) suffixes, making up 5,275 entries in total. As a further enhancement, we enriched the subword dictionary with a simple semantic relation. The EQ relations links subwords which stand in a semantic *equivalence* relation to each other. This extension is particularly directed at foreign-language (mostly Greek or Latin) translates of source language terms, e.g., German *'nier'* EQ Latin *'ren'* (EQ English *'kidney'*).

The morphological segmentation engine builds *all* possible morphological segmentations for an input word using the above-mentioned resources and concatenation regularities. If ambiguous morphological segmentations of an input word are encountered,[3] they are ranked according to the following ordering of preference criteria:

1. longest match from the left,
2. minimal number of stems per word,
3. minimal number of consecutive affixes (this criterion penalizes utterly formal segmentations), and
4. relative weight — more specifically, a semantic weight factor
   $w = 2$ is assigned to all subwords and some semantically important suffixes, such as *'-tomie'* [*'-tomy'*] or *'-itis'* (cf. Pacak et al. [11]);
   $w = 1$ is assigned to prefixes and derivational suffixes; and
   $w = 0$ holds for inflectional suffixes and infixes.

The MEDSEARCH system implements the subword segmentation model outlined above within the programming environment of Visual Basic^{TM} 6.0. We

---

[3]   Roughly on the order of 5 to 10 readings can be expected for typical medical compounds, depending on the saturation of the subword dictionary.

also provide a graphical interface for the interactive maintenance of the subword dictionary. Besides the graphical interface, the entire system can be run as a COM (Microsoft Component Object Model) server application, thus exporting its methods within a network.

We started our work with an empty subword dictionary and a fully specified affix list. In each training cycle, complex text tokens were imported from various medical corpora, while segmentation was performed on the basis of the subwords and affixes already specified. The weighted segmentation results were ordered according to the above mentioned criteria. It was left to the system manager, a medical expert, to include additional subwords into the dictionary — mainly to avoid over-stemming. As a heuristic guideline to identify possible segmentation errors in the wealth of data we used the following identification criteria: no segmentation at all, word segments contain only three characters or even less (the main source of errors), word segment size exceeds ten characters.

Figure 1 gives a flavor of the user interface and illustrates the segmentation of the nominal compound 'Postgastrektomiesymptomatik'. Note that lots of ambiguous analyses are displayed, since 'gastr' at this update stage has not yet been specified as a valid subword. Upper case characters mark segments with a weight factor of $w = 2$. Each relevant subword that is found to be missing is manually inserted in the subword dictionary such as it is shown in Figure 1. By adding new entries the segmentation capability of our tool continuously improved. Prior to storing a new subword, attributes concerning its subword class and language (viz. German, Latin, Greek) have to be specified.

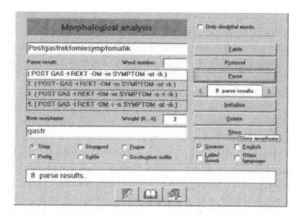

Fig. 1. MEDSEARCH User Interface

3 The Retrieval Environment

The document collection underlying our experiments consists of a bilingual source, viz. the "MSD - Manual der Diagnostik und Therapie", a close though not fully parallel translation of *"The Merck Manual of Diagnosis and Therapy"*. We chose a bilingual corpus, since we intend to extend our experiments from the mono- to various multilingual cases. The German edition is a standard textbook for students and practitioners in the field of clinical medicine. It contains 5,517 handbook-style articles (about 2.4 million text tokens) on a broad range of clinical medical knowledge, which includes general introductions to each topic, detailed information about diagnosis and therapy, as well as epidemiological data.

We envisage the routine application of our approach in a nonexperimental, highly standardized system framework. Hence, we chose the *AltaVista*^{TM}*Search Engine 3.0* as our testbed (`http://solutions.altavista.com/downloads/downloads.html`), a widely distributed, easy to install off-the-shelf IR system. The system manual is not fully conclusive about the details of index term processing but the following criteria are mentioned. *"The relevancy of a document is determined by the frequency of words, the position of words in documents, whether the words appear in the document title, whether the complete phrase exists for multi-word query, and the proximity of words to each other within documents."* [p. 6, Software Product Description 052000].[4] All text types from the document collection are assembled in an inverted term *index* accessible for retrieval. Given these criteria, the search engine produces a ranked output of documents (in our experiments, we chose a cut-off value for the top 200 documents retrieved).

4   Retrieval Experiments

In our retrieval experiments we tried to cover a wide range of topics from clinical medicine. We therefore decided to use the collection of multiple choice questions from the nationally standardized year 5 examination questionnaire for medical students in Germany as the basis of our queries. Questions from general medicine, psychosomatic medicine and psychiatry, as well as public health, were excluded, because these fields use standard medical terminology only to a limited degree. From a total of 580 questions, 210 questions remained for further investigation.

We then asked 63 students (between the 3rd and 5th study year) from our university's Medical School during regular classroom hours to formulate free-form natural language queries in order to retrieve documents that would help to answer these questions, assuming an ideal search engine. Acronyms and abbreviations were allowed, but the length of each query was restricted to a maximum of ten terms. Each student was assigned ten topics at random, so we ended up with 630 queries, from which 25 were randomly chosen for further consideration (the set contained no duplicate queries).

The relevance judgments were done by three medical experts (none of them was involved in the system development), identifying relevant documents in the whole test collection for each of the 25 queries prior to the conduct of morphological analysis. We ran the following experiments:

---

[4]   Additional term processing tools (spelling correction, phrase recognition, thesaurus, stemming, etc.) were disabled except for Test 2 (cf. Section 4).

- **Test 1: Token Search.** In this scenario, a direct match between text tokens in the documents and those in the query is attempted, i.e., no term processing is done before indexing or before submitting the query. The search was run on the index covering the entire MSD document collection (182,306 index terms). This scenario serves as the baseline for determining the benefits of our approach.
- **Test 2: Token Search with Stemming.** In the second scenario, text tokens in the documents and in the queries were submitted to the operation of the (language-specific) stemmer included as an add-on feature in the *AltaVista^{TM} Search Engine*.
- **Test 3: Morphological Segmentation.** The third scenario consists of the subword-based retrieval approach described in Section 2. Morphological segmentation yielded a shrunk index, with 39,315 index terms remaining. This yields a reduction rate of 78% compared with the number of text types in MSD.[5]
- **Test 4: Morphological Segmentation and Synonym Expansion.** In the fourth scenario, we augmented the plain subword model by introducing the EQ semantic relation between suitable subwords and exploited these relations for subsequent retrieval. In the documents, as well as in the queries, each known word form was substituted by an alphabetic code (here referred to as *pseudo word*) identifying the thesaurus class.

The assessment of the experimental results is based on the aggregation of all 25 selected queries. In particular, we calculated the average interpolated precision values at fixed recall levels (we chose a continuous increment of 10%) based on the consideration of the top 200 documents retrieved by the *AltaVista^{TM} Search Engine*. The corresponding P/R values for all four test scenarios are summarized in Figure 2 and visualized in Figure 3.[6]

For our baseline, *Test 1*, the direct match between query terms and document terms, precision is already poor at low recall points ($R \leq 30$), ranging in an interval from 54% to 35%. At high recall points ($R \geq 70$), precision drops from 18% to 5%.

Adding the (German) stemming procedure of the *AltaVista^{TM} Search Engine* in *Test 2* (surprisingly) increases noise in the system, since precision values drop by a factor 10% for real low recall values, while for high ones the precision curve almost overlaps with that for searches without stemming, showing no significant improvement.

---

[5]   The data for the English version, 50,934 text types with 24,539 index entries remaining after segmentation, indicates a significantly lower reduction rate of 52%. The size of the English subword dictionary (only 300 entries less than the German one, which contains 4,630 subwords) does not explain the data. Rather this finding demonstrates the obvious tendency in English to have fewer single-word compounds than German.

[6]   We are unable to quantify the degree of inter-rater reliability, since the process of relevance assignments was driven by qualitative interaction between the raters in order to arrive at some reasonable consensus.

	Test 1	Test 2	Test 3	Test 4
Recall(%)	Precision(%) over 25 queries of top r=200 retr. documents			
0	53.6	43.3	69.4	66.9
10	51.7	42.1	65.5	60.5
20	45.4	37.6	61.4	54.9
30	34.9	33.3	55.4	51.6
40	29.5	30.5	51.4	46.7
50	27.8	29.7	49.7	44.1
60	26.2	27.1	40.7	39.2
70	18.1	19.7	32.6	31.7
80	15.2	17.4	26.3	22.4
90	5.6	5.4	20.1	11.4
100	5.4	5.3	16.3	11.0
3pt avrg	29.5	28.2	45.8	40.5
11pt avrg	28.5	26.5	44.4	40.0

Fig. 2. Evaluation Results — Precision/Recall Table

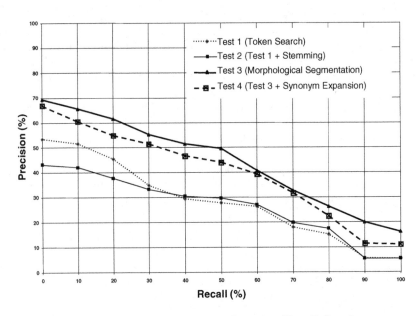

Fig. 3. Evaluation Results — Precision/Recall Graph

The subword approach in *Test 3* clearly outperforms the results achieved for *Test 1* and *Test 2*. For low recall values the gain in precision ranges from 14% to 21%, while for high recall values the gain is still in the range of 11% to 15%.

Adding equivalent terms slightly decreases the performance of MEDSEARCH, roughly on the order of 5%. This indicates that truly equivalent terms are hard to determine, even in the medical domain. Since the addition of equivalent terms produced no advantage over simple segmentation into subwords, we cannot recommend their inclusion into the search process on the basis of our data.

In order to estimate the statistical significance of this result, we compared relevant test pairs for each fixed recall level, using the two-tailed sign test (for a description and its applicability for the interpretation of P/R graphs, cf. [14]), and obtained the following results, the details of which are summarized in Figure 4:

Recall Level	Test2 vs. Test1	Test3 vs. Test1	Test4 vs. Test1	Test2 vs. Test3	Test4 vs. Test3
0%	< 0.05	n.s.	n.s..	< 0.005	n.s.
10%	n.s.	< 0.05	n.s.	< 0.005	n.s.
20%	< 0.05	< 0.05	n.s.	< 0.05	n.s.
30%	n.s.	< 0.05	n.s.	< 0.05	n.s.
40%	n.s.	< 0.05	< 0.05	< 0.05	n.s.
50%	n.s.	< 0.005	< 0.05	n.s.	n.s.
60%	n.s.	n.s.	n.s.	n.s.	n.s.
70%	n.s.	n.s.	n.s.	< 0.05	n.s.
80%	n.s.	n.s.	n.s.	n.s.	n.s.
90%	n.s.	< 0.005	< 0.05	< 0.005	n.s.
100%	n.s.	< 0.005	< 0.05	< 0.005	n.s.

Fig. 4. Significance Judgments for Relevant Test Pairs on Fixed Recall Levels

- *Test 2* vs. *Test 1* — Token search with *AltaVista*^{TM} stemming performs significantly poorer than token search without at recall points 0% and 20% with a significance level of .05 (otherwise, no significance).
- *Test 3* vs. *Test 1* — Morphological segmentation outperforms simple token search with a significance level of .05 at recall points 10% to 40%, while at the recall points 50%, 90% and 100% a .005 significance level is determined. No significant difference was found at recall points 0% and 60% to 80%.
- *Test 3* vs. *Test 2* — Morphological segmentation outperforms token search with *AltaVista*^{TM} stemming with a significance level of .005 at the recall points 0%, 10%, 90% and 100%, and with a significance level of .05 at the recall points 30%, 40% and 70%. No significant difference was found at 50%, 60% and 80%.
- *Test 4* vs. *Test 1* — Morphological segmentation with thesaurus expansion (including the equivalence relation) improves over simple token search with a significance level of .05 at the recall points 40%, 50%, 90% and 100% (otherwise, no significance).
- *Test 4* vs. *Test 3* — No statistically significant difference at all was found for the comparison between morphological segmentation with and without thesaurus expansion.

Generalizing the interpretation of our data in the light of these findings, we recognize a substantial increase of retrieval performance when query and text tokens are segmented according to the principles of the subword model. The benefit we achieve is not dependent on whether we aim at maximizing precision or recall.

5 Conclusions

There has been some controversy, at least for simple stemmers [9,13], about the effectiveness of morphological analysis for document retrieval [4,5,8]. The key issue for quality improvement seems to be rooted mainly in the presence or absence of some form of dictionary, i.e., a list of content items in some agreed-upon basic lexical format plus, possibly, additional linguistic information concerning parts of speech, gender, number, tense, semantic relations, etc. Empirical evidence has been brought forward that inflectional and/or derivational stemmers augmented by dictionaries indeed perform substantially better than those without access to such lexical repositories [8,7,15].

This result is particularly valid for natural languages, such as German, Dutch, Russian, Finnish, Hebrew, with a rich morphology — both in terms of derivation and (single-word) composition. Document retrieval on these languages suffers from serious performance degradation with the direct query-term-to-text-word matching paradigm that underlies most current search engines. Therefore, we proposed a dictionary-based approach in which morphologically complex word forms, which appear in the query as well as in the documents, are segmented into relevant subwords and are subsequently submitted to the matching procedure. This way, the impact of word form alterations can be eliminated from the retrieval procedure. We evaluated our hypothesis with the $AltaVista^{TM}Search$ $Engine$ on a large medical document collection. Our experiments lent (mostly) statistically significant support to the subword hypothesis.

References

1. R. H. Baud, C. Lovis, A.-M. Rassinoux, and J.-R. Scherrer. Morpho-semantic parsing of medical expressions. In *AMIA'98 – Proceedings of the 1998 AMIA Annual Fall Symposium*, pages 760–764. Orlando, FL, November 7-11, 1998.

2. Y. Choueka. RESPONSA: An operational full-text retrieval system with linguistic components for large corpora. In A. Zampolli, editor, *Computational Lexicology and Lexicography: A Volume in Honor of B. Quemada*. Pisa: Giardini Press, 1992.

3. P. Dujols, P. Aubas, C. Baylon, and F. Grémy. Morphosemantic analysis and translation of medical compound terms. *Methods of Information in Medicine*, 30(1):30–35, 1991.

4. D. Harman. How effective is suffixing? *Journal of the American Society for Information Science*, 42(1):7–15, 1991.

5. D. A. Hull. Stemming algorithms: A case study for detailed evaluation. *Journal of the American Society for Information Science*, 47(1):70–84, 1996.

6. H. Jäppinen and J. Niemistö. Inflections and compounds: Some linguistic problems for automatic indexing. In *RIAO 88 – Proceedings of the RIAO 88 Conference,* volume 1, pages 333–342. Cambridge, MA, March 21-24, 1988.

7. W. Kraaij and R. Pohlmann. Viewing stemming as recall enhancement. In *SIGIR'96 – Proceedings of the 19th Annual International ACM SIGIR Conference on Research and Development in Information Retrieval,* pages 40–48. Zurich, Switzerland, August 18-22, 1996.

8. R. Krovetz. Viewing morphology as an inference process. In *SIGIR'93 – Proceedings of the 16th Annual International ACM SIGIR Conference on Research and Development in Information Retrieval,* pages 191–203. Pittsburgh, PA, USA, June 27 - July 1, 1993.

9. J. B. Lovins. Development of a stemming algorithm. *Mechanical Translation and Computational Linguistics,* 11(1/2):22–31, 1968.

10. L. M. Norton and M. G. Pacak. Morphosemantic analysis of compound word forms denoting surgical procedures. *Methods of Information in Medicine,* 22(1):29–36, 1983.

11. M. G. Pacak, L. M. Norton, and G. S. Dunham. Morphosemantic analysis of *-itis* forms in medical language. *Methods of Information in Medicine,* 19(2):99–105, 1980.

12. M. Popovic and P. Willett. The effectiveness of stemming for natural language access to Slovene textual data. *Journal of the American Society for Information Science,* 43(5):384–390, 1992.

13. M. F. Porter. An algorithm for suffix stripping. *Program,* 14(3):130–137, 1980.

14. C. J. van Rijsbergen. *Information Retrieval.* London: Butterworths, 2nd edition, 1979.

15. E. Tzoukermann, J. L. Klavans, and C. Jacquemin. Effective use of natural language processing techniques for automatic conflation of multi-word terms: The role of derivational morphology, part of speech tagging, and shallow parsing. In *SIGIR'97 – Proceedings of the 20th Annual International ACM SIGIR Conference on Research and Development in Information Retrieval,* pages 148–155. Philadelphia, PA, USA, July 27-31, 1997.

16. F. Wingert. Morphologic analysis of compound words. *Methods of Information in Medicine,* 24(3):155–162, 1985.

17. S. Wolff. The use of morphosemantic regularities in the medical vocabulary for automatic lexical coding. *Methods of Information in Medicine,* 23(4):195–203, 1984.

Errors Detection and Correction
in Large Scale Data Collecting

Renato Bruni and Antonio Sassano

Dipartimento di Informatica e Sistemistica,
Università di Roma "La Sapienza", Via Buonarroti 12 - 00185 Roma, Italy,
{bruni,sassano}@dis.uniroma1.it

Abstract. The paper is concerned with the problem of automatic detection and correction of inconsistent or out of range data in a general
process of statistical data collecting. Under such circumstances, errors
are usually detected by formulating a set of rules which the data records
must respect in order to be declared correct. As a first relevant point, the
set of rules itself is checked for inconsistency or redundancy, by encoding
it into a propositional logic formula, and solving a sequence of Satisfiability problems. This set of rules is then used to detect erroneous data.
In the subsequent phase of error correction, the above set of rules must
be satisfied, but the erroneous records should be altered as little as possible, and frequency distributions of correct data should be preserved. As
a second relevant point, error correction is modeled by encoding the rules
with linear inequalities, and solving a sequence of set covering problems.
The proposed procedure is tested on a real-world case of Census.

1 Introduction

When dealing with a large amount of collected information, a relevant problem
arises: perform the requested elaboration considering only correct data. Examples of data collecting are cases of statistical investigations, marketing analysis,
experimental measures, etc. Our attention will be focused on the problem of
statistic projections carried out by processing answers to questionnaires. We
will consider, in particular, the case of a census of population. A data record
is a set of values v_i for a set of fields f_i. In our case, a record is the set of the
answers given to one questionnaire $Q = \{f_1 = v_1, f_2 = v_2, \ldots, f_p = v_p\}$.

Examples of fields f_i are `age` or `marital status`, corresponding examples
of values v_i are `18` or `single`. Fields can be distinguished in quantitative and
qualitative ones. A quantitative field is a field on whose values are applied (at
least some) mathematical operators (e.g. $>$, $+$), hence such operators should be
defined. Examples of quantitative field are numbers (real or integer), or even
the elements of an ordered set. A qualitative field simply requires its value to
be member of a discrete set with finite number of elements. Errors, or, more
precisely, inconsistencies between answers or out of range answers, can be due
to the original compilation of the questionnaire, or introduced during any later
phase of information processing. The problem of *error detection* is generally

F. Hoffmann et al. (Eds.): IDA 2001, LNCS 2189, pp. 84–94, 2001.
© Springer-Verlag Berlin Heidelberg 2001

approached by formulating a set of rules that the records must respect in order to be *correct*. Rules are generally written in form of *edits*. An edit expresses the error condition, as a conjunction of expressions ($f_i < relation > v_{f_i}$).

Example 1.1. An inconsistent answer can be to declare

marital status as married and age as 10 years old.

The rule to detect this kind of errors could be: if marital status is married, age must be not less than, say, 14. Hence, there is an error if marital status = married and age < 14, and the edit, that is the error condition, is

(marital status = married) ∧ (age < 14)

Questionnaires which verify the condition defined in at least one edit are declared *erroneous*. Obviously, the set of edits must be free from *inconsistency* (i.e. edits must not contradict each other), and, preferably, from *redundancy* (i.e. edits must not be logically implied by other edits). In the case of real questionnaires, edits can be very numerous, since a high number of edits allows a better quality error detection. Many commercial software systems deal with the problem of questionnaires correction, by using a variety of different edits encoding and solution algorithm (e.g. [1,11,12]). In practical case, however, they suffer from severe limitations, due to the inherent computational complexity of the problem. Some methods ignore edit testing, and just divide erroneous questionnaires from correct ones. In such cases, since results are incorrect if edits contains contradictions, the number of edits must be small enough to be validated by inspection by a human operator. Moreover, edits updating turns out to be very difficult. Other methods try to check for contradiction and redundancy by generating all implied edits, such as the 'Fellegi Holt' procedure [6]. Their limitation is that, as the number of edits slightly increases, they produce very poor performance, since the number of implied edits exponentially grows with the number of original edits. The above limitations prevented to now the use of a set of edits whose cardinality is above a certain value. Another serious drawback is that simultaneous processing of quantitative and qualitative fields is seldom allowed.

By encoding the rules in clauses, the above problem of checking the set of rules against inconsistencies and redundancies is here transformed into a *propositional logic* problem (Sect. 2). A sequence of *propositional Satisfiability* problems is therefore solved (Sect. 3). Since generally information collecting has a cost, we would like to utilize erroneous records as well, by performing an *error correction*. During such phase, erroneous records are changed in order to satisfy the above rules. This should be done by keeping as much as possible the correct information contained in the erroneous records (Sect. 4). The above problem is modeled by encoding the rules in linear inequalities, and solving a sequence of *set covering* problems (Sect. 5). The proposed procedure is tested on a real world Census. The application and part of the data were kindly provided by the Italian National Statistic Institute (Istat). Additional low-level details are in [5].

2   A Logical Representation of the Set of Edits

The usefulness of logic or Boolean techniques is proved by many approaches to similar problems of information representation (e.g. [3]). A representation of the set of edit by means of first-order logic is not new, with consequent computational limitations. In this paper we propose an edit encoding by means of the easier-to-solve propositional logic. A propositional logic formula $\mathcal{F}$ in *conjunctive normal form* (CNF) is a conjunction of clauses C_j, each clause being a disjunction of literals, each literal being either a positive (α_i) or a negative ($\neg\alpha_i$) logic variable. By denoting the possible presence of $\neg$ by $[\neg]$, this is

$$\bigwedge_{j=1..m} (\bigvee_{i=1..|C_j|} [\neg]\alpha_i) \tag{1}$$

Given truth values (*True* or *False*) to the logical variables, we have a truth value for the formula. A formula $\mathcal{F}$ is *satisfiable* if and only if there exists a truth assignment that makes $\mathcal{F}$ *True* (i.e. a *model*). If this does not exist, $\mathcal{F}$ is *unsatisfiable*. The problem of testing satisfiability of propositional formulae in conjunctive normal form, named SAT, is well-known to be NP-complete [7], and plays a protagonist role in mathematical logic and computing theory. In the case of questionnaires, every edit can be encoded in a propositional logic clause. Moreover, since edits have a very precise syntax, encoding could be performed by means of the following automatic procedure.

Edit propositional encoding procedure

1. *Identification of the domains D_f for each one of the p fields f, considering that we are dealing with errors.*
2. *Identification of k_f subsets $S_f^1, S_f^2, \ldots, S_f^{k_f}$ in every domain D_f, by using breakpoints, or cut points, b_f^j obtained from the edits, and by merging (possible) equivalent subsets within each domain D_f.*
3. *Definition of n_f logical variables $\alpha_f^1, \alpha_f^2, \ldots, \alpha_f^{n_f}$ to encode the k_f subsets S_f^j of each field f.*
4. *Expression of each edit by means of clauses defined over the introduced logical variables α_f^j.*
5. *Generation of congruency clauses to supply information not present in edits.*

Example 2.1. For the qualitative field `marital status`, answer can vary on a discrete set of possibilities in mutual exclusion, or, due to errors, be missing or not meaningful. Both latter cases are expressed with the value `blank`.

$$D_{\text{marital status}} = \{\text{single}, \text{married}, \text{separate}, \text{divorced}, \text{widow}, \text{blank}\}$$

For the quantitative field `age`, due to errors, the domain is

$$D_{\text{age}} = (-\infty, +\infty) \cup \{\text{blank}\}$$

Values v appearing in the edits are called *breakpoints*, or *cut points*, for the domains. They represent the logical *watershed* between values of the domain, and will be indicated with b_f^j. Such breakpoints are used to split every domain D_f into subsets S_f^j representing values of the domain which are *equivalent* from the edits' point of view. We congruently have $D_f = \bigcup_j S_f^j$.

Example 2.2. For the field `age` we have the following breakpoints

$$b_{\text{age}}^1 = 0, \ b_{\text{age}}^2 = 14, \ b_{\text{age}}^3 = 18, \ b_{\text{age}}^4 = 26, \ b_{\text{age}}^5 = 120, \ b_{\text{age}}^6 = \text{blank}$$

and, by using the breakpoints and the edits to cut D_{age}, we have the subsets

$$S_{\text{age}}^1 = (-\infty, 0), \ S_{\text{age}}^2 = [0, 14), \quad S_{\text{age}}^3 = [14, 18), \quad S_{\text{age}}^4 = \{18\},$$
$$S_{\text{age}}^5 = (18, 26), \ S_{\text{age}}^6 = [26, 120], \ S_{\text{age}}^7 = (120, +\infty), \ S_{\text{age}}^8 = \{\text{blank}\}$$

Subsets $(-\infty, 0), (120, +\infty), \{\text{blank}\}$, representing *out of range* values, are equivalent (can be automatically detected) and collapse in to the same subset S_{age}^1.

$$S_{\text{age}}^1 = (-\infty, 0) \cup (120, +\infty) \cup \{\text{blank}\}, S_{\text{age}}^2 = [0, 14), S_{\text{age}}^3 = [14, 18),$$
$$S_{\text{age}}^4 = \{18\}, S_{\text{age}}^5 = (18, 26), S_{\text{age}}^6 = [26, 120]$$

So far, subsets can be encoded with logic variables in several ways (for instance, k_f subsets can be encoded by $\lceil \log_2 k_f \rceil$ logic variables). We choose to encode the k_f subsets of every domain with $n_f = k_f - 1$ variables, with the aim to produce an easier-to-solve CNF. When the value v of field f belongs to subset S_f^j, this means $\alpha_f^j = True$ and $\alpha_f^h = False$, for $h = 1, \ldots, n_f, h \neq j$. The same holds for the other subsets of f, except for the *out of range* subset (present for every field), which is encoded by putting all variables α_f^h at $False$, for $h = 1, \ldots, n_f$.

Example 2.3. The field `marital status` is divided in 6 subsets, hence we have the $6\text{-}1 = 5$ logical variables $\alpha_{[\text{single}]}, \alpha_{[\text{married}]}, \alpha_{[\text{separate}]}, \alpha_{[\text{divorced}]}, \alpha_{[\text{widow}]}$.

Now, every expressions $(f_i < relation > v_{f_i})$ can be substituted by the corresponding logical variable, obtaining a conjunction of logic variables. Since we are interested in clauses satisfied by correct records, and being edits the error condition, we negate such conjunction, obtaining a disjunction, hence a clause.

Example 2.4. Consider the edit (`marital status = married`) $\land$ (`age < 14`). By substituting the logical variables, we have the logic formula $\alpha_{[\text{married}]} \land \alpha_{[0,14)}$. By negating it, and applying De Morgan's law, we obtain the following clause

$$\neg \alpha_{[\text{married}]} \lor \neg \alpha_{[0,14)}$$

In addition to information given by edits, there is other information that a human operator would consider obvious, but which must be provided. With our choice for variables, we need to express that fields must have one and only one value, and therefore $\binom{n_f}{2}$ (number of combinations of class 2 of n_f objects) clauses, named congruency clauses, are added. Altogether, the set of edits produces a set of m clauses with n logical variables, hence a CNF formula $\mathcal{E}$. The set of answers to a questionnaire Q produces a truth assignment for such logical variables. We say, briefly, that Q must satisfy $\mathcal{E}$ to be correct.

3   Edits Validation

In order to check the set of edits against inconsistency and redundancy, we study the models of $\mathcal{E}$. When every possible set of answers to the questionnaire is declared incorrect, we have the situation called *complete inconsistency* of the set of edits. When the edit inconsistency appears only for particular values of particular fields, we have the (even more insidious) situation of *partial inconsistency*.

Example 3.1. A very simple complete inconsistency, with edits meaning: (a) everybody must have a seaside house, (b) everybody must have a mountain house, (c) it is not allowed to have both seaside and mountain house.

$$\texttt{seaside house = no} \qquad \text{(a)}$$
$$\texttt{mountain house = no} \qquad \text{(b)}$$
$$\texttt{(seaside house = yes)} \wedge \texttt{(mountain house = yes)} \quad \text{(c)}$$

Example 3.2. A very simple partial inconsistency, with edits meaning: (a) one must have a seaside house if and only if annual income is greater then or equal to 1000, (b) one must have a mountain house if and only if annual income is greater then or equal to 2000, (c) it is not allowed to have both seaside and mountain house. For annual income < 2000, this partial inconsistence does not show any effect, but every questionnaires where the subject has an annual income ≥ 2000 is declared erroneous, even if it should not. We have a partial inconsistency with respect to the subset annual income ≥ 2000.

$$\texttt{(annual income} \geq 1000\texttt{)} \wedge \texttt{(seaside house = no)} \quad \text{(a)}$$
$$\texttt{(annual income} \geq 2000\texttt{)} \wedge \texttt{(mountain house = no)} \quad \text{(b)}$$
$$\texttt{(mountain house = yes)} \wedge \texttt{(seaside house = yes)} \quad \text{(c)}$$

In large sets of edits, or after edit updating, inconsistencies may easily occur. Due to the following result, they are detected by solving a series of SAT problems.

Theorem 3.1. *By encoding the set of edits in a CNF formula $\mathcal{E}$, complete inconsistency occurs if and only if $\mathcal{E}$ is unsatisfiable. A partial inconsistency with respect to a subset S_f^j occurs if and only if the formula obtained from $\mathcal{E}$ by fixing $\alpha_{S_f^j} = True$ is unsatisfiable.*

Moreover, in the case of inconsistency, we are interested in restoring consistency. The approach of deleting edits corresponding to clauses that we could not satisfy is not useful. In fact, every edit has its function, and cannot be deleted, but only modified by the human expert who writes the edits. On the contrary, the selection of the set of conflicting edits can guide the human expert in modifying them. This corresponds to selecting which part of the unsatisfiable CNF causes the unsolvability, i.e. a minimal unsatisfiable subformula (MUS). Therefore, we used a SAT solver which, in the case of unsatisfiable instances, is able to select a MUS or at least an unsatisfiable subformula approximating a MUS [4].

Some edits could be logically implied by others, being therefore redundant. It would be preferable to remove them, because decreasing the number of edits while maintaining the same power of error detection can simplify the whole process and make it less error prone.

Example 3.3. A very simple redundancy, with edits meaning: (a) head of the house must have an annual income greater then or equal to 100, (b) everybody must have an annual income greater then or equal to 100. (a) is clearly redundant.

$$(\text{role} = \text{head of the house}) \wedge (\text{annual income} < 100) \quad \text{(a)}$$
$$\text{annual income} < 100 \quad \text{(b)}$$

A SAT formulation is used to solve the problem of logical implication. Given a set of statements S and a single statement s, $S \Rightarrow s$ if and only if $S \cup \neg s$ is an unsatisfiable formula [8,9]. Therefore, the following holds.

Theorem 3.2. *The clausal representation of an edit e_j is implied by the clausal representation of a set of edits E if and only if $E \cup \neg e_j$ is unsatisfiable.*

We check if an edit with clausal representation e_j is redundant by testing if the formula $(\mathcal{E} \setminus e_j) \cup \neg e_j$ is unsatisfiable. Redundancy of every edit is checked by iterating the above operation.

Detection of erroneous questionnaires Q^e trivially becomes the problem of checking if the truth assignment corresponding to Q satisfies the formula $\mathcal{E}$.

4   The Problem of Imputation

After detection of erroneous records, if information collecting has no cost, we could just cancel erroneous records and collect new information until we have enough correct records. Since usually information collecting has a cost, we would like to use also the correct part of information contained in the erroneous records. Given an *erroneous questionnaire* Q^e, the *imputation* process consists in changing some of his values, obtaining a *corrected questionnaire* Q^c which satisfies the formula $\mathcal{E}$ and is as close as possible to the (unknown) *original questionnaire* Q^o (the one we would have if we had no errors). Two principles should be followed [6]: to apply the minimum changes to erroneous data, and to modify as less as possible the original frequency distribution of the data. Generally, a cost for changing each field is given, based on the reliability of the field. It is assumed that, when error is something unintentional, the erroneous fields are the minimum-cost set of fields that, if changed, can restore consistency.

The problem of *error localization* is to find a set W of fields of minimum total cost such that Q^c can be obtained from Q^e by changing (only and all) the values of W. Imputation of actual values of W can then be performed in a deterministic or probabilistic way. This cause the minimum changes to erroneous data, but has little respect for the original frequency distributions.

A *donor questionnaire* Q^d is a correct questionnaire which, according to some distance function $d(Q^e, Q^d) \in \mathbb{R}_+$, is the nearest one to the erroneous questionnaire Q^e, hence it represents a record with similar characteristics. The problem of *imputation trough a donor* is to find a set D of fields of minimum total cost such that Q^c can be obtained from Q^e by copying from the donor Q^d (only and all) the values of D. This is generally recognized to cause low alteration of the original frequency distributions, although changes caused to erroneous data are not minimum. We are interested in solving both of the above problems.

Example 4.1. We have the following erroneous questionnaire Q^e

{... age = 17, car = no, city of residence = A, city of work = B ...}

Suppose we restore consistency when city of work = A. The solution of error localization is $W = \{$city of work$\}$. However, we have the following donor Q^d

{... age = 18, car = yes, city of residence = A, city of work = B ...}

The solution of imputation through donor is $D = \{$age, car$\}$, having higher cost.

5   A Set Covering Formulation

Given a ground set S of n elements s_i, each one with a cost $c_i \in \mathbb{R}_+$, and a collection $\mathcal{A}$ of m sets A_j of elements of S, the weighted set covering problem [10] is the problem of taking the set of elements s_i of minimum total weight such that at least one element for every A_j is taken. Let a^j be the incidence vector of A_j, i.e. a vector in $\{0, 1\}^n$ whose i-th component a_i^j is 1 if $s_i \in A_j$, 0 if $s_i \notin A_j$. By using a vector of variables $x \in \{0, 1\}^n$ representing the incidence vector of the set of elements s_i we take, we have

$$\min \sum_{i=1}^{n} c_i x_i \quad \text{s.t.} \quad \sum_{i=1}^{n} a_i^j x_i \geq 1 \quad j = 1 \ldots m, \quad x \in \{0, 1\}^n \qquad (2)$$

This problem is well-known NP-complete [7], and is of great relevance in many applied fields. In order to work here with binary optimization, a positive literal α_i becomes a binary variable x_i, and a negative literal $\neg \alpha_i$ becomes a negated binary variable $\bar{x}_i$. A questionnaire Q, which mapped to a truth assignment in $\{True, False\}^n$, will now map to a binary vector in $\{0, 1\}^n$. A clause c_j

$$(\alpha_i \vee \ldots \vee \alpha_j \vee \neg \alpha_k \vee \ldots \vee \neg \alpha_n)$$

becomes now the following linear inequality, by defining the set A_π of the logical variables appearing positive in c_j, and the set A_ν of the logical variables appearing negated in c_j, together with their incidence vectors a^π and a^ν

$$\sum_{i=1}^{n} a_i^\pi x_i + \sum_{i=1}^{n} a_i^\nu \bar{x}_i \geq 1$$

Example 5.1. Suppose that the (correct) questionnaire Q maps to the truth assignment $\{\alpha_1 = False, \alpha_2 = False, \alpha_3 = True\}$, and that $\mathcal{E}$ is

$$(\neg \alpha_1 \vee \alpha_2 \vee \neg \alpha_3) \wedge (\neg \alpha_1 \vee \neg \alpha_2) \wedge (\alpha_2 \vee \alpha_3)$$

The binary vector corresponding to Q is $\{x_1 = 0, x_2 = 0, x_3 = 1\}$, and the system of linear inequalities corresponding to $\mathcal{E}$ is

$$\begin{pmatrix} 0 & 1 & 0 \\ 0 & 0 & 0 \\ 0 & 1 & 1 \end{pmatrix} \begin{pmatrix} x_1 \\ x_2 \\ x_3 \end{pmatrix} + \begin{pmatrix} 1 & 0 & 1 \\ 1 & 1 & 0 \\ 0 & 0 & 0 \end{pmatrix} \begin{pmatrix} \bar{x}_1 \\ \bar{x}_2 \\ \bar{x}_3 \end{pmatrix} \geq \begin{pmatrix} 1 \\ 1 \\ 1 \end{pmatrix}$$

We can now model the two above imputation problems as follows. We have

- The binary vector $e = \{e_1, \ldots, e_n\} \in \{0,1\}^n$ corresponding to the erroneous questionnaire Q^e.
- In the case of *imputation through a donor* only, the binary vector $d = \{d_1, \ldots, d_n\} \in \{0,1\}^n$ corresponding to the donor questionnaire Q^d.
- The binary variables $x = \{x_1, \ldots, x_n\} \in \{0,1\}^n$ and their complements $\bar{x} = \{\bar{x}_1, \ldots, \bar{x}_n\} \in \{0,1\}^n$, with the coupling constraints $x_i + \bar{x}_i = 1$. They correspond to the corrected questionnaire Q^c that we want to find.
- The system of linear inequalities $A^\pi x + A^\nu \bar{x} \geq 1$, with $A^\pi, A^\nu \in \{0,1\}^{m \times n}$, that e does not satisfy. We know that such system has binary solutions, since $\mathcal{E}$ is satisfiable and has more than one solution.
- The vector $c = \{c_1, \ldots, c_n\} \in \mathbb{R}_+^n$ of costs that we pay for changing e. We pay c_i for changing e_i.

We furthermore introduce a vector of binary variables $y = \{y_1, \ldots, y_n\} \in \{0,1\}^n$ representing the changes we introduce in e.

$$y_i = \begin{cases} 1 & \text{if we change } e_i \\ 0 & \text{if we keep } e_i \end{cases}$$

The minimization of the total cost of the changes can be expressed with

$$\min_{y_i \in \{0,1\}} \sum_{i=1}^{n} c_i y_i = \min_{y \in \{0,1\}^n} c'y \tag{3}$$

However, the constraints are expressed for x. A key issue is that there is a relation between variables y and x (and consequently $\bar{x}$). In the case of error localization, this depends on the values of e, as follows:

$$y_i = \begin{cases} x_i & (= 1 - \bar{x}_i) \text{ if } e_i = 0 \\ 1 - x_i & (= \bar{x}_i) \text{ if } e_i = 1 \end{cases} \tag{4}$$

In the case of imputation through donor, this depends on the values of e and d.

$$y_i = \begin{cases} x_i & (= 1 - \bar{x}_i) \text{ if } e_i = 0 \text{ and } d_i = 1 \\ 1 - x_i & (= \bar{x}_i) \text{ if } e_i = 1 \text{ and } d_i = 0 \\ 0 & \text{if } e_i = d_i \end{cases} \tag{5}$$

By using the above results, we can express the two imputation problems with the following set covering formulation. In the case of error localization, (3) becomes

$$\min_{x_i, \bar{x}_i \in \{0,1\}} \sum_{i=1}^{n} (1 - e_i) c_i x_i + \sum_{i=1}^{n} e_i c_i \bar{x}_i \tag{6}$$

Conversely, in the case of imputation through a donor, our objective (3) becomes

$$\min_{x_i, \bar{x}_i \in \{0,1\}} \sum_{i=1}^{n} (1 - e_i) d_i c_i x_i + \sum_{i=1}^{n} e_i (1 - d_i) c_i \bar{x}_i \tag{7}$$

Subject, in both cases (6) and (7), to the following set of constraints

$$\begin{aligned} A^\pi x + A^\nu \bar{x} &\geq 1 \\ x_i + \bar{x}_i &= 1 \end{aligned} \qquad x, \bar{x} \in \{0,1\}^n$$

6   Implementation and Results

Satisfiability problems are solved by means of the efficient enumerative solver Adaptive Core Search (ACS) [4]. In the case of unsatisfiable instances, ACS is able to select an unsatisfiable subset of clauses. Set covering are solved by means of a solver based on the Volume Algorithm (VA), effective procedure recently presented by Barahona [2] as an extension of the subgradient algorithm. We added a simple heuristic in order to obtain an integer solution. Moreover, since this is an approximate procedure, we compared its results with the commercial branch-and-bound (B&B) solver *Xpress*. The process of edits validation and data imputation in the case of a Census of Population is performed. Edits are provided by the Italian National Statistic Institute (Istat). We checked different sets of real edits, and (in order to test the limits of the procedure) artificially generated CNF instances of larger size (up to 15000 var. and 75000 clauses) representing simulated sets of edits. When performing the whole inconsistency and redundancy checking, every CNF with n variables and m clauses, produces about $1+n+n/10+m$ SAT problems. Total times for solving such entire sequence of SAT problems (on a Pentium II 450MHz PC) are reported. Inconsistency or redundancy present in the set of edits were detected in the totality of the cases. Detection of erroneous questionnaires was performed, as a trivial task. In the cases of error localization and imputation through a donor, for each set of edits we considered various simulated erroneous answers with different percentage of activated edits. Note that a CNF with n variables and m clauses corresponds to a set covering problem with $2n$ variables and $m + n$ constraints. In the case of error localization, the heuristic VA can solve problems of size not solvable by B&B within the time limit of 6 hours. Further occasional tests with higher error percentage shows that VA does not increase its running time, but it is often unable to find a feasible integer solution, while B&B would reach such solution but in an a prohibitive amount of time. In the case of imputation through a donor, the heuristic VA is unable to find a feasible integer solution for problems with large error percentage, while B&B solves all problems in very short times. This holds because, when using a donor, many variables are fixed if $e_i = d_i$ (see (5)). Therefore, such problems become similar to error localization problems with a smaller number of variables but a higher error percentage.

Real sets of edits			
n	m	# of problems	time
315	650	975	0.99
350	710	1090	1.35
380	806	1219	1.72
402	884	1321	2.21
425	960	1428	2.52
450	1103	1599	3.24

Simulated sets of edits			
n	m	# of problems	time
1000	5000	6101	15
3000	15000	18301	415
5000	25000	30501	1908
8000	40000	48801	7843
10000	50000	61001	16889
15000	75000	91501	>36000

Tables 1 & 2: Edit validation procedure on real and simulated sets of edits.

Real problems 480 var. and 1880 const.				
	time		value	
error	VA	B&B	VA	B&B
0.4%	0.04	1.91	58.6	58.6
0.7%	0.10	1.91	108.6	108.6
1.0%	0.11	2.54	140.1	140.1
1.6%	0.16	1.90	506.7	506.7
2.5%	0.20	2.50	1490.1	1490.1

Simulated problems 30000 var. and 90000 const.				
	time		value	
error	VA	B&B	VA	B&B
0.4%	35.74	>21600	6754.7	-
0.9%	47.33	>21600	12751.3	-
1.3%	107.97	>21600	20135.4	-
2.0%	94.42	>21600	31063.6	-
4.0%	186.92	>21600	66847.4	-

Tables 3 & 4: Error localization procedure on real and simulated sets of edits.

Real problems 480 var. and 1880 const.				
	time		value	
error	VA	B&B	VA	B&B
0.4%	0.04	0.02	63.6	63.6
0.7%	0.06	0.02	144.7	144.7
1.0%	0.06	0.02	264.5	264.5
1.6%	0.06	0.02	643.5	643.1
2.5%	0.07	0.02	1774.3	1774.1

Simulated problems 30000 var. and 90000 const.				
	time		value	
error	VA	B&B	VA	B&B
0.4%	3.58	1.27	6788.1	6788.1
0.9%	3.22	1.27	12760.0	12759.5
1.3%	0.97	0.9	20140.1	20140.0
2.0%	2.89	1.27	31258.1	31082.2
4.0%	-	1.59	-	67764.94

Tables 5 & 6: Imputation through a donor on real and simulated sets of edits.

7 Conclusions

A binary encoding is the more direct and effective representation both for records and for the set of edit rules, allowing automatic detection of inconsistencies and redundancies in the set of edit rules. Erroneous records detection is carried out with an inexpensive procedure. The proposed encoding allows, moreover, to automatically perform error localization and data imputation. Related computational problems are overcome by using state-of-the-art solvers. Approached real problems have been solved in extremely short times. Artificially generated problems are effectively solved until sizes which are orders-of-magnitude larger than the above real-world problems. Hence, noteworthily qualitative improvements in a general process of data collecting are made possible.

References

1. M. Bankier. Experience with the New Imputation Methodology used in the 1996 Canadian Census with Extensions for future Census. *UN/ECE Work Session on Statistical Data Editing*, Working Paper n.24, Rome, Italy, 2-4 June 1999.
2. F. Barahona and R. Anbil. The Volume Algorithm: producing primal solutions with a subgradient method. *IBM Research Report* RC21103, 1998.
3. E. Boros, P.L. Hammer, T. Ibaraki and A. Kogan. Logical analysis of numerical data. *Mathematical Programming*, 79:163–190, 1997.

4. R. Bruni and A. Sassano. Finding Minimal Unsatisfiable Subformulae in Satisfiability Instances. *in proc. of 6th Internat. Conf. on Principles and Practice of Constraint Programming*, Lecture Notes in Computer Science 1894, Springer, 500–505, 2000.

5. R. Bruni and A. Sassano. CLAS: a Complete Learning Algorithm for Satisfiability. *Dip. di Inf. e Sist., Univ. di Roma "La Sapienza"* , Technical Report 01-01, 1999.

6. P. Fellegi and D. Holt. A Systematic Approach to Automatic edit and Imputation. *Journal of the American Statistical Association*, 17:35–71(353), 1976.

7. M.R. Garey and D.S. Johnson. *Computers and Intractability: A Guide to the Theory of NP-Completeness*. W.H. Freeman and Company, San Francisco, 1979.

8. R. Kowalski. *Logic for Problem solving*. North Holland, 1978.

9. D.W.Loveland. *Automated Theorem Proving: a Logical Basis*. North Holland 1978.

10. G.L. Nemhauser and L.A. Wolsey. *Integer and Combinatorial Optimization*. J. Wiley, New York, 1988.

11. C. Poirier. A Functional Evaluation of Edit and Imputation Tools. *UN/ECE Work Session on Statistical Data Editing*, Working Paper n.12, Rome, Italy, 2-4 June 1999.

12. W.E. Winkler. State of Statistical Data Editing and current Research Problems. *UN/ECE Work Session on Stat. Data Edit.*, W. P. n.29, Rome, Italy, 2-4 June 1999.

A New Framework to Assess Association Rules

Fernando Berzal, Ignacio Blanco, Daniel Sánchez*, and María-Amparo Vila

Department of Computer Science and Artificial Intelligence, University of Granada,
E.T.S.I.I., Avda. Andalucia 38, 18071 Granada, Spain

Abstract. The usual support/confidence framework to assess association rules has several drawbacks that lead to obtain many misleading
rules, even in the order of 95% of the discovered rules in some of our
experiments. In this paper we introduce a different framework, based on
Shortliffe and Buchanan's certainty factors and the new concept of *very
strong rules*. The new framework has several good properties, and our
experiments have shown that it can avoid the discovery of misleading
rules.

1   Introduction

One of the main problems in the field of Data Mining is how to assess the patterns
that are found in data, such as association rules in T-sets [1]. We call T-set a set
of transactions, where each transaction is a subset of items. Association rules are
" implications" that relate the presence of items in the transactions of a T-set.
More formally, given a set of items I and a T-set R on I, an association rule is
an expression of the form $A \Rightarrow C$, with $A, C \subset I$, $A \cap C = \emptyset$, where A and C
are called *antecedent* and *consequent* of the rule respectively.

The usual measures to assess association rules are support and confidence,
both based on the concept of support of an *itemset* (a subset of items). Given a
set of items I and a T-set R on I, the support of an itemset $I_0 \subseteq I$ is

$$supp(I_0) = \frac{|\{\tau \in R \mid I_0 \subseteq \tau\}|}{|R|} \qquad (1)$$

i.e., the probability that the itemset appears in a transaction of R. The support
of the association rule $A \Rightarrow C$ in R is

$$Supp(A \Rightarrow C) = supp(A \cup C) \qquad (2)$$

and its confidence is

$$Conf(A \Rightarrow C) = \frac{supp(A \cup C)}{supp(A)} = \frac{Supp(A \Rightarrow C)}{supp(A)}. \qquad (3)$$

Support is the percentage of transactions where the rule holds. Confidence is
the conditional probability of C with respect to A or, in other words, the relative

---

* Corresponding author. E-mail: daniel@decsai.ugr.es, Phone: +34 958 246397
  Fax: +34 958 243317 .

F. Hoffmann et al. (Eds.): IDA 2001, LNCS 2189, pp. 95–104, 2001.

cardinality of C with respect to A. The techniques for mining association rules attempt to discover rules whose support and confidence are greater than user-defined thresholds called *minsupp* and *minconf* respectively. These are called *strong rules*.

However, several authors have pointed out some drawbacks of this framework that lead to find many more rules than it should [3,6,9]. The following example is (from [3]): in the CENSUS database of 1990, the rule "past active duty in military $\Rightarrow$ no service in Vietnam" has a very high confidence of 0.9. This rule suggests that knowing that a person served in military we should believe that he/she did not serve in Vietnam. However, the itemset "no service in Vietnam" has a support over 95%, so in fact the probability that a person did not serve in Vietnam *decreases* (from 95% to 90%) when we know he/she served in military, and hence the association is negative. Clearly, this rule is misleading.

In this paper we introduce a new framework to assess association rules in order to avoid to obtain misleading rules. In section 2 we describe some drawbacks of the support/confidence framework. Section 3 contains some related work. Section 4 is devoted to describe our new proposal. Experiments and conclusions are summarized in sections 5 and 6, respectively.

2 Drawbacks of the Support/Confidence Framework

2.1 Confidence

Confidence is an accuracy measure of a rule. In [5], Piatetsky-Shapiro suggested that any accuracy measure ACC should verify three specific properties in order to separate strong and weak rules (in the sense of assigning them high and low values respectively). The properties are the following:

P1 $ACC(A \Rightarrow C) = 0$ when $Supp(A \Rightarrow C) = supp(A)\, supp(C)$. This property claims that any accuracy measure must test the independence (though values other than 0 could be used, depending on the range of ACC).
P2 $ACC(A \Rightarrow C)$ monotonically increases with $Supp(A \Rightarrow C)$ when other parameters remain the same.
P3 $ACC(A \Rightarrow C)$ monotonically decreases with $supp(A)$ (or $supp(C)$) when other parameters remain the same.

Now we show that confidence does not verify all the properties:

Proposition 1. *Confidence does not verify the property* **P1**.

Proof. Here is a counterexample: let $I_1 = \{i_1, i_2, i_3, i_4\}$ be a set of items, and let R_1 be the T-set on I_1 of table 1.A. Rows represent transactions, and columns represent items. A cell containing "1" means that the item/column is present in the transaction/row. Table 1.B shows the support of three itemsets with items in I_1. Since $supp(\{i_1\})supp(\{i_2\}) = 1/3 = supp(\{i_1, i_2\})$, i_1 and i_2 are statistically independent and hence the confidence should be 0. However, $Conf(\{i_1\} \Rightarrow \{i_2\}) = = \frac{1/3}{1/2} = 2/3 \neq 0$.

i_1	i_2	i_3	i_4
1	0	1	0
0	0	0	1
0	1	1	1
0	1	1	1
1	1	1	1
1	1	1	1

A

Itemset	Support
$\{i_1\}$	1/2
$\{i_2\}$	2/3
$\{i_1, i_2\}$	1/3

B

Table 1. (A) The T-set R_1. **(B)** Support of several itemsets in R_1

Proposition 2. *Confidence verifies the property* **P2**.

Proof. Trivial regarding (3).

Proposition 3. *Confidence verifies the property* **P3** *only for* $supp(A)$.

Proof. It is easy to see for $supp(A)$ regarding (3). It is also trivial to see that **P3** does not hold with respect to $supp(C)$ since $supp(C)$ does not appear in (3), and $supp(A \cup C)$ and $supp(A)$ remain the same by the conditions of **P3**.

In summary, confidence is not able to detect statistical independence (**P1**) nor negative dependence between items (the examples in the introduction), because it does not take into account the support of the consequent.

2.2 Support

A common principle in association rule mining is "the greater the support, the better the itemset", but we think this is only true to some extent. Indeed, itemsets with very high support are a source of misleading rules because they appear in most of the transactions, and hence any itemset (despite its meaning) seems to be a good predictor of the presence of the high-support itemset.

An example is $\{i_3\}$ in table 1.A. It is easy to verify that any itemset involving only i_1 and i_2 is a perfect predictor of $\{i_3\}$ (any rule with $\{i_3\}$ in the consequent has total accuracy, that is, confidence is 1 for all such rules). Also, $Conf(\{i_4\} \Rightarrow \{i_3\}) = 0.8$, that is pretty high. But we cannot be sure that these associations hold in real world. In fact what holds most times is negative dependence or independence, as the examples in the introduction showed.

As we have seen, an accuracy measure verifying **P1 - P3** can solve the problem when $Conf(A \Rightarrow C) \le supp(C)$ (i.e., negative dependence or independence). But when $supp(C)$ is very high and $Conf(A \Rightarrow C) > supp(C)$, we can obtain a high accuracy. However, there is a lack of variability in the presence of C in data that does not allow us to be sure about the rule. Fortunately, this situation can be detected by checking that $supp(C)$ is not very high, but no method to check this has been incorporated into the existing techniques to find association rules.

These problems lead to obtain much more rules than it should. Suppose we have a T-set R on a set of items I, from where a set of reliable rules S has been obtained. Think of adding an item i_{vf} to I and to include it in the transactions of R so that i_{vf} has a very high support. It is very likely that adding i_{vf} to the consequent of any rule, both support and accuracy of the rules don't change. The same can be expected if we add the item to the antecedent, so we can obtain in the order of three times more rules (the original set, and those obtained by adding i_{vf} to the antecedent, or to the consequent). But we must also consider that since i_{vf} is very frequent, almost any itemset could be a good predictor of the presence of i_{vf} in a transaction, so we may obtain in the order of $2^{|I|}$ more rules. For example, if $|I| = 10$ (without i_{vf}) and $|S| = 50$ (a modest case), by adding i_{vf} we could obtain 1124 misleading rules in the worst case! Even if we restrict ourselves to find rules with only one item in the consequent, we are talking about 1074 rules in the worst case.

The problem is clearly that the user is overwhelmed with a big amount of misleading rules. The situation gets worse exponentially if we add two or more items with very high support. Now think of mining a real database such as the CENSUS data employed in [3], where $|I| = 2166$ and there are many items with support above 95%. It is clear that a new framework to assess association rules is needed.

3   Related Work

Several authors have proposed alternatives to confidence, see [5,3,9,10,8] among others. In this section we briefly describe two of them.

3.1   Conviction

Conviction was introduced in [3] to be

$$Conv(A \Rightarrow C) = \frac{supp(A)\, supp(\neg C)}{supp(A \cup \neg C)} \tag{4}$$

where $\neg C$ means the absence of C. Its domain is $(0, \infty)$, 1 meaning independence. Values in $(0, 1)$ mean negative dependence. In our opinion, the main drawback of this measure is that its range is not bounded, so it is not easy to compare the conviction of rules because differences between them are not meaningful and, much more important, to define a conviction threshold (for example, some rules considered interesting in [3] have conviction values of 1.28, 2.94, 50 and ∞, this last meaning total accuracy). Also, from (4) it is easy to see that conviction does not verify property **P3** for supp(A).

3.2   Interest

In [9], the χ^2 test is used to find dependencies between items. However, the value of the χ^2 statistic is not suitable to measure the degree of dependence, so interest is used instead. The interest is defined as

$$Int(A \Rightarrow C) = \frac{Supp(A \Rightarrow C)}{supp(A)\, supp(C)} \tag{5}$$

Interest verifies **P1-P3**, the value 1 meaning independence. But as conviction, its range is not bounded so it has the same drawbacks. Moreover, interest is symmetric (i.e. the interest of $A \Rightarrow C$ and $C \Rightarrow A$ is the same), and this is not intuitive in most of the cases. Association rules require to measure the strength of implication in both directions, not only the degree of dependence.

4   A New Framework to Assess Association Rules

4.1   Measuring Accuracy

To assess the accuracy of association rules we use Shortliffe and Buchanan's *certainty factors* [7] instead of confidence. Certainty factors were developed to represent uncertainty in the rules of the MICYN expert system, and they have been recognized as one of the best models in the development of rule-based expert systems (however, they have been also used in data mining [4,6]).

Definition 1. *We name certainty factor of $A \Rightarrow C$ to the value*

$$CF(A \Rightarrow C) = \frac{Conf(A \Rightarrow C) - supp(C)}{1 - supp(C)} \tag{6}$$

if $Conf(A \Rightarrow C) > supp(C)$, and

$$CF(A \Rightarrow C) = \frac{Conf(A \Rightarrow C) - supp(C)}{supp(C)} \tag{7}$$

if $Conf(A \Rightarrow C) < supp(C)$, and 0 otherwise.

The certainty factor is interpreted as a measure of *variation* of the probability that C is in a transaction when we consider only those transactions where A is. More specifically, a positive CF measures the decrease of the probability that C is not in a transaction, given that A is. A similar interpretation can be done for negative CFs.

By (6) and (7) it is clear that CFs take into account both the confidence of the rule and the support of C. Moreover, they verify properties **P1-P3**, as the following propositions show:

Proposition 4. *Certainty factors verify* **P1**.

Proof. If $Supp(A \Rightarrow C) = supp(A)\, supp(C)$ then $Conf(A \Rightarrow C) = supp(C)$ and then by definition $CF(A \Rightarrow C) = 0$.

By this property, CFs are an independence test. But CFs can also detect the kind of dependence. If there is a positive dependence between A and C then $Supp(A \Rightarrow C) > supp(A)\, supp(C)$, so $Conf(A \Rightarrow C) > supp(C)$ and hence $CF(A \Rightarrow C) > 0$. If there is a negative dependence then $Conf(A \Rightarrow C) < supp(C)$ and hence $CF(A \Rightarrow C) < 0$.

Proposition 5. *Certainty factors verify* **P2**.

Proof. Confidence verifies **P2** and, when confidence increases and $supp(C)$ remains the same, CF increases (see (6) and (7)). Hence, CF increases with $Sup(A \Rightarrow C)$ when other parameters remain the same.

Proposition 6. *Certainty factors verify* **P3**.

Proof. CF verifies **P3** for $supp(A)$ since confidence does (see the proof of **P2**). Now we shall prove **P3** for $supp(C)$.

- Suppose $CF(A \Rightarrow C) < 0$. Then, by (7) it is clear that if $supp(C)$ increases then $CF(A \Rightarrow C)$ decreases.
- Suppose $CF(A \Rightarrow C) > 0$. CF is a function of confidence and $supp(C)$. By the conditions of **P3** we assume that confidence remain the same (i.e., it is a constant). If we derive with respect to $supp(C)$ we obtain

$$CF'(A \Rightarrow C) = \frac{Conf(A \Rightarrow C) - 1}{(1 - supp(C))^2}$$

  so $CF'(A \Rightarrow C) \leq 0$, and hence $CF(A \Rightarrow C)$ monotonically decreases with $supp(C)$.
- Let $Conf(A \Rightarrow C) = c_0$ with $0 < c_0 < 1$. Let us increase monotonically $supp(C)$ from 0 to 1. While $supp(C) < c_0$ it holds that $CF(A \Rightarrow C) > 0$ and monotonically decreases, as we have shown. When $supp(C)$ reaches c_0 then $CF(A \Rightarrow C) = 0$, so it keeps decreasing. Finally, when $supp(C) > c_0$ it holds that $CF(A \Rightarrow C) < 0$, so it has decreased, and it keeps decreasing as $supp(C)$ increases, as we have shown.

Further properties of CFs can be found in [2]. We omit them here for lack of space. From now on, we will call a rule *strong* if its support and CF are greater than user-specified thresholds *minsupp* and *minCF* respectively. Let us remark that we are interested only in rules with positive CF, meaning positive dependence among items, and hence we shall assume $minCF > 0$.

4.2   Solving the Support Drawback

A simple solution would be to use a maximum support threshold *maxsupp* to solve the support drawback, and to avoid reporting those rules involving itemsets with support above *maxsupp*. However, the user should provide the value for *maxsupp*. In order to avoid this we introduce the concept of *very strong rule*.

Definition 2. *The rule* $A \Rightarrow C$ *is very strong if both* $A \Rightarrow C$ *and* $\neg C \Rightarrow \neg A$ *are strong rules.*

The rationale behind this definition is that $A \Rightarrow C$ and $\neg C \Rightarrow \neg A$ are logically equivalent, so we should look for strong evidence of both rules to believe that they are interesting. This definition can help us to solve the support drawback since when $supp(C)$ (or $supp(A)$) is very high, $Supp(\neg C \Rightarrow \neg A)$ is very low, and hence the rule $\neg C \Rightarrow \neg A$ won't be strong and $A \Rightarrow C$ won't be very strong.

By definition, a very strong rule must verify:

1. Support conditions:
   (a) $Supp(A \Rightarrow C) > minsupp$
   (b) $Supp(\neg C \Rightarrow \neg A) > minsupp$
2. CF conditions:
   (a) $CF(A \Rightarrow C) > minCF$
   (b) $CF(\neg C \Rightarrow \neg A) > minCF$

So there are two new conditions for a rule to be interesting, 1.b and 2.b. But, in practice, only one CF condition must be checked, as a result of the following proposition:

Proposition 7. *If* $CF(A \Rightarrow C) > 0$ *then* $CF(A \Rightarrow C) = CF(\neg C \Rightarrow \neg A)$.

Proof. We shall use the usual probability notation, i.e., $Conf(X \Rightarrow Y) = p(Y|X)$ and $supp(X) = p(X)$. By Bayes' Rule

$$p(A|\neg C) = \frac{p(\neg C|A)p(A)}{p(\neg C)} = \frac{(1 - p(C|A))\,p(A)}{p(\neg C)}$$

and

$$p(\neg A|\neg C) = 1 - p(A|\neg C) = 1 - \frac{(1 - p(C|A))\,p(A)}{p(\neg C)}$$

Let us use (6) to obtain $CF(\neg C \Rightarrow \neg A)$. If we obtain a positive value, that will be the CF of the rule, since in that case $Conf(\neg C \Rightarrow \neg A) > supp(\neg A)$ (i.e., $p(\neg A|\neg C) > p(\neg A)$). Otherwise, we should have used (7).

$$CF(\neg C \Rightarrow \neg A) = \frac{p(\neg A|\neg C) - p(\neg A)}{1 - p(\neg A)} = \frac{1 - \frac{(1-p(C|A))p(A)}{p(\neg C)} - (1 - p(A))}{p(A)} =$$

$$= \frac{p(A) - \frac{(1-p(C|A))p(A)}{p(\neg C)}}{p(A)} = 1 - \frac{1 - p(C|A)}{p(\neg C)} = \frac{(1 - p(C)) - (1 - p(C|A))}{(1 - p(C))} =$$

$$= \frac{p(C|A) - p(C)}{1 - p(C)} = CF(A \Rightarrow C).$$

Since $CF(A \Rightarrow C) > 0$, we have used the correct expression and we have shown that $CF(\neg C \Rightarrow \neg A) = CF(A \Rightarrow C)$.

The last property is not only useful, but also intuitive. Conviction also verifies it, but confidence and interest don't. Details can be found in [2].

4.3   Implementation

One of the advantages of our new framework is that it is easy to incorporate it into existing algorithms. Most of them work in two steps:

Step 1. Find the itemsets whose support is greater than *minsupp* (called *frequent itemsets*). This step is the most computationally expensive.

Step 2. Obtain rules with accuracy greater than a given threshold from the frequent itemsets obtained in step 1, specifically the rule $A \Rightarrow C$ is obtained from the itemsets $A \cup C$ and A.

To find very strong rules, step 1 remains the same. In step 2 we obtain the CF of the rule from the rule confidence and $supp(C)$, both calculated in step 1 (since $A \cup C$ is frequent, A and C also are), and we verify the CF condition. Support condition 1.a is ensured because $A \cup C$ is a frequent itemset. Support condition 1.b is also easy to verify since

$$supp(\neg C \cup \neg A) = 1 - supp(C) - supp(A) + supp(A \cup C) \qquad (8)$$

and $supp(C)$, $supp(A)$ and $supp(A \cup C)$ are available. An important feature of these modifications is that they keep both the time and space complexity of the algorithms.

Finally, let us remark that support condition 1.a is usually employed to bound the search for frequent itemsets in step 1 (hence reducing both time and space complexity). Further reduction can be obtained by also using 1.b. This can benefit from the following property:

Proposition 8. *If $supp(A \cup C) > 1 - minsupp$ then $supp(\neg C \cup \neg A) < minsupp$.*

Proof. If $supp(A \cup C) > 1 - minsupp$ then $1 - supp(A \cup C) < minsupp$. Clearly $supp(A \cup C) + supp(\neg C \cup \neg A) \le 1$, so $supp(\neg C \cup \neg A) \le 1 - supp(A \cup C) < minsupp$.

The last proposition also suggests very strong rules implicitly use a value *maxsupp* of at most 1-*minsupp*. A detailed description about how to use 1.b. to reduce the implementation complexity, together with some algorithms, can be found in [2].

5   Experiments

Table 2 and figure 1 show the results of one of our experiments on a T-set, containing more than $225 \cdot 10^6$ transactions and 10 items, obtained from data about surgical operations in the University Hospital of Granada [6]. By using CFs, rules with negative dependence or independence are discarded, and hence much fewer rules are obtained (we used a value *minsupp*=0.01).

We also detected the influence of the items with very high support "blood" and "prosthesis". Table 3 and figure 2 show how the number of rules is reduced if "blood" and "prosthesis" are not considered.

α	0.1	0.2	0.3	0.4	0.5	0.6	0.7	0.8	0.98	1
minC=α	1185	1078	1066	1034	971	839	706	494	96	0
minFC=α	795	633	549	431	304	190	104	96	20	0

Table 2. Number of rules obtained by using confidence and CF

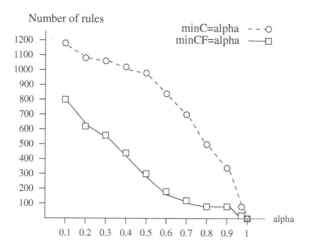

Fig. 1. Graphical representation of table 2

For a minimum accuracy of 0.8, the number of rules has been reduced from 494 (using confidence) to 8. Among the discarded rules, 398 are rules with negative dependence or independence, 32 have "blood" in the consequent, 32 have "prosthesis" in the consequent, and 24 are obtained from the 8 very strong rules by adding to the antecedent "blood", or "prosthesis", or both. They were all misleading rules. Other experiments, involving the CENSUS database, are detailed in [2] and show similar results.

6   Conclusions

Very strong rules based on CFs are a suitable framework to discard misleading rules. The concept of very strong rule is very intuitive. CFs are successfully used in expert systems where the task is predictive, and it is well-known that CFs of rules can be obtained from humans, so CFs are meaningful and that improves the understanding and comparison of rules, and the definition of $minCF$.

minFC	0.1	0.2	0.3	0.4	0.5	0.6	0.7	0.8	0.9	1
With B/P	795	633	549	431	304	190	104	96	96	0
Without B/P	140	108	95	76	54	28	10	8	8	0

Table 3. Number of rules obtained with and without items "blood" and "prosthesis"

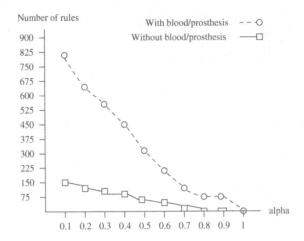

Fig. 2. Graphical representation of table 3

References

1. R. Agrawal, T. Imielinski, and A. Swami. Mining association rules between sets of items in large databases. In *Proc. Of the 1993 ACM SIGMOD Conference*, pages 207–216, 1993.
2. F. Berzal, M. Delgado, D. Sánchez, and M.A. Vila. Measuring the accuracy and importance of association rules. Technical Report CCIA-00-01-16, Department of Computer Science and Artificial Intelligence, University of Granada, 2000.
3. S. Brin, R. Motwani, J.D. Ullman, and S. Tsur. Dynamic itemset counting and implication rules for market basket data. *SIGMOD Record*, 26(2):255–264, 1997.
4. L.M.Fu and E.H.Shortliffe. The application of certainty factors to neural computing for rule discovery. *IEEE Transactions on Neural Networks*, 11(3):647–657, 2000.
5. G. Piatetsky-Shapiro. Discovery, analysis, and presentation of strong rules. In G. Piatetsky-Shapiro and W. Frawley, editors, *Knowledge Discovery in Databases*, pages 229–238. AAAI/MIT Press, 1991.
6. D. Sánchez. *Adquisición de Relaciones Entre Atributos En Bases de Datos Relacionales (Translates to: Acquisition of Relationships Between Attributes in Relational Databases) (in Spanish)*. PhD thesis, Department of Computer Science and Artificial Intelligence, University of Granada, December 1999.
7. E. Shortliffe and B. Buchanan. A model of inexact reasoning in medicine. *Mathematical Biosciences*, 23:351–379, 1975.
8. A. Silberschatz and A. Tuzhilin. On subjective measure of interestingness in knowledge discovery. In *Proc. First Int'l Conf. Knowledge Discovery and Data Mining (KDD'95)*, pages 275–281, August 1995.
9. C. Silverstein, S. Brin, and R. Motwani. Beyond market baskets: Generalizing association rules to dependence rules. *Data Mining and Knowledge Discovery*, 2:39–68, 1998.
10. P. Smyth and R.M. Goodman. Rule induction using information theory. In G. Piatetsky-Shapiro and W. J. Frawley, editors, *Knowledge Discovery in Databases*. AAAI/MIT Press, 1991.

Communities of Interest

Corinna Cortes, Daryl Pregibon, and Chris Volinsky

AT&T Shannon Research Labs
Florham Park, New Jersey, USA

Abstract. We consider problems that can be characterized by large dynamic graphs. Communication networks provide the prototypical example of such problems where nodes in the graph are network IDs and the edges represent communication between pairs of network IDs. In such graphs, nodes and edges appear and disappear through time so that methods that apply to static graphs are not sufficient. We introduce a data structure that captures, in an approximate sense, the graph and its evolution through time. The data structure arises from a bottom-up representation of the large graph as the union of small subgraphs, called Communities of Interest (COI), centered on every node. These subgraphs are interesting in their own right and we discuss two applications in the area of telecommunications fraud detection to help motivate the ideas.

1 Introduction

Transactional data consists of records of interactions between pairs of entities occurring over time. For example, a sequence of credit card transactions consists of purchases of retail goods by individual consumers from individual merchants. Transactional data can be represented by a graph where the nodes represent the transactors and the edges represent the interactions between pairs of transactors. Viewed in this way, interesting new questions can be posed concerning the connectivity of nodes, the presence of atomic subgraphs, or whether the graph structure leads to the identification and characterization of "interesting" nodes. For example, Kleinberg (1998) introduces the notion of "hubs" and "authorities" as interesting nodes on the internet. The data used by Kleinberg differ significantly from the data we consider in that he uses static links to induce a graph over web pages. In our case, we use actual network traffic, as captured by interactions between pairs of transactors, to define our graph. Thus in a very real sense, the graph we consider is dynamic since nodes and edges appear and disappear from the graph through time.

There are many challenging issues that arise for dynamic graphs and we have used a specific application to focus our research, namely the graph induced by calls carried on a large telecommunications network. This application is interesting, both because of its size (i.e., hundreds of millions of nodes) and its rate of change (i.e., hundreds of millions of new edges each day). Like all networks, it is also diverse in the sense that some nodes are relatively inactive while others are superactive.

F. Hoffmann et al. (Eds.): IDA 2001, LNCS 2189, pp. 105–114, 2001.

In thinking about dynamic graphs, the first question that arises concerns the definition of $\mathcal{G}_t$, namely the graph $\mathcal{G}$ at time t. The intuitive notion is that $\mathcal{G}_t$ consists of the nodes and edges active at time t, or in a small interval around t. We consider discrete time applications where new sets of nodes and edges corresponding to the transactions from time step t to $t+1$ only become available at the end of the time step, for example once a day. Associated with every edge is a weight that is derived from an aggregation function applied to all (directed) transactions between a pair of nodes at time step t. For example, the aggregation function can be the "total duration of calls" or the "number of calls" from one node to another.

Let the graph corresponding to the transactions during time step t be g_t. We can define $\mathcal{G}_t$ from g_i where $i = 1, \ldots, t$ in several ways. Let us first define the sum of two graphs g and h

$$G = \alpha g \oplus \beta h$$

where α and β are scalars. The nodes and edges in G are obtained from the union of the nodes and edges in g and h. The weight of an edge in G is

$$\mathrm{w}(G) = \alpha \mathrm{w}(g) + \beta \mathrm{w}(h)$$

where the weight of an edge is set to zero if the edge is absent from the graph.

If one defines $\mathcal{G}_t = g_t$, $\mathcal{G}_t$ is very unstable as it might change dramatically at each time step. On the other hand, if one defines

$$\mathcal{G}_t = g_1 \oplus g_2 \oplus \ldots \oplus g_t = \bigoplus_{i=1}^{t} g_i,$$

$\mathcal{G}_t$ is perhaps too stable, as it includes all historic transactions from the beginning of time. To allow the graph $\mathcal{G}_t$ to track the dynamics of the transactional data stream, one can define $\mathcal{G}_t$ as a moving window over n time steps:

$$\mathcal{G}_t = g_{t-n} \oplus g_{t-n+1} \oplus \ldots \oplus g_t = \bigoplus_{i=t-n}^{t} g_i.$$

However, this definition suffers from two problems: first, one needs to store the graphs g_i corresponding to the last n time steps to compute $\mathcal{G}_t$, and second, it gives equal weight to all time steps.

To allow for a smooth dynamic evolution of $\mathcal{G}_t$ without incurring the storage problems of the moving window approach, we adopt the recursive definition:

$$\mathcal{G}_t = \theta \mathcal{G}_{t-1} \oplus (1 - \theta)g_t \tag{1}$$

where $0 \leq \theta \leq 1$ is a parameter that allows more (θ near 1) or less (θ near 0) history to influence the current graph. An alternative representation of (1) is obtained by expanding the recursion:

$$\mathcal{G}_t = \omega_1 g_1 \oplus \omega_2 g_2 \oplus \ldots \oplus \omega_t g_t = \bigoplus_{i=1}^{t} \omega_i g_i \tag{2}$$

where $\omega_i = \theta^{t-i}(1 - \theta)$. This representation highlights the fact that more recent data contributes more heavily to $\mathcal{G}_t$ than older data. Figure 1 displays this graphically. If processing occurs daily, then a value of $\theta = 0.85$ roughly corresponds to $\mathcal{G}_t$ capturing a moving month of network activity. Finally note that the convex combination defined in (1) is "smoother" than a simple 30 day moving window as network peaks and troughs, induced say by holiday traffic patterns, are effectively moderated by θ.

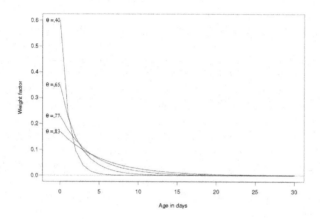

Fig. 1. *Damping factor of edge weights as a function of time steps (days) in the recursive definition (2) of $\mathcal{G}_t$. The values of θ correspond to effectively no influence from an edge after 1 ($\theta = 0.40$), 2 ($\theta = 0.65$), 3 ($\theta = 0.77$), or 4 weeks ($\theta = 0.83$).*

The paper is organized as follows. Section 2 describes the data structure that we use to capture network activity and to evolve it through time. We also briefly discuss the strategy we employ to traverse this data structure to quickly and efficiently build graphs around individual nodes. Section 3 introduces a pair of examples that illustrate how these subgraphs are used in practice. Section 4 summarizes the findings and discusses future work.

2   Data Structure

We propose a constructive approach to evolving a large time-varying graph. Consider a node in the graph, its associated directed edges, and weights associated with each edge. A data structure that consists of these weighted directed edge sets for each node is a representation of the complete graph. This data structure is redundant since it is indexed by nodes so that edges must be stored twice, once for the originating node and once for the terminating node. In contrast, a data structure that stores each edge once must be doubly indexed by nodes. The cost of edge duplication is often mitigated by gains in processing speed when subgraphs around nodes are expanded. For this reason we have chosen to represent our graphs as a singly indexed list of nodes, each with an associated array of weighted directed edges.

The data structure outlined above is complete in the sense that it captures the entire graph. However in the applications that we are familiar with, the computational horsepower to maintain complete graphs with hundreds of millions and nodes and billions of edges is neither feasible nor desirable. Instead we define a new graph where the atomic unit is the subgraph consisting of a node and its directed top-k edges to other nodes. The meaning of "top" is relative to the aggregation function applied to transactions associated with each edge, so it might be the top-k edges in terms of the "number of calls." In addition to the top-k inbound and top-k outbound edges, we also define an overflow node, called "other", for aggregating traffic to/from nodes not contained in the top-k slots. While the value of k determines how well our data structure approximates the true network graph, it is worth noting that larger is not necessarily better when network graphs are used to study connectivity of the nodes. For example, assuming that a few percent of all calls are misdialed, do we really want the corresponding edges reflected in the graph? In our experience we have found the answer to this question to be "no" and have used a value of k that balances computational complexity (e.g., as regards speed and storage) with empirically determined accuracy (see below).

In practice, the edges may not always be retained in a symmetric fashion. If one node receives calls from many nodes, like 800CALLATT, it overflows its top-k slots so most callers will find their edges absorbed by the "other" node. However, any single node making calls to 800CALLATT may not have edges to more than k other nodes, and the edge will be contained in its top-k edge set. Duplication of edges becomes complete only in the limit as $k \to \infty$. For all finite k, the only invariance over the set of subgraphs is that the sum of the outbound edge weights equals the sum of the inbound edge weights, where the sum includes node "other". For most of our applications, we use $k = 9$ since most residential long distance accounts do not exceed this number of edges (see Figure 2).

To accommodate the time evolution of the network graph, the data structures are updated at fixed time steps. Between updating steps, transactions are collected and temporarily stored. At the end of that time period, the transactions are aggregated and the subgraph updated. The length of the time period represents another trade-off in accuracy: the longer the time period, the better an estimate of the top-k edge set, but the more outdated the resulting subgraph. In the applications discussed in Section 3, we perform daily updates, thereby maintaining reasonable accuracy while requiring temporary disk space for only one day of data. For a more detailed discussion see Cortes and Pregibon (1999).

Let $\hat{\mathcal{G}}_{t-1}$ denote the top-k approximation to $\mathcal{G}_{t-1}$ at time $t-1$ and let g_t denote the graph derived from the new transactions at time step t. The ap-

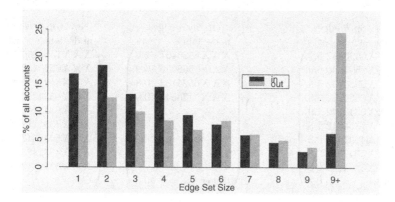

Fig. 2. *Edge set sizes for a random sample of residential accounts. Black bars indicate the size of the inbound edge set and gray bars indicate the size of the outbound edge set.*

proximation to $\mathcal{G}_t$ is formed from $\hat{\mathcal{G}}_{t-1}$ and g_t, node by node, using a top-k approximation to Eq. 1:

$$\hat{\mathcal{G}}_t = \text{top-}k\{\theta\hat{\mathcal{G}}_{t-1} \oplus (1-\theta)g_t\} \tag{3}$$

Thus, we first calculate the edge weights for all the edges of $\theta\hat{\mathcal{G}}_{t-1} \oplus (1-\theta)g_t$. The overflow node "other" is treated as any other node in the graph. Then for each node we sort the edges according to their weight. The top-k are preserved, and if there are more than k edges in the edge set for that node, the weights of the remaining edges are added to the weight of the edge going from the node to node "other". These operations are displayed pictorially in Figure 3 using $\theta = .85$.

The subgraph consisting of the top-k inbound and the top-k outbound edges of a node is ideal for fast extraction of larger subgraphs centered on the node. The data structures can be queried recursively for each node in the top-k edge sets of the center node. We grow such subgraphs in a breadth-first traversal of the data structure. For notational purposes, we denote the edge set of the node itself by d_1 (depth of 1). Similarly let d_2 denote the edge set formed by the union of d_1 and the edge sets of all nodes contained in d_1. We rarely explore edge sets greater than d_2 in our applications as they become unmanageably large and remarkably uninformative. Indeed we often apply a thresholding function to the edge weights, even for d_2, to further reduce "clutter" in a subgraph. Thus any edge with weight less than ϵ need not be expanded if one feels that such edges are inconsequential for the application at hand.

In the next section we introduce two applications that exploit the index structure of our representation. This is critical since we often need to compute and compare many subgraphs on a daily basis. We have tuned our algorithms so

Old top-k edges		Today's edges		New top-k edges	
node–labels	wts	node–labels	wts	node–labels	wts

$$
\theta
\begin{pmatrix}
XXX6525467 & 5.2 \\
XXX7562656 & 5.0 \\
XXX6524132 & 4.5 \\
XXX6534231 & 2.3 \\
XXX6243142 & 1.9 \\
XXX7354212 & 1.8 \\
XXX4231423 & 0.8 \\
XXX5342312 & 0.5 \\
XXX5264532 & 0.2 \\
\text{Other} & 0.1
\end{pmatrix}
+ (1-\theta)
\begin{pmatrix}
XXX6525467 & 2.0 \\
XXX7562656 & 6.2 \\
XXX6524132 & 0.8 \\
XXX5436547 & 10.0 \\
\\
\\
\\
\\
\\
\text{Other} & 0.0
\end{pmatrix}
=
\begin{pmatrix}
XXX7562656 & 5.2 \\
XXX6525467 & 4.6 \\
XXX6524132 & 3.9 \\
XXX6534231 & 2.0 \\
XXX6243142 & 1.6 \\
XXX7354212 & 1.5 \\
XXX5436547 & 1.5 \\
XXX4231423 & 0.7 \\
XXX5342312 & 0.4 \\
\text{Other} & 0.3
\end{pmatrix}
$$

Fig. 3. *Computing a new top-k edge set from the old top-k edge set and today's edges. Note how a new edge enters the top-k edge set, forcing an old edge to be added to Other.*

that the average time for retrieving and rendering a d_2 edge set from our data structure of close to 400M nodes is just under one second (on a single processor).

3   Applications

In the telecommunications industry, there are many different types of fraudulent behavior. Subscription fraud is a type of fraud that occurs when an account is set up by an individual who has no intention of paying any bills. The enabler in such cases involves either flawed processes for accepting and verifying customer supplied information, or identity-theft where an individual impersonates another person. In either case, if left undetected, the fraud is typically only discovered when the bill is returned to sender, often after thousands of dollars have been lost. In the first example we use COI and a "guilt by association" argument to detect new cases of fraud in the network. The second example uses a distance metric between COI to suggest that a fraudster has assumed a new network identity.

3.1   Guilt by Association

When a new account is activated on the network, is there any way to label it according to its inherent riskiness? In this section we explore this possibility by defining a procedure that assesses risk on the basis of a node's connectivity to other nodes. To illustrate the idea we use data on subscribers from a large metropolitan city for a one month period. We have labeled the nodes as fraudulent or legitimate according to the disposition determined by network security.

Figure 4 displays the distribution of the number of edges from a node to the closest node that is labeled as fraudulent. The figure shows that fraudsters tend

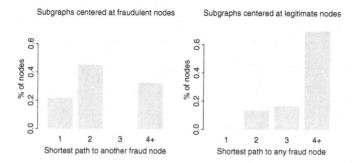

Fig. 4. *Guilt by association - what is the shortest path to a fraudulent node?*

to be closer to other fraudsters than random accounts are to fraud. Specifically we see that relatively few legitimate accounts are directly adjacent to fraudulent accounts. Indeed network security investigators have found that fraudsters seldom work in isolation from each other. There are brokers who compromise a legit customer's service and then sell this service to their own set of customers. Or even if there is no organized criminal element operating, fraudsters often cannot stop themselves from sharing their tricks with their friends and family. Because fraudsters seem to form an affinity group, we label the network connectivity process to catch fraud as "guilt by association."

Our process consists of the following steps:

- compute the d_2 edge sets for all new accounts one week after they are activated on the network
- label each of the nodes in the resulting COI as fraudulent or legitimate based on the most recent information from security associates
- rank the new accounts according to how much fraud appears in their COI

The left panel in Figure 5 provides an illustrative example where five nodes surrounding a new suspect (labeled XXX8667665) were recently deactivated for fraudulent behavior. The right panel summarizes the performance of the methodology for 105 cases presented to network security. While there is noise in the plot, it clearly shows that the probability that an account is fraudulent is an increasing function of the number of fraudulent nodes in its COI.

3.2 Record Linkage Using COI-based Matching

Consider the case where we have information on an account that was recently disconnected for fraud and we are looking for a new account that has the same individual behind it. Assuming that identity-theft was the root cause of the prior fraudulent account, it is likely that the new account is under a different name and address than the old one (i.e., the fraudster has now assumed the identity of a new victim). We attack this problem with the intuition that while the subscription information is not useful for matching network IDs to the same individual, the

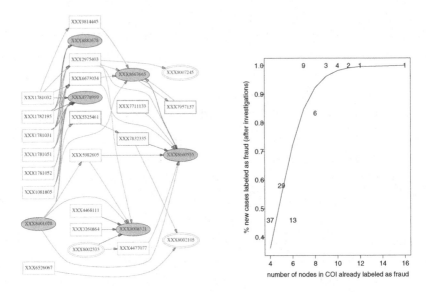

Fig. 5. Left Panel. *A guilt by association plot. Circular nodes correspond to wireless service accounts while rectangular nodes are conventional land line accounts. Shaded nodes have been previously labeled as fraudulent by network security associates.* Right Panel. *Calibration plot. The per cent fraud is plotted against the number of fraudsters in a COI for 105 cases presented to security. The plotting symbol is the number of these cases at each level of fraud infection. The curve superimposed on the points is the fit of a simple logistic model.*

calling patterns of the new account, as characterized by its COI, should not change very much from the previous account. The left panel of Figure 6 shows a convincing case where two nodes appear to belong to the same individual. We now have a problem of matching COI, with the underlying problem of deriving a reasonable distance function to quantify the closeness of a pair of COI.

The matching problem is computationally difficult because of the size of our network – each day we see tens of thousands of new accounts. For each of these, we need to compute their COI, and then the distance from each of these to the COI of all recently confirmed fraudulent accounts. Assuming for these purposes that we maintain a library of the most recent 1000 fraudulent accounts, tens of millions of pairwise distances need to be computed. To carry out the computations we use a d_2 COI for all accounts in our "fraud library" and d_1 COI for all new accounts.

The distance between two COI depends on both the quantity and the quality of the overlapping nodes. The quantity of the overlap is measured by counting the number of overlapping nodes and calculating the percentage of a COI which consists of overlapping nodes. However all overlapping nodes are not equally informative, so we need a measure of quality as well. Many graphs will inter-

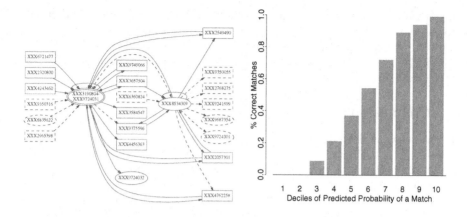

Fig. 6. Left Panel. *Visualization of linking accounts by their COI. The two individual COI are superimposed (in the double-lined oval) to show their similarity. Solid lines indicate edges common to both COI, while dashed lines and nodes indicate subgraphs belonging to only one COI.* Right Panel. *Success of COI matching. Observed proportions of matching node-pairs versus decile of predicted matching probability.*

sect at high-use nodes, such as large telemarketing shops or widely advertised customer service numbers. An informative overlapping node is one that has relatively low in- and out-degree, and in the best case, is shared only by the nodes under consideration for a match. We now describe a measure that captures these notions.

Given two COI for nodes a and b with a non-zero overlap, we define

$$\text{Overlap}(\text{COI}_a, \text{COI}_b) = \sum_{o \in O} \frac{w_{ao} w_{bo}}{w_o} \frac{1}{d_{ao}} \frac{1}{d_{bo}},$$

where O is the set of all overlapping nodes in the two COI, w_{ao} is the weight of edges between node a and node o in a's d_1 edge set, w_o is the overall weight of node o (the sum of all edges in the d_1 edge set for node o), and d_{ao} is the minimal distance from node a to node o in COI_a. [In the case where $d_{ao} > 1$, it is not clear what the weight w_{ao} should be, since there is no direct edge between the two. For this application we elected to set such a weight at $w_{ao} = .01$ in order to minimize the effect of these overlaps]. Intuitively, the numerator measures the strength of the connection from a and b to the overlap, while the denominator corrects for an overlap node which is either common to many nodes or is further in the graph from a or b. This measure scores high for overlap nodes that have strong links to the nodes of interest, but otherwise have low overall volume.

A decision tree built with this score, along with several covariates obtained from the information provided by the subscriber, produces a "matching" probability for any list of node pairs. For some of the node pairs that we scored, investigators were able to determine whether the old and new accounts belonged to the same individual. The right panel of Figure 6 shows the performance of our COI matching model. For the sample of pairs that we validated ($n = 1537$), we display the observed proportion of matching node-pairs for each decile of predicted matching probability.

4   Conclusions

In this paper we introduced the concept of a dynamic graph and our definition as an exponentially weighted average of the previous graph and a new edge set. We introduced a data structure that could be used to capture the evolution of a graph through time that was amenable to the exponential weighting scheme. This data structure allows the subgraph around any particular node to be quickly and efficiently expanded to an arbitrary diameter. Several applications were introduced that capitalized on this feature.

We have concentrated on the computational aspects of building and evolving the data structure for real applications. We have not explored the statistical aspects of treating our data structure and the associated algorithm for traversal as an approximation $\hat{\mathcal{G}}_t^k(d)$ to the true graph $\mathcal{G}_t$ where k denotes the size of the top-k edge set maintained in the data structure and d the diameter employed by the traversal algorithm. We hope to initiate research to explore these ideas in the near future.

Another topic for further research is how to prune an extracted subgraph so that only informative edges and nodes are retained. A common approach from (static) graph theory is to extract the strongly connected component. However, we feel that certain features inherent to telecommunication networks such as asymmetric edges (due to some customers subscribing to a competitor), sinks (toll-free calling) and sources (large corporations), makes strongly connected components an inferior choice for pruning back a COI.

The updating and storage of the data structures are facilitated by the programming language Hancock, [Cortes *et al* (2000)]. Hancock is a domain-specific C-based language for efficient and reliable programming with transactional data. Hancock is publicly available for non-commercial use at

http://www.research.att.com/~kfisher/hancock/.

References

Kleinberg (1998) Authoritative sources in a hyperlinked environment. *J. Kleinberg.* Proceedings 9th ACM-SIAM Symposium on Discrete Algorithms, 1998.

Cortes and Pregibon (1999) An information Mining Platform. *C. Cortes & D. Pregibon.* Proceedings of KDD99, San Diego, CA.

Cortes *et al* (2000) Hancock: A language for extracting signatures from data streams. *C. Cortes, K. Fisher, D. Pregibon, A. Rogers, & F. Smith.* Proceedings of KDD2000, Boston, MA.

An Evaluation of Grading Classifiers

Alexander K. Seewald and Johannes Fürnkranz

Austrian Research Institute for Artificial Intelligence, Schottengasse 3, A-1010 Wien
{alexsee, juffi}@oefai.at

Abstract. In this paper, we discuss grading, a meta-classification technique that tries to identify and correct incorrect predictions at the base level. While stacking uses the predictions of the base classifiers as meta-level attributes, we use "graded" predictions (i.e., predictions that have been marked as correct or incorrect) as meta-level classes. For each base classifier, one meta classifier is learned whose task is to predict when the base classifier will err. Hence, just like stacking may be viewed as a generalization of voting, grading may be viewed as a generalization of selection by cross-validation and therefore fills a conceptual gap in the space of meta-classification schemes. Our experimental evaluation shows that this technique results in a performance gain that is quite comparable to that achieved by stacking, while both, grading and stacking outperform their simpler counter-parts voting and selection by cross-validation.

1 Introduction

When faced with the decision "Which algorithm will be most accurate on my classification problem?", the predominant approach is to estimate the accuracy of the candidate algorithms on the problem and select the one that appears to be most accurate. [13] has investigated this approach in a small study with three learning algorithms on five UCI datasets. His conclusions are that on the one hand this procedure is on average better than working with a single learning algorithm, but, on the other hand, the cross-validation procedure often picks the wrong base algorithm on individual problems. This problem is expected to become more severe with an increasing number of classifiers.

As a cross-validation basically computes a prediction for each example in the training set, it was soon realized that this information could be used in more elaborate ways than simply counting the number of correct and incorrect predictions. One such meta-classification scheme is the family of *stacking* algorithms [19]. The basic idea of stacking is to use the predictions of the original classifiers as attributes in a new training set that keeps the original class labels.

In this paper, we investigate another technique, which we call *grading*. The basic idea is to learn to predict for each of the original learning algorithms whether its prediction for a particular example is correct or not. We therefore train one classifier for each of the original learning algorithms on a training set that consists of the original examples with class labels that encode whether

F. Hoffmann et al. (Eds.): IDA 2001, LNCS 2189, pp. 115–124, 2001.

Attributes	Class
x_{11} ... x_{1n_a}	t
x_{21} ... x_{2n_a}	f
... ...	...
x_{n_e1} ... $x_{n_e n_a}$	t

(a) training set

C_1 C_2 ... C_{n_c}
t t ... f
f t ... t
....
f f ... t

(b) predictions

C_1 C_2 ... C_{n_c}
$+$ $+$ ... $-$
$+$ $-$ ... $-$
....
$-$ $-$ ... $+$

(c) "graded" predictions

C_1 C_2 ... C_{n_c}	Cl
t t ... f	t
f t ... t	f
....	...
f f ... t	t

(d) t. s. for stacking

Attributes	Cl
x_{11} ... x_{1n_a}	$+$
x_{21} ... x_{2n_a}	$+$
... ...	...
x_{n_e1} ... $x_{n_e n_a}$	$-$

...

Attributes	Cl
x_{11} ... x_{1n_a}	$-$
x_{21} ... x_{2n_a}	$-$
... ...	...
x_{n_e1} ... $x_{n_e n_a}$	$+$

(e) n_c training sets for grading

Fig. 1. Illustration of stacking and grading. In this hypothetical situation, n_c classifiers are tried on a problem with n_a attributes, n_e examples and $n_l = 2$ classes (t, f).

the prediction of this learner was correct on this particular example. Hence—in contrast to stacking—we leave the original examples unchanged, but instead modify the class labels.

The idea of grading is not entirely new. [9] and [8] independently introduce algorithms based on the same basic idea, but provide only a preliminary evaluation of the approach. We will describe this idea in more detail and compare it to cross-validation, stacking, and a voting technique.

2   Grading

Figure 1(a) shows a hypothetical learning problem with n_e training examples, each of them encoded using n_a attributes x_{ij} and a class label cl_i. In our example, the number of different class labels n_l is 2 (the values t and f) but this is no principal restriction. We are now assuming that we have n_c base classifiers C_k, which were evaluated using some cross-validation scheme. As cross-validation ensures that each example is used as a test example exactly once, we have obtained one prediction for each classifier and for each training example (Figure 1(b)).

A straight-forward use of this prediction matrix is to let each classifier vote for a class, and predict the class that receives the most votes. Stacking, as originally described by [19], makes a more elaborate use of the prediction matrix. It adds the original class labels cl_i to it and uses this new data set—shown in Figure 1(d)—for training another classifier[1]. Examples are classified by submit-

---

[1] In fact, we followed [17] in using probability distributions for stacking. Instead of merely adding the classes that are considered to be most likely by each base classifier, they suggest to add the entire n_l-dimensional class probability vector P_{ik}, yielding a meta dataset with $n_l \times n_c$ attributes instead of only n_c for conventional stacking.

ting them to each base classifier (trained on the entire training set) and using the predicted labels as input for the meta classifier learned from the prediction matrix. Its prediction is then used as the final prediction for the example.

By providing the true class labels as the target function, stacking provides its meta-learner with indirect feedback about the correctness of its base classifiers. This feedback can be made more explicit. Figure 1(c) shows an evaluation of the base classifiers' predictions. Each entry c_{ik} in the prediction matrix is compared to the corresponding label cl_i. Correct predictions ($c_{ik} = cl_i$) are "graded" with +, incorrect predictions with $-$.

Grading makes use of these graded predictions for training a set of meta classifiers that learn to predict when the base classifier is correct. The training set for each of these meta classifiers is constructed using the graded predictions of the corresponding base classifier as new class labels for the original attributes. Thus we have n_c two-class training sets (classes + and $-$), one for each base classifier (Figure 1(e)). We now train n_c level 1 classifiers, everyone of which gets exactly one of these training sets, based on the assumption that different base classifiers make different errors. Thus, every one of these level 1 classifiers tries to predict when its base classifier will err.

Note that the proportion of negatively graded examples in these datasets is simply the error rate of the corresponding base classifier as estimated by the cross-validation procedure. Hence, while selection by cross-validation [13] simply picks the classifier corresponding to the dataset with the fewest examples of class $-$ as the classifier to be used for all test examples, grading tries to make this decision for each example separately by focussing on those base classifiers that are predicted to be correct on this example. In this sense, grading may be viewed as a generalization of selection by cross-validation.

At classification time, each base classifier makes a prediction for the current example. The final prediction is derived from the predictions of those base classifiers that are predicted to be correct by the meta-classification schemes. Conflicts (several classifiers with different base-level predictions are predicted to be correct) may be resolved by voting or by making use of the confidence estimates of the base classifiers.

In our implementation, the confidence[2] of the level 1 classifier will be summed per class and afterwards normalized to yield a proper class probability distribution. In the rare case that no base classifier is predicted to be correct, all base classifiers are used with $(1 - confidence)$ as the new confidence, thus preferring those base classifiers for which the level 1 classifier is more unsure of its decision. The most probable class from the final class distribution is chosen as the final prediction. Ties are cut by choosing among all most probable classes the class which occurs more frequently in the training data.

More formally, let p_{ikl} be the class probability calculated by base classifier k for class l and example i. For simplification, we write P_{ik} to mean $(p_{ik1}, p_{ik2},$

---

[2] We measure confidence with the meta classifiers' estimate of the probability $p(+)$ that the example is classified correctly by the corresponding base classifier. Since meta datasets as defined are two-class, confidence for $-$ is $p(-) = 1 - p(+)$.

$\ldots p_{ikn_l}$), i.e., the vector of all class probabilities for example i and classifier k. The prediction of the base classifier k for example i is the class l with maximum probability p_{ikl}, more formally $c_{ik} = \arg\max_l\{p_{ikl}\}$.

Grading then constructs n_c training sets, one for each base classifier k, by adding the graded predictions g_{ik} as the new class information to the original dataset (g_{ik} is 1 if the base classifier k's prediction for example i was correct (graded $+$) and 0 otherwise). $prMeta_{ik}$ is the probability that base classifier k will correctly predict the class of example i as estimated by meta classifier k.

From this information we compute the final probability estimate for class l and example i. In case at least one meta classifier grades its base classifier as $+$ (i.e., $prMeta_{ik} > 0.5$), we use the following formula:[3]

$$prGrading_{il} = \sum \{prMeta_{ik}|c_{ik} = l \wedge prMeta_{ik} > 0.5\}$$

Otherwise, if no base classifiers are presumed correct by the meta classifiers, we use all base classifiers in our voting. The class with highest probability is then chosen as the final prediction.

3 Empirical Evaluation

In this section, we compare our implementation of grading to stacking, voting, and selection by cross-validation. We implemented Grading in Java within the Waikato Environment for Knowledge Analysis (WEKA).[4] All other algorithms at the base and meta-level were already available within WEKA.

For an empirical evaluation we chose twenty-six datasets from the UCI Machine Learning Repository [2]. The datasets were selected arbitrarily before the start of the experiments and include both two-class and multi-class problems with up to 2310 examples. All reported accuracy estimates are the average of ten ten-fold stratified cross validations, except when stated otherwise.

We evaluated each meta-classification scheme using the following six base learners, which were chosen to cover a variety of different biases.

- DecisionTable: a decision table learner
- J48: a Java port of C4.5 Release 8 [12]
- NaiveBayes: the Naive Bayes classifier using kernel density estimation
- KernelDensity: a simple kernel density classifier
- MultiLinearRegression: a multi-class learner which tries to separate each class from all other classes by linear regression (*multi-response linear regression*)
- KStar: the K* instance-based learner [5]

All algorithms are implemented in WEKA Release 3.1.8. They return a class probability distribution, i.e., they do not predict a single class, but give probability estimates for each possible class.

---

[3] Penalizing cases where the base classifier predicts class l but the meta classifier considers this prediction wrong ($c_{ik} = l \wedge prMeta_{ik} \le 0.5$) did not work as well.

[4] The Java source code of WEKA has been made available at www.cs.waikato.ac.nz

Table 1. Accuracy (%) for all meta-classification schemes.

Dataset	Grading	X-Val	Stacking	Voting
audiology	83.36	77.61	76.02	84.56
autos	80.93	80.83	82.20	83.51
balance-scale	89.89	91.54	89.50	86.16
breast-cancer	73.99	71.64	72.06	74.86
breast-w	96.70	97.47	97.41	96.82
colic	84.38	84.48	84.78	85.08
credit-a	86.01	84.87	86.09	86.04
credit-g	75.64	75.48	76.17	75.23
diabetes	75.53	76.86	76.32	76.25
glass	74.35	74.44	76.45	75.70
heart-c	82.74	84.09	84.26	81.55
heart-h	83.64	85.78	85.14	83.16
heart-statlog	84.22	83.56	84.04	83.30

Dataset	Grading	X-Val	Stacking	Voting
hepatitis	83.42	83.03	83.29	82.77
ionosphere	91.85	91.34	92.82	92.42
iris	95.13	95.20	94.93	94.93
labor	93.68	90.35	91.58	93.86
lymph	83.45	81.69	80.20	84.05
primary-t.	49.47	49.23	42.63	46.02
segment	98.03	97.05	98.08	98.14
sonar	85.05	85.05	85.58	84.23
soybean	93.91	93.69	92.90	93.84
vehicle	74.46	73.90	79.89	72.91
vote	95.93	95.95	96.32	95.33
vowel	98.74	99.06	99.00	98.80
zoo	96.44	95.05	93.96	97.23
Avg	85.04	84.59	84.68	84.88

All base algorithms have their respective strengths and weaknesses and perform well on some datasets and badly on others. Judging by the average accuracy (a somewhat problematic measure, see below), KernelDensity and KStar seem to have the competitive edge. Since space restrictions prevent us from showing all the details, these and other experimental results can be found in [15]. On these base algorithms, we tested the following four meta-classification schemes:

- Grading is our implementation of the grading algorithms. It uses the instance-based classifier IBk with ten nearest neighbors as the meta-level classifier. Our implementation made use of the class probability distributions returned by the meta classifiers. IBk estimates the class probabilities with a Laplace-estimate of the proportion of the neighbors in each class. These estimates are then normalized to yield a proper probability distribution.[5]
- X-Val chooses the best base classifier on each fold by an internal ten-fold CV.
- Stacking is the stacking algorithm as implemented in WEKA, which follows [17]. It constructs the meta dataset by adding the entire predicted class probability distribution instead of only the most likely class. Following [17], we also used MultiLinearRegression as the level 1 learner.[6]
- Voting is a straight-forward adaptation of voting for distribution classifiers. Instead of giving its entire vote to the class it considers to be most likely, each classifier is allowed to split its vote according to its estimate of the class probability distribution for the example. It is mainly included as a benchmark of the performance that could be obtained without resorting to the expensive CV of every other algorithm.

---

[5] $confidence = p(+) = \frac{p + \frac{1}{n_e}}{k + \frac{2}{n_e}}$, n_e is training set size, p of k neighbors graded as $+$.

[6] Relatively global and smooth level-1 generalizers should perform well [19,17].

Table 2. Significant wins/losses for the four meta-classification schemes against themselves and against all meta learning algorithms. The first number shows significant wins for the algorithm in the column, and the second number for the algorithm in the row.

	Grading	X-Val	Stacking	Voting
Grading	—	5/6	7/7	2/4
X-Val	6/5	—	6/3	9/7
Stacking	7/7	3/6	—	7/8
Voting	4/2	7/9	8/7	—
∑ meta	17/14	15/21	21/17	18/19

	Grading	X-Val	Stacking	Voting
DecisionTable	24/0	22/0	24/0	26/0
J48	20/2	17/3	18/1	20/1
KernelDensity	21/1	18/2	20/2	23/2
KStar	19/0	17/4	17/4	19/1
MLR	17/3	12/2	13/7	14/5
NaiveBayes	13/5	14/1	14/5	17/7
∑ base	114/11	100/12	106/19	119/16

As noted above, all algorithms made use of class probability distributions. This choice was partly motivated by the existing implementation of **Stacking** within WEKA, and partly because of the experimental evidence that shows that the use of class probability distributions gives a slightly better performance in combining classifiers with stacking [17] and ensemble methods [1]. We did some preliminary studies which seemed to indicate that this is also the case for **Grading**, but we did not yet attempt a thorough empirical verification of this matter.

Table 1 shows the accuracies for all meta-classification schemes w.r.t. each dataset. Not surprisingly, no individual algorithm is a clear winner over all datasets; each algorithm wins on some datasets and loses on others. On average, **Grading** seems to be slightly more accurate than **Stacking** (85.04% vs. 84.68%). Somewhat surprising is the performance of **Voting**: although it does not use the expensive predictions obtained from the internal cross-validation, it seems to perform no worse than the other algorithms which do use this information. Of course, a comparison of algorithms with their average accuracy over a selection of datasets has to be interpreted very cautiously because of the different baseline accuracies and variances on the different problems. In the following, we take a closer look at the performance differences.

Table 2 shows significant wins/losses of the meta-classification schemes versus themselves and versus the base classifiers. Significant differences were calculated by a t-test with 99% significance level. Positive and negative differences between classifiers were counted as wins and losses respectively. Among the meta-classification schemes, **Stacking** and **Grading** are best with **Grading** slightly behind ($wins − losses = 4$ vs. 3) while **X-Val** and **Voting** lag far behind (more losses than wins). Although this evaluation shows that—contrary to the average performance discussed above—**Voting** does not get close to **Grading** and **Stacking**, it nevertheless outperforms **X-Val**. Hence, adding up the predicted class probabilities of different classifiers seems to be a better decision procedure than selecting the best algorithm by cross-validation, which is consistent with other results on ensembles of classifiers [6]. Table 2 shows that all meta-classification schemes are almost always an improvement over the six base learners. Measured in terms of the differences between wins and losses, Voting and Grading are both on first place (*wins* − *losses* = 103). Interestingly, Stacking's performance seems to be the worst: it has fewer wins than Voting and Grading, and the most losses of all meta-classification schemes. Still, meta-classification schemes are clearly an improvement over base classifiers in any case.

Grading is not worse than any base classifiers on seventeen datasets and better than all base classifiers on two datasets, among them the largest (*segment*). Nine times, Grading is worse than at least one classifier, two times it is worse than two, and never worse than three or more. Our meta-classification scheme can thus be considered an improvement over the base learners in the majority of cases.

However, while our results show no significant differences across the datasets we studied, they also seem to indicate that the performance of stacking and grading varies considerably across different domains. For example, our results seemed to indicate that Grading seems to be better than Stacking on smaller datasets, while it performs worse on on larger datasets.

In summary, the overall performance of Grading is comparable to that of Stacking. Although Grading has a higher average performance over all datasets and has more significant wins against base classifiers, its performance in a head-to-head comparison is equal to that of stacking.

In a second series of experiments, we wanted to test whether our intuitions were correct, and tested our original choice of IBk and the six base learners as level 1 learners. For these experiments, we used only a one-time ten-fold cross-validation all datasets (as opposed to the average of ten cross-validations shown in the previous table).

It turns out that IBk with ten nearest neighbors[7] is in fact one of the best level 1 learners among the seven tested.

The only algorithm that appears to be slightly better is NaiveBayes. However, all relative performances are within 1% of that of IBk, i.e., all algorithms perform about equally good. This came as a little surprise to us, because we had expected that there would be more differences in this wide range of algorithms. Some of the algorithms learn local models (IBk), while others always consider all examples for deriving a prediction (KStar). Likewise, some of the algorithms always use all of the features (NaiveBayes), while others try to focus on important ones (J48). Apparently, the particular type of meta-learning problem encountered in grading has some properties that make it uniformly hard for a wide variety of learning algorithms. We plan further investigations of this matter in the future.

4   Related Work

Our original motivation for the investigation of grading was to evaluate the potential of using qualitative error-characterization techniques proposed by [9] and [8] as an ensemble technique. There are some minor differences to these workshop papers with respect to how we compute the predictions at the meta

---

[7] After choosing IBk as the level 1 classifier, prior to the evaluation described here, we had determined the number of neighbors (10) using a ten-fold CV on all datasets.

level (e.g. the approach of [8] can only handle 2-class problems), but the idea underlying these approaches is more or less the same. This paper provides a thorough empirical evaluation of grading, and compares it to stacking, selection by cross-validation, and voting.

[4] propose the use of *arbiters* and *combiners*. A combiner is more or less identical to stacking. [4] also investigate a related form, which they call an *attribute-combiner*. In this architecture, the original attributes are not replaced with the class predictions, but instead they are added to them. As [14] shows in his paper about bi-level stacking, this may result in worse performance. On the other hand, [8] compared this approach to the above-mentioned 2-class version of grading, and found that grading performs significantly better on the KRK problem. These results have to be reconciled in future work.

An arbiter [4] is a separate, single classifier, which is trained on a subset of the original data. This subset consists of examples on which the base classifiers disagree. They also investigate *arbiter trees*, in which arbiters that specialize in arbiting between pairs of classifiers are organized in a binary decision tree. Arbiters are quite similar in spirit to grading. The main difference is that arbiters use information about the disagreement of classifiers for selecting a training set, while grading uses disagreement with the target function (estimated by a cross-validation) to produce a new training set.

Quite related to grading is also the work by [16], who proposed to use the predictions of base classifiers for learning a function that maps the algorithms' internal confidence measure (e.g., instance typicality for nearest neighbor classifiers or a posteriori probabilities for Naive Bayes classifiers) to an estimate of its accuracy on the output. The main differences to grading is that we use the original feature vectors as inputs, and select the best prediction based on the class probabilities returned by those meta classifiers that predict that their corresponding base classifiers are correct on the example.

A third approach with a similar goal, *meta decision trees* [18], aims at directly predicting which classifier is best to classify an individual example. To this end, it uses information about statistical properties of the predicted class distribution as attributes and predicts the right algorithm from this information. The approach is not modular (in the sense that any algorithm could be used at the meta-level) but implemented as a modification to the decision tree learner C4.5. Grading differs from meta decision trees in that respect, and by the fact that we use the original attributes in the datasets for learning an ensemble of classifiers which learn the errors of each base classifier.

There are many approaches to combine multiple models without resorting to elaborate meta-classification schemes. Best known are ensemble methods such as bagging and boosting, which rely on learning a set of diverse base classifiers (typically via different subsamples of the training set), whose predictions are then combined by simple voting [6]. Another group of techniques, *meta-learning*, focuses on predicting the right algorithm for a particular problem based on characteristics of the dataset [3] or based on the performance of other, simpler learning algorithms [11]. Finally, another common decision procedure (especially with larger datasets) is to take a subsample of the entire dataset and try each algorithm on this sample. This approach was analyzed by [10].

5   Conclusions

We have examined the meta-classification scheme grading. Essentially, the idea behind grading is to train a new classifier that predicts which base classifier is correct on a given example. To that end, the original data set is transformed into a two-class dataset with new class labels that encode whether the the base classifier was able to correctly predict this example in an internal cross-validation or not. This approach may be viewed as a direct generalization of selection by cross-validation, which would always select the base classifier that corresponds to the meta dataset with the highest default accuracy.

The experimental evaluation showed that grading is slightly better than stacking according to some performance measures (like the average performance on a selection of UCI datasets or an indirect comparison to the base classifiers), but in a head-to-head comparison the differences are not statistically significant. Both algorithms, stacking and grading, perform better than voting and selection by cross-validation. We believe that both approaches are valid alternatives that should be considered when working with multiple models.

Our results also show that the method seems to be quite insensitive to the choice of a meta-learning algorithm. This is quite surprising, as we investigated a variety of algorithms with very different biases. We are yet unsure why this is the case.

A possible reason may lie in the fact that for reasonable performances of the base learning algorithms, the two-class meta data set consists of far more correct than incorrectly predicted examples. Hence the learner should be able to deal with imbalanced training sets, which none of the algorithms we tested specializes in. We have not yet investigated the influence of this issue upon the performance of the learning system.

We remain hopeful that our approach may in time become complementary to stacking, in particular if the respective strengths and weaknesses of the two approaches are better understood. To this end, we plan to investigate these issues by performing a strict empirical evaluation of the diversity of both classification schemes as well as study the influence of the diversity of the base classifiers on these meta classification schemes. Our current work concentrates on the definition of a common framework for meta-classification schemes, which allows a thorough experimental and theoretical comparison of the different approaches that have been proposed in the literature.

Acknowledgments This research is supported by the Austrian *Fonds zur Förderung der Wissenschaftlichen Forschung (FWF)* under grant no. P12645-INF and the ESPRIT LTR project METAL (26.357). The Austrian Research Institute for Artificial Intelligence is supported by the Austrian Federal Ministry of Education, Science and Culture. We want to thank Johann Petrak and Gerhard Widmer for valuable discussions and comments.

References

1. Bauer, E., & Kohavi, R. (1999). An empirical comparison of voting classification algorithms: Bagging, boosting, and variants. *Machine Learning*, *36*, 105–169.
2. Blake, C. L., & Merz, C. J. (1998). UCI repository of machine learning databases. http://www.ics.uci.edu/~mlearn/MLRepository.html. Department of Information and Computer Science, Unifirsty of California at Irvine, Irvine CA.
3. Brazdil, P. B., Gama, J., & Henery, B. (1994). Characterizing the applicability of classification algorithms using meta-level learning. *Proceedings of the 7th European Conference on Machine Learning (ECML-94)* (pp. 83–102). Catania, Italy: Springer-Verlag.
4. Chan, P. K., & Stolfo, S. J. (1995). A comparative evaluation of voting and meta-learning on partitioned data. *Proceedings of the 12th International Conference on Machine Learning (ICML-95)* (pp. 90–98). Morgan Kaufmann.
5. Cleary, J. G., & Trigg, L. E. (1995). K*: An instance-based learner using an entropic distance measure. *Proceedings of the 12th International Conference on Machine Learning* (pp. 108–114). Lake Tahoe, CA.
6. Dietterich, T. G. (2000a). Ensemble methods in machine learning. *First International Workshop on Multiple Classifier Systems* (pp. 1–15). Springer-Verlag.
7. Kononenko, I., & Bratko, I. (1991). Information-based evaluation criterion for classifier's performance. *Machine Learning*, *6*, 67–80.
8. Koppel, M., & Engelson, S. P. (1996). Integrating Multiple Classifiers By Finding Their Areas of Expertise. *Proceedings of the AAAI-96 Workshop on Integrating Multiple Models* (pp. 53–58).
9. Ortega, J. (1996). Exploiting Multiple Existing Models and Learning Algorithms *Proceedings of the AAAI-96 Workshop on Integrating Multiple Models* (pp. 101–106).
10. Petrak, J. (2000). Fast subsampling performance estimates for classification algorithm selection. *Proceedings of the ECML-00 Workshop on Meta-Learning: Building Automatic Advice Strategies for Model Selection and Method Combination* (pp. 3–14). Barcelona, Spain.
11. Pfahringer, B., Bensusan, H., & Giraud-Carrier, C. (2000). Meta-learning by landmarking various learning algorithms. *Proceedings of the 17th International Conference on Machine Learning (ICML-2000)*. Stanford, CA.
12. Quinlan, J. R. (1993). *C4.5: Programs for Machine Learning*. San Mateo, CA: Morgan Kaufmann.
13. Schaffer, C. (1993). Selecting a classification method by cross-validation. *Machine Learning*, *13*, 135–143.
14. Schaffer, C. (1994). Cross-validation, stacking and bi-level stacking: Meta-methods for classification learning. In P. Cheeseman and R. W. Oldford (Eds.), *Selecting models from data: Artificial Intelligence and Statistics IV*, 51–59. Springer-Verlag.
15. Seewald, A. K., & Fürnkranz, J. (2001). *Grading classifiers* (Technical Report OEFAI-TR-2001-01). Austrian Research Institute for Artificial Intelligence, Wien.
16. Ting, K. M. (1997). Decision combination based on the characterisation of predictive accuracy. *Intelligent Data Analysis*, *1*, 181–206.
17. Ting, K. M., & Witten, I. H. (1999). Issues in stacked generalization. *Journal of Artificial Intelligence Research*, *10*, 271–289.
18. Todorovski, L., & Džeroski, S. (2000). Combining multiple models with meta decision trees. *Proceedings of the 4th European Conference on Principles of Data Mining and Knowledge Discovery (PKDD-2000)* (pp. 54–64). Lyon, France: Springer-Verlag.
19. Wolpert, D. H. (1992). Stacked generalization. *Neural Networks*, *5*, 241–260.

Finding Informative Rules in Interval Sequences[*]

Frank Höppner and Frank Klawonn

Department of Electrical Engineering and Computer Science
University of Applied Sciences, Emden
Constantiaplatz 4
D-26723 Emden, Germany
frank.hoeppner@ieee.org

Abstract. Observing a binary feature over a period of time yields a sequence of observation intervals. To ease the access to continuous features (like time series), they are often broken down into attributed intervals, such that the attribute describes the series' behaviour within the segment (e.g. increasing, high-value, highly convex, etc.). In both cases, we obtain a sequence of interval data, in which temporal patterns and rules can be identified. A temporal pattern is defined as a set of labeled intervals together with their interval relationships described in terms of Allen's interval logic. In this paper, we consider the evaluation of such rules in order to find the most informative rules. We discuss rule semantics and outline deficiencies of the previously used rule evaluation. We apply the J-measure to rules with a modified semantics in order to better cope with different lengths of the temporal patterns. We also consider the problem of specializing temporal rules by additional attributes of the state intervals.

1 Introduction

Most of the data analysis methods assume static data, that is, they do not consider time explicitly. The value of attributes is provided for a single point in time, like "patient A has disease B". If we observe the attributes over a period of time, we have to attach a time interval in which the attribute holds, for example "patient A has had disease B from 1st to 7th of July". It may happen that patient A gets disease B a second time, therefore sequences of labeled state intervals (state sequences) can be viewed as a natural generalization of static attributes to time-varying domains. Compared to static data, little work has been done to analyse interval data.

Even in continuous domains discretization into intervals can be helpful. As an example, the problem of finding common characteristics of multiple time series or different parts of the same series requires a notion of similarity. If a process is subject to variation in time (translation or dilation), those measures used traditionally for estimating similarity (e.g. pointwise Euclidean norm) will fail

---

[*] This work has been supported by the Deutsche Forschungsgemeinschaft (DFG) under grant no. Kl 648/1.

F. Hoffmann et al. (Eds.): IDA 2001, LNCS 2189, pp. 125–134, 2001.

in providing useful hints about the time series similarity in terms of the cognitive perception of a human. This problem has been addressed by many authors in the literature [1,4,6]. In [8] we have used qualitative descriptions to divide up the time series in small segments (like increasing, high-value, convexly decreasing, etc.), each of it easy to grasp and understand by the human. Then, matching of time series reduces to the identification of patterns in interval sequences.

Motivated by association rule mining [2], and more specific the discovery of frequent episodes in event sequences [10], we have proposed a method to discover frequent temporal patterns from a single state sequence in [8]. From the patterns rules can be formed that identify dependencies between such patterns. In this paper, we reconsider rule semantics and the problem of rule evaluation. We use the J-measure [11] to rank rules by their information content. And we discuss how to specialize rules by incorporating additional information about the state intervals to further improve the rules.

The outline of the paper is as follows: In Sect. 2 we define our notion of a state sequence and temporal patterns. We briefly summarize the process of mining frequent patterns in Sect. 3. We discuss the problems in measuring how often a pattern occurs in a sequence in section 4 and concentrate on rule evaluation in section 5. Section 6 deals with specializing rules to increase their usefulness (information content).

2   Temporal Patterns in State Sequences

Let S denote the set of all possible trends, properties, or states that we want to distinguish. A state $s \in S$ holds during a period of time $[b, f)$ where b and f denote the *initial point* in time when we enter the state and the *final point* in time when the state no longer holds. A state sequence on S is a series of triples defining state intervals

$$(b_1, s_1, f_1), (b_2, s_2, f_2), (b_3, s_3, f_3), (b_4, s_4, f_4), \ldots$$

where $b_i \leq b_{i+1}$ and $b_i < f_i$ holds. We do not require that one state interval has ended before another state interval starts. This enables us to mix up several state sequences (possibly obtained from different sources) into a single state sequences.

We use Allen's temporal interval logic [3] to describe the relation between state intervals. For any pair of intervals we have a set $\mathcal{I}$ of 13 possible relationships. For example, we say "A meets B" if interval A terminates at the same point in time at which B starts. The inverse relationship is "B is-met-by A". Given n state intervals (b_i, s_i, f_i), $1 \leq i \leq n$, we can capture their relative positions to each other by an $n \times n$ matrix R whose elements $R[i, j]$ describe the relationship between state interval i and j. As an example, consider the state sequence in Fig. 1. Obviously state A is always followed by B. And the lag between A and B is covered by state C. Below the state interval sequence both of these patterns are written as a matrix of interval relations. Formally, a temporal pattern P of size n is defined by a pair (s, R), where $s : \{1, .., n\} \to S$ maps index

i to the corresponding state, and $R \in \mathcal{I}^{n \times n}$ denotes the relationship between $[b_i, f_i]$ and $[b_j, f_j]$[1]. By $\dim(P)$ we denote the dimension (number n of intervals) of the pattern P. If $\dim(P) = k$, we say that P is a k-pattern. Of course, many sets of state intervals map to the same temporal pattern. We say that the set of intervals $\{(b_i, s_i, f_i) \mid 1 \leq i \leq n\}$ is an *instance* of its temporal pattern (s, R). If we remove some states (and the corresponding relationships) from a pattern, we obtain a *subpattern* of the original pattern.

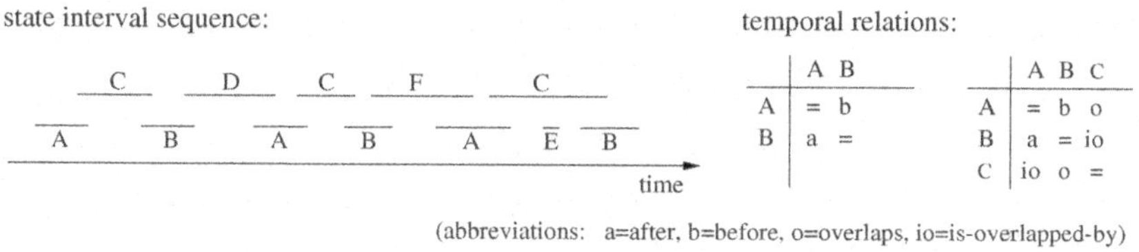

Fig. 1. Example for state interval patterns expressed as temporal relationships.

3   Pattern Discovery

In this section we briefly review the process of pattern discovery and rule generation, which is on a coarse view roughly the same with many kinds of patterns. For a more detailed treatment, see [2,10].

As already mentioned, we intend to search for frequent temporal patterns. The support of a pattern denotes how often a pattern occurs. Postponing the exact definition of support for the moment, a pattern is called *frequent*, if its support exceeds a threshold supp_{min}. To find all frequent patterns we start in a first database pass with the estimation of the support of every single state (also called candidate 1-patterns). After the kth run, we remove all candidates that have missed the minimum support and create out of the remaining frequent k-patterns a set of candidate $(k+1)$-patterns whose support will be estimated in the next pass. This procedure is repeated until no more frequent patterns can be found. The fact that the support of a pattern is always less than or equal to the support of any of its subpatterns guarantees that we do not miss any frequent patterns.

After having determined all frequent patterns, we can construct rules $X \mapsto Y$ from every pair (X, Y) of frequent patterns where X is a subpattern of Y. If the confidence of the rule $\text{conf}(A \rightarrow B) = \frac{\text{supp}(A \rightarrow B)}{\text{supp}(A)}$ is greater than a minimal confidence, the rule is printed as an interesting candidate for an important rule.

The space of temporal patterns is even larger than the space of itemsets in association rule mining [2] or episodes in event sequences [10], since for every pair of objects (intervals) we have to maintain a number of possible relationships.

---

[1] To determine the interval relationships we assume closed intervals $[b_i, f_i]$

Thus, efficient pruning techniques are a must to overcome the combinatorial explosion of possible patterns. We refer the reader to [8] for the details of the frequent pattern discovery process.

4 Counting Temporal Patterns in State Series

What is a suitable definition of support in the context of temporal patterns? Perhaps the most intuitive definition is the following: The support of a temporal pattern is the number of temporal patterns in the state series. Let us examine this definition in the context of the following example:

$$
\text{if } \begin{array}{c|cc} & A & B \\ \hline A & = & b \\ B & a & = \end{array} \quad \text{then} \quad \begin{array}{c|cccc} & A & B & A & B \\ \hline A & = & b & b & b \\ B & a & = & b & b \\ A & a & a & = & m \\ B & a & a & im & = \end{array} \quad \text{with probability } p \tag{1}
$$

We call the pattern in the premise the *premise pattern P* and the pattern in the conclusion the *rule pattern R*. The rule pattern also comprises the premise pattern. If we remove the premise from the rule pattern we obtain the *conclusion pattern C*. In (1), the pattern is depicted below the relation matrix (a=after, b=before, m=meets, im=is-met-by).

How often does the pattern in the conclusion occur in the state series in Fig. 2(a)? We can easily find 3 occurrences as shown in Fig. 2(b). The remaining (unused) states do not form a 4th pattern. How often does the premise pattern occur? By pairing states (1,4), (2,6), (3,7), etc. we obtain a total number of 7. So we have $p = \frac{3}{7}$. This may correspond to our intuitive understanding of the rule, but we can improve p to $\frac{4}{7}$ when using the rule pattern assignment in Fig. 2(c). The latter assignment is perhaps less intuitive than the first, because the pattern's extension in time has increased. But now we have a state series that is assembled completely out of rule patterns, there is no superfluous state. Then, would it not be more natural to have a rule probability near 1 instead of $\frac{4}{7}$?

The purpose of the example is to alert the reader that the rule semantics is not that clear as might be expected. Furthermore, determining the maximum number of pattern occurrences is a complex task and does not necessarily correspond to our intuitive counting. We therefore define the total time in which (one or more) instances of the patterns can be observed in a sliding window of width w as the support of the pattern. If we divide the support of a pattern by the length of the state sequence plus the window width w we obtain the relative frequency p of the pattern: If we randomly select a window position we can observe the pattern with probability p.

We note in passing, that the rule probability $p = \frac{3}{7}$ is obtained by using the concept of minimal occurrences [10], as used by Mannila et al. for the discovery of frequent episodes in event sequences. An instance of a pattern P in a time interval

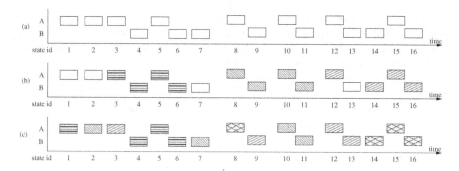

Fig. 2. Counting the occurrences of temporal patterns. (States with different labels (A and B) are drawn on different levels. Note that the pattern of interest (1) requires a *meets* relation in the conclusion.)

$[t_0, t_3]$ is a minimal occurrence, if there is no $[t_1, t_2] \subset [t_0, t_3]$ such that there is also an instance of P within $[t_1, t_2]$. We do not follow this idea, since we consider the rule discovery to be less robust when using minimal occurrences. Consider a pattern "A before B", where the length of the intervals is characteristic for the pattern. If the interval sequence is noisy, that is, there may be additional short B intervals in the gap of the original pattern, the minimal occurrence of A and noisy B would prevent the detection of A and original B. Rule specialization as we will discuss in Sect. 6 would not have a chance to recover the original pattern. Such a situation can easily occur in an automatically generated state sequence which describes the local trend of a time series, where noise in the time series will cause noise in the trend sequence.

5   Rule Evaluation

Let us consider the case when two patterns perfectly correlate in a state sequence. Using again our example rule (1), let us assume that whenever we observe "A before B", we find another two states A and B such that in combination they form the rule pattern in (1). Usually, the support and confidence value of the rule are used to decide about its usefulness [2]. If a sequence consists of rule patterns only, we should expect a confidence value near 1, however, this is not necessarily the case.

5.1   Modified Rule Semantics

There are two possible reasons for a low rule probability (or confidence). The greater the (temporal) extent of the pattern, the lesser the probability of observing the pattern in the sliding window. Consequently, the confidence of a rule decreases as the extent of the rule pattern increases. Secondly, if there are more premise patterns and less rule patterns, rule confidence also decreases. The latter is what we usually associate with rule confidence, whereas the first seems a bit counterintuitive. To reduce the effect of pattern extension, we define a different rule semantics: *Given a randomly selected sliding window that contains an instance of the premise pattern, then with probability p this window overlaps a sliding window that contains the rule pattern.* Loosely speaking, the effect of this redefinition is an increase in the support of the rule, since we substitute "number of windows that contain rule pattern" by "number of windows that contain the premise and overlap a window with a rule pattern".

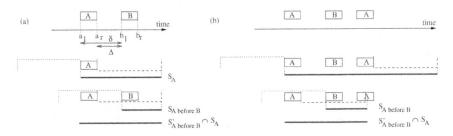

Fig. 3. Support sets of "A" and "A before B", determined by the sliding window positions when the pattern is observed for the first time (dotted window position) and for the last time (dashed window position).

Figure 3(a) illustrates the problem. We consider the premise pattern $P = $ "A", the conclusion pattern $C = $ "B", and the rule pattern $R = $ "A before B", w denotes the window width. For any pattern Q, let S_Q be the support set of a pattern Q, that is, a set of sliding window positions for which Q is observable. Then we have $\text{supp}(Q) = \text{card}(S_Q)$ (cardinality). In the example we have $\text{supp}(P) = \text{card}([a_l, a_r + w])$ and $\text{supp}(C) = \text{card}([b_l, b_r + w])$. Here $\text{supp}(R) = \text{supp}(P \cap C)$ holds and hence $\text{supp}(R) = \text{card}(S_A \cap S_B) = \text{card}([b_l, a_r + w])$. Thus, defining $\Delta := b_l - a_r < w$ and denoting the length of A by l_A, the rule confidence is

$$\text{conf}(A \to A \text{ before } B) = \frac{\text{card}(S_A \cap S_B)}{\text{card}(S_A)} = \frac{a_r + w - b_l}{a_r + w - a_l} = \frac{w - \Delta}{w + l_A}$$

Obviously, as the gap (Δ) between A and B increases, the confidence approaches zero. Now for any pattern Q, let $S'_Q := S_Q \cup \{t - w \mid t \in S_P\}$. S'_Q can be interpreted as the support of a pseudo-pattern "pattern Q is visible or will be visible within time w". If we now replace the cardinality of "windows that contain rule patterns" by the cardinality of "windows that contain premise and overlap a window that contains rule patterns" as required by the new semantics, we obtain[2]

$$\text{conf}(A \to A \text{ before } B) = \frac{\text{card}(S_A \cap S'_B)}{\text{card}(S_A)} = \frac{a_r + w - a_l}{a_r + w - a_l} = 1$$

---

[2] Here we have $S_R = S_A \cap S_B$ and therefore $S'_R \cap S_P = S'_A \cap S'_B \cap S_A = S'_B \cap S_A$.

Thus, as long as we can see A and the beginning of B within the sliding window, we obtain a confidence value of 1, no matter how far A and B are apart. Cases where no conclusion pattern occurs are not affected by this modification (see Fig. 3(b)). Thus, this modification helps to recover the usual semantics of confidence values.

The sets S_P have been determined while searching for frequent patterns anyway [8], they can be handled easily as sorted lists of intervals. Therefore the operations discussed above can be implemented efficiently without looking at the data again: We replace every interval $[l, r] \in S_Q$ by $[l - w, r]$ to obtain S'_Q. In general, rule confidence is then given by

$$\text{conf}(P \to R) = \frac{\text{card}(S_P \cap S'_R)}{\text{card}(S_P)}$$

5.2   Information Content of a Rule

Usually one obtains a large number of frequent patterns and thus a large number of rules. Considerable efforts have been undertaken in the literature to make the vast amount of rules more amenable. We use the J-measure [11] to rank the rule by their information content. It is considered as one of the most promising measures for rule evaluation [5], however, it is still not widely used. Given a rule "if $Y = y$ then $X = x$" on random variables X and Y, the J-measure compares the a priori distribution of X with the a posteriori distribution of X given that $Y = y$. In the context of a rule, we are only interested in two cases, given that $Y = y$, either the rule was right ($X = x$) or not ($X = \bar{x}$), that is, we only consider the distribution of X over $\{x, \bar{x}\}$. Then, the relative information

$$j(X|Y = y) = \sum_{z \in \{x, \bar{x}\}} Pr(X = z | Y = y) \log_2 \left(\frac{Pr(X = z | Y = y)}{Pr(X = z)} \right)$$

yields the *instantaneous* information that $Y = y$ provides about X (j is also known as the Kullbach-Leibler distance or cross-entropy). When applying the rule multiple times, on average we have the information $J(X|Y = y) = Pr(Y = y) \cdot j(X|Y = y)$. The value of J is bounded by ≈ 0.53 bit.

In our context, the random variable Y indicates whether the premise occurred in the sliding window $\mathcal{W}$ or not. The probability $Pr(P \in \mathcal{W})$ when choosing a sliding window position at random is $\text{supp}(y)/T$ where T is the support of the whole sequence. The random variable X indicates whether the rule pattern has occurred. The a priori probability for $R \in \mathcal{W}$ is $\text{supp}(S_R)/T$, the a posteriori probability is given by $\text{supp}(S_R)/\text{supp}(S_P) = \text{conf}(P \to R)$. When using the modified rule semantics, we have to replace S_R by $S'_R \cap S_P$.

5.3   From Rules to Correlations

We have investigated rules $P \to R$ so far, what about $C \to R$ (where C is the conclusion pattern that is determined uniquely by P and R)? If $P \to R$ and

$C \rightarrow R$ hold, then we have a correlation or equivalence $P \leftrightarrow_R C$, that is, the premise is an indication for the conclusion and also vice versa. We can easily extend the rule evaluation to consider correlations. Then, Y denotes a random variable that indicates whether the conclusion has been found in the sliding window (or in a window overlapped by it, if we use modified rule semantics), thus $Pr(C \in \mathcal{W}) = \text{card}(S_C)$ (or $\text{card}(S'_C)$). The random variable X is left unchanged. So we obtain two J-values for $P \rightarrow R$ and $C \rightarrow R$; if one of them is much higher than the other, we can print the rule $P \rightarrow R$ or $C \rightarrow R$, if both values are similar we print $P \leftrightarrow_R C$.

6 Rule Specialisation

The rule evaluation considers only the interval relationships of temporal patterns, but often there is additional information available for each interval. For example, we have not yet evaluated the length of the intervals, or the size of a gap between two states, etc. These lengths are always available when dealing with interval data, but there might be additional information attached to the intervals. For instance, if the intervals denotes ingredients in a chemical process, an additional attribute might denote the intensity or dose of the admixture. A rule that seems interesting to an expert might not have reached the desired confidence value or information content, unless this additional information is incorporated into the rule. For instance, the desired product quality might be achieved only if admixture D has been supplemented to the process at a dose greater than x. In this section we consider the problem of improving rule confidence using such additional state information.

Given a rule $P \rightarrow R$ with temporal patterns P and R and a real-valued attribute a attached to one of the states used in R. Besides some notational differences, we do not make a distinction between attributes of states that occur in the premise (e.g. "if $A \wedge \text{length}(A) < 3$ then A before B") or in the conclusion (e.g. "if A then A before $B \wedge \text{length}(B) > 1$"). Potentially it is possible to improve the information content of a rule in both cases. For notational convenience, however, let us consider the first case, where we examine an attribute of a state in the premise. Now, we run once through the database and store for each instance i of the rule pattern a triple $(a_i, I_i^{(P)}, I_i^{(R)})$, where a_i denotes the value of the attribute, $I_i^{(P)}$ the support interval of the premise instance, and $I_i^{(R)}$ the support interval of the rule instance. For all instances i of the premise pattern that cannot be completed to a rule pattern we store $(a_i, I_i^{(P)}, I_i^{(R)} = \emptyset)$. In contrast to the frequent pattern mining process, now we are not satisfied if we know that there is an occurrence of the rule pattern in the sliding window, but this time we are interested in *all* occurrences with all possible state combinations. This is computationally more expensive, therefore only selected rules should be considered (e.g., best 100 rules without specialisation).

Next, we have to find a threshold α such that the J-value of either $P \wedge (a > \alpha) \rightarrow R$ or $P \wedge (a < \alpha) \rightarrow R$ is maximized. This can be accomplished by sweeping α once through $[\min_i a_i, \max_i a_i]$ and calculating the J-value each time. When J

becomes maximal, we have found the α value that yields the most informative rule. Having done this for all available attributes, we specialise the rule with the most informative attribute. Then, we can refine the specialised rule again, or use the bounds on the J-value [11] to stop when no improvement is possible.

Sweeping through the range of possible attribute values is done incrementally. Let us assume that the indices are chosen such that $a_{i+1} > a_i$, that is, we sort by attribute values. Furthermore, without loss of generality we assume that for no i we have $a_{i+1} = a_i$. If there are i and j with $a_i = a_j$, we substitute $(a_i, I_i^{(P)}, I_i^{(R)})$ and $(a_j, I_j^{(P)}, I_j^{(R)})$ by $(a_i, I_j^{(P)} \cup I_i^{(P)}, I_j^{(R)} \cup I_i^{(R)})$. Now, we run once through the indices and set $\alpha = \frac{a_i + a_{i+1}}{2}$. We start with empty sets for the support of the rule pattern S_R and premise pattern S_P. After the incrementation of i and α, we incrementally update S_P to $S_P \cup I_i^{(P)}$ and S_R to $S_R \cup I_i^{(R)}$. Given the support sets S_P and S_R we can now calculate the J-value for this α. If we want to check for correlations rather than just rules, we additionally maintain the support of the conclusion pattern S_C.

7   Example

We have applied our technique to various real data sets (weather data, music) that would require more detailed background information in order to understand the results. Due to the lack of space, we consider only a small artificial example. We have generated a test data set where we have randomly toggled three states A, B, and C at discrete time points in $\{1, 2, ..., 9000\}$ with probability 0.2, yielding a sequence with 2838 states. Whenever we have encountered a situation where only A is active during the sequence generation, we generate with probability 0.3 a 4-pattern A meets B, B before C, and C overlaps a second B instance. The length and gaps in the pattern were chosen randomly out of $\{1, 2, 3\}$. We have executed the pattern discovery ($\text{supp}_{min} = 2\%$) and rule generation process several times, using the old and new rule semantics and different sliding window widths (8,10,12). We consider the artificially embedded pattern and any subpattern consisting of at least 3 states as interesting. As expected, using the old rule semantics the confidence value is not very helpful in finding interesting rules. Most of the top-ranking rules were not interesting. Among the top 10 rules, we have found 1/2/3 interesting patterns for $w = 8/10/12$, they all had 2-3 states in the premise and 1 in the conclusion pattern. The J-measure yields much better results, even when using the old semantics. When using the modified semantics, we obtain higher confidence values and J-values. The top 5 rules rated by J-values were identical, regardless of the window width, among them all 3 possible rules with 4 states. It is interesting to note that the rule $A \rightarrow BCB$ ranges among the top 5 although its confidence value is still clearly below 0.5 (and the best interesting rule has a confidence of 0.96).

In a second dataset, we have created the described pattern whenever the length of the A interval is greater or equal to 5. For this dataset the rule "$AB \rightarrow CB$" obtained conf $= 0.75$ and $J = 0.26$ bit. We have searched for a threshold α to specialize the rule with "length(A) $\geq \alpha$". When comparing the rules with different α values we obtain a single rule where J becomes maximal with the correct value ($\alpha = 5$). The confidence increases to 0.85 and the information content to 0.34 bit. The confidence value for $\alpha = 5$ represents only a local maximum, beyond $\alpha = 8$ confidence increases monotonically with α. The run time for the last example was slightly above 1 minute on a Mobile Pentium II, 64 MB Linux computer (40s pattern discovery (see also [8]); 28s rule evaluation (naive implementation); 2s rule specialization).

8   Conclusion

Compared to static data, the development of attributes over time is much more difficult to understand by a human. Rather than explaining the temporal process as a whole, which is usually very difficult or even impossible, finding local rules or correlations between temporal patterns can help a human to understand interdependencies and to develop a mental model of the data [9]. In this paper, we have discussed means to find out the most promising temporal rules, which have been generated out of a set of frequent patterns in a state sequence [8]. In combination, modified rule semantics, J-measure, and rule specialisation are much better suited to rank informative rules than support and confidence values alone.

References

1. R. Agrawal, C. Faloutsos, and A. Swami. Efficient similarity search in sequence databases. In *Proc. of the 4th Int. Conf. on Foundations of Data Organizations and Algorithms*, pages 69–84, Chicago, 1993.
2. R. Agrawal, H. Mannila, R. Srikant, H. Toivonen, and A. I. Verkamo. Fast discovery of association rules. In [7], chapter 12, pages 307–328. MIT Press, 1996.
3. J. F. Allen. Maintaing knowledge about temporal intervals. *Comm. ACM*, 26(11):832–843, 1983.
4. D. J. Berndt and J. Clifford. Finding patterns in time series: A dynamic programming approach. In [7], chapter 9, pages 229–248. MIT Press, 1996.
5. M. Berthold and D. J. Hand, editors. *Intelligent Data Analysis*. Springer, 1999.
6. C. Faloutsos, M. Ranganathan, and Y. Manolopoulos. Fast subsequence matching in time-series databases. In *Proc. of ACM SIGMOD Int. Conf. on Data Management*, May 1994.
7. U. M. Fayyad, G. Piatetsky-Shapiro, P. Smyth, and R. Uthurusamy, editors. *Advances in Knowledge Discovery and Data Mining*. MIT Press, 1996.
8. F. Höppner. Discovery of temporal patterns – learning rules about the qualitative behaviour of time series. In *Proc. of the 5th Europ. Conf. on Princ. and Pract. of Knowl. Discovery in Databases*, Freiburg, Germany, Sept. 2001. Springer.
9. K. B. Konstantinov and T. Yoshida. Real-time qualitative analysis of the temporal shapes of (bio)process variables. *Artificial Intelligence in Chemistry*, 38(11):1703–1715, Nov. 1992.
10. H. Mannila, H. Toivonen, and A. I. Verkamo. Discovery of frequent episodes in event sequences. Technical Report 15, University of Helsinki, Finland, Feb. 1997.
11. P. Smyth and R. M. Goodman. Rule induction using information theory. In *Knowledge Discovery in Databases*, chapter 9, pages 159–176. MIT Press, 1991.

Correlation-Based and Contextual Merit-Based Ensemble Feature Selection

Seppo Puuronen, Alexey Tsymbal, and Iryna Skrypnyk

Department of Computer Science and Information Systems, University of Jyväskylä,
P.O. Box 35, FIN-40351 Jyväskylä, Finland
{sepi, alexey, iryna }@cs.jyu.fi

Abstract. Recent research has proved the benefits of using an ensemble of diverse and accurate base classifiers for classification problems. In this paper the focus is on producing diverse ensembles with the aid of three feature selection heuristics based on two approaches: correlation and contextual merit -based ones. We have developed an algorithm and experimented with it to evaluate and compare the three feature selection heuristics on ten data sets from UCI Repository. On average, simple correlation-based ensemble has the superiority in accuracy. The contextual merit -based heuristics seem to include too many features in the initial ensembles and iterations were most successful with it.

1 Introduction

Recent research has proved the benefits of using an ensemble of diverse and accurate base classifiers for classification problems [6]. One way to construct a base classifier is to use supervised machine learning. In this approach a learning algorithm is given a set $\mathbf{T}$ of training instances of the form $\{(\mathbf{x}_1, y_1), \ldots, (\mathbf{x}_M, y_M)\}$, where the vectors $\mathbf{x}_i = < x_{i,1}, \ldots, x_{i,N} >$ are composed of the values $x_{i,j}$ of the features $f_j \in \mathbf{F}$. Commonly, a diverse ensemble of base classifiers is obtained using different training sets $\mathbf{T}_k \subseteq \mathbf{T}$ for the construction of different base classifiers h_k. Well known bagging and boosting methods build training sets $\mathbf{T}_k$ by selecting different subsets of instances or changing weights of instances.

We apply another approach to form the training sets $\mathbf{T}_k$. Instead of selecting subsets of instances we use all instances and select for each $\mathbf{T}_k$ a subset of features $\mathbf{F}_k \subseteq \mathbf{F}$ that defines which features are taken into account in the construction of a base classifier h_k. The subsets of features are selected using feature selection heuristics. In this paper we describe and experiment with three feature selection heuristics: the basic (SCR) and advanced (CR) correlation-based heuristics, and the contextual merit -based (CM) heuristics [1], [2], [3], [8]. In the context of an ensemble, an additional goal for feature selection is to find multiple feature subsets that produce an ensemble of base classifiers promoting productive disagreement among the base classifiers [7].

Section 2 describes the heuristics. In Section 3 we present our iterative algorithm. In Section 4 the results of our experimental study are described. We summarize with conclusions in Section 5.

F. Hoffmann et al. (Eds.): IDA 2001, LNCS 2189, pp. 135–144, 2001.

2   Feature Selection Heuristics

Correlation-based approach is a widespread approach used to estimate interrelations between two features, or between a feature and the class variable. The approach that takes into account only feature-class Pearson's correlation coefficient r_{xy} is called Simple Correlation-Based (SCR) feature selection in this paper (see for example [8]. The advanced correlation-based heuristic (CR) takes into account the correlations between the pairs of features, too. It is calculated as in [2] by means of the formula (1):

$$CR_{F_k} = \frac{n|\bar{r}_{f_i c}|}{\sqrt{n + n(n-1)|\bar{r}_{f_i f_j}|}}. \tag{1}$$

where F_k is a feature subset containing n features, c is a class, $\bar{r}_{f_i c}$ is the mean feature-class correlation ($f_i \in F$), and $\bar{r}_{f_i f_j}$ is the average feature-feature intercorrelation ($f_i, f_j \in F$). For categorical variables we apply binarization as in [2].

The main assumption of the CM measure, which was developed in [3] is that features important for classification should differ significantly in their values to predict instances from different classes. The CM measure assigns a merit value to a given feature taking into account the degree to which the other features are capable to discriminate between the same instances as the given feature. We calculate the value of CM measure CM_{f_i} of a feature f_i as it has been presented in [3] according the formula (2):

$$CM_{f_i} = \sum_{r=1}^{M} \sum_{\mathbf{x}_k \in \overline{C(\mathbf{x}_i)}} w_{\mathbf{x}_i \mathbf{x}_k}^{(f_i)} d_{\mathbf{x}_i \mathbf{x}_j}^{(f_i)}. \tag{2}$$

where M is the number of instances, $\overline{C(\mathbf{x}_i)}$ is the set of vectors which are not from the same class as the vector $\mathbf{x}_i$, and $w_{\mathbf{x}_i \mathbf{x}_k}^{(f_i)}$ is a weight chosen so that vectors that are close to each other, i.e., they differ only in a few features, have greater influence in determining each feature's CM measure. In [3] weights $w_{\mathbf{x}_i \mathbf{x}_k}^{(f_i)} = \frac{1}{D_{\mathbf{x}_i \mathbf{x}_k}^2}$ were used when $\mathbf{x}_k$ is one of the K nearest neighbors of $\mathbf{x}_i$ in terms of the distance $D_{\mathbf{x}_i \mathbf{x}_k}$ between the vectors $\mathbf{x}_i$ and $\mathbf{x}_k$, in the set $\overline{C(\mathbf{x}_i)}$, and $w_{\mathbf{x}_i \mathbf{x}_k}^{(f_i)} = 0$ otherwise. K is the binary logarithm of the number of vectors in the set $\overline{C(\mathbf{x}_i)}$ as in [3].

3   An Iterative Algorithm for Ensemble Feature Selection

In this chapter an overview of our algorithm EFS_ref for ensemble feature selection using the three heuristics above is presented (a previous version using only CM-based heuristic was presented in [10]). The algorithm constructs three ensembles each including as many base classifiers as there exist different classes

among the instances of the training set. The objective is to build each base classifier on the feature subset that includes features most relevant for distinguishing
the corresponding class from the other classes according to each heuristic. It
is necessary to note that in this algorithm the base classifiers are not binary,
but each of them distinguishes all the classes present in the data set. First,
the EFS_ref algorithm (see below) generates the initial ensemble. Second, the
algorithm generates new candidate ensembles inside the iterative loop using a
different procedure.

```
Algorithm EFS_ref(DS)
  DS     the whole data set
  TRS    training set
  VS     validation set used during the iterations
  TS     test set
  Ccurr        the current ensemble of base classifiers
  Accu   the accuracy of the final ensemble
  FS     set of feature subsets for base classifiers
  Threshold Threshold value used to select the features for the
    initial ensemble
  Begin
    divide_instances(DS,TRS,VS,TS) {divides DS into TRS, VS,
    and TS using stratified random sampling}
    for Heuristics{CM,Corr,Scorr}
      Ccurr=build_initial_ensemble(Heuristic,TRS,FS,Threshold)
      loop
        cycle(Delete,Heuristic,TRS,Ccurr,FS,VS)
        {developes candidate ensembles trying to delete features
        and updates Accu,Ccurr,and FS when necessary}
        cycle(Add,Heuristic,TRS,Ccurr,FS,VS)
      until no_changes
      Accu=accuracy(TS,Ccurr)
    end for
  end algorithm EFS_ref
```

With each heuristic, the number of features for the initial ensemble is fixed for
each data set separately. The features with the highest merit values are selected.
The selection is based on a predefined threshold value for each heuristics and
different data sets. The threshold is fixed ($\in \{0.1, 0.2, ..., 0.9\}$) in advance so that
the accuracy of the ensemble over the training set is the highest one among all
the accuracies obtained with the threshold values for the weighted voting. After
the features are selected the base classifiers are built using the C4.5 learning
algorithm over all instances of the training set taking into account only the
selected features. The main outline of the algorithm forming the initial ensemble
is presented below.

The initial ensemble is modified by suggesting replacement of one base classifier during each iteration. First, the least diverse base classifier is recognized.
Then, the algorithm forms a new candidate base classifier to replace the selected

one. The new candidate base classifier is learned using a modified feature subset that is formed either excluding or including one feature from the corresponding feature subset. With the correlation-based heuristic the feature subset of the candidate base classifier is formed using forward inclusion procedure.

```
build_initial_ensemble(Heuristic,TRS,FS,Threshold)
  L     number of classes and number of base classifiers
  Begin
    Set Ensemble and FS empty
    for i from 1 to L
      MERITS[i]=calculate_merits(Heuristic,TRS,i)
      FS[i]=select_features(MERITS,Threshold)
      C[i]=C4.5(TRS,FS[i]){learns classifier i}
      Add C[i] to Ensemble;Add FS[i] to FS
    end for
  end build_initial_ensemble and return Ensemble and FS
```

The diversity measure used to select the least diverse base classifier is calculated as an average difference in predictions between all pairs of the classifiers. We modified the formula presented in [9] to calculate the approximated diversity Div_i of a classifier h_i as it is presented by the formula (3)

$$Div_i = \frac{\sum_{j=1}^{M} \sum_{k=1,k\neq i}^{H} Dif(h_i(\mathbf{x}_j), h_k(\mathbf{x}_j))}{M(H-1)}. \tag{3}$$

where H denotes the number of the base classifiers, $h_i(x_j)$ denotes the classification of the vector $\mathbf{x}_j$ by the classifier h_i, and $Dif(a,b)$ is 0 if the classifications a and b are same and 1 if they are different, and M is the number of instances in the validation set. The diversity of an ensemble is calculated as the average diversity of all the base classifiers.

```
cycle(Operation,Heuristic,TRS,Ccurr,FS,VS)
  begin
    loop
      for i from 1 to L DIV[i]=calculate_diversities(TRS,Ccurr)
      Cmin=argmin DIV[i];NewCcurr=Ccurr\C[Cmin];NewFS=FS
      MERITS[Cmin]=calculate_merits(Heuristic,TRS,Cmin)
      if Operation is delete
        then NewFS[Cmin]=FS[Cmin]-feature with min MERITS[Cmin]
            included in FS[Cmin]
        else NewFS[Cmin]=FS[Cmin]+feature with max MERITS[Cmin]
            included in FS[Cmin]
      Add new base classifier C4.5(TRS,NewFS[Cmin]) to NewCcurr
      if accuracy      (VS,NewCcurr)>=accuracy(VS,Ccurr)
        then Ccurr=NewCcurr; FS=NewFS
        else no changes to Ccurr and FS
    until no change
  end cycle
```

4   Experiments

In this chapter, we present our experiments with the EFS_ref algorithm for ten data sets taken from the UCI machine learning repository [5]. These data sets were chosen so as to provide a variety of application areas, sizes, combinations of feature types, and difficulty as measured by the accuracy achieved on them by current algorithms.

For each data set 30 test runs are made. In each run the data set is first split up into the training set and two test sets by stratified random sampling keeping the class distribution of instances in each set approximately the same as in the initial data set. The training set (TRS) includes 60 percent of instances and the test sets (VS and TS) both 20 percent of instances. The first test set, VS (validation set) is used for tuning the ensemble of classifiers, adjusting the initial feature subsets so that the ensemble accuracy with weighted voting becomes as high as possible for each heuristics separately. The other test set, TS is used for the final estimation of the ensemble accuracy. The base classifiers themselves are learnt using the C4.5 decision tree algorithm with pruning [12] and the test environment is implemented within the MLC++ framework [4].

Table 1. Accuracies (%) of the CM-, CR-, and SCR-based ensembles, and C4.5

Data set	CM-based			CR-based			SCR-based			C4.5
	Bef.	Aft.	Threshold	Bef.	Aft.	Threshold	Bef.	Aft.	Threshold	
Car	86.6	87.9	0.5	86.0	86.8	0.6	**88.1**	88.0	0.1	87.9
Glass	64.8	65.2	0.1	65.7	66.0	0.3	65.8	**66.4**	0.4	62.6
Iris	**94.6**	94.3	0.9	94.1	94.2	0.7	94.1	94.3	0.8	93.9
LED_17	64.4	64.6	0.1	59.5	62.1	0.4	66.5	**66.9**	0.2	65.0
Lymph	**76.4**	75.3	0.5	74.0	73.6	0.6	74.2	74.7	0.4	74.2
Thyroid	91.4	92.7	0.6	**93.4**	93.1	0.3	**93.4**	93.3	0.5	92.8
Vehicle	67.0	67.3	0.1	67.6	68.1	0.5	70.3	**70.8**	0.2	68.7
Wavef.	73.0	74.2	0.2	74.3	74.2	0.5	**75.8**	**75.8**	0.4	72.2
Wine	93.5	93.5	0.2	92.3	93.4	0.7	**93.6**	**93.6**	0.7	92.9
Zoo	92.2	**93.5**	0.1	87.0	88.1	0.5	92.5	92.7	0.1	92.4

Table 1 summarizes the accuracies of the three heuristics-based ensembles on each data set before (Bef.) and after (Aft.) iterations. The predefined values of thresholds used for each dataset are presented in the corresponding columns (Threshold) for each heuristic. The last column presents corresponding C4.5 accuracies without any preceding feature selection. The highest accuracy value achieved for each data set is boldfaced in Table 1. As can be seen SCR produces the highest final accuracies for seven data sets (one shared with CR) and CM produces the highest final accuracies for the remaining three data sets.

In order to make conclusions we calculated statistics of Tables 2 and 3 over the 30 runs. For each comparison (as for example SCR/CM) there are two columns: the first titled "w/t/l" includes the number of runs when the first heuristic results higher accuracy (w-value), same accuracy (t-value), and smaller accuracy (l-value) than the second heuristic; the second column titled "P,2-t." includes the corresponding p-value of paired two-tailed t-test between the accuracies resulted by the two heuristics. The content of a cell is boldfaced when the result is statistically significant using the 95% confidence interval. In the column "w/t/l" according to the sign test and in the column "P,2-t." according to the paired two-tailed t-test. When the p-value is underlined then the corresponding distribution test gives negative result.[1] Table 2 represents statistics for comparison of SCR to CM and CR between the accuracies of the initial ensembles and the final ensembles.

Table 2. Comparisons of the accuracies of three heuristic-based ensembles by the sign test and the paired two-tail t-test

| Data set | Initial ensembles | | | | Final ensembles | | | |
| | SCR/CM | | SCR/CR | | SCR/CM | | SCR/CR | |
	w/t/l	P,2-t.	w/t/l	P,2-t.	w/t/l	P,2-t.	w/t/l	P,2-t.
Car	**21/8/1**	**0.00**	**29/0/1**	<u>**0.00**</u>	14/8/8	0.58	**21/8/1**	**0.00**
Glass	14/4/12	0.43	13/8/9	0.79	15/5/10	0.32	10/10/10	0.63
Iris	4/21/5	<u>0.30</u>	1/28/1	<u>1.00</u>	5/20/5	<u>0.99</u>	1/28/1	<u>0.66</u>
LED_17	16/6/8	**0.04**	**25/2/3**	**0.00**	18/5/7	**0.01**	**22/6/2**	**0.00**
Lymph	10/5/15	0.12	12/10/8	0.77	10/8/12	0.53	9/18/3	<u>0.10</u>
Thyroid	14/8/8	**0.02**	2/24/4	<u>0.59</u>	7/13/10	<u>0.42</u>	2/24/4	<u>0.79</u>
Vehicle	**25/1/4**	**0.00**	**22/1/7**	**0.00**	**25/1/4**	**0.00**	**21/1/8**	**0.00**
Wavef.	**26/2/2**	**0.00**	**24/0/6**	**0.00**	20/1/9	**0.00**	**21/0/9**	**0.01**
Wine	7/10/13	<u>0.94</u>	12/11/7	<u>0.75</u>	13/7/10	<u>0.93</u>	10/13/7	0.76
Zoo	7/19/4	0.60	**24/4/2**	**0.00**	4/18/8	0.40	**22/5/3**	**0.00**
Aver	8/0/2	0.08	**8/2/0**	**0.03**	7/1/2	0.09	**10/0/0**	**0.01**

Comparisons between CM and CR have been presented in our earlier paper [11] where we reported that CM initial ensemble results in higher average accuracy with 6 data sets (2 statistically significant). CR initial ensemble resulted in higher accuracy with 4 data sets but non of them were statistically significant according to the sign test and one was statistically significant according to the t-test. CM final ensemble resulted in higher average accuracy also with the same 6 data sets (1 statistically significant according to the sign test and 3 according

---

[1] The normality of distribution is tested using standardized skewness and standardized kurtosis. The values outside the range of -2 to +2 indicate significant departures from normality, which would tend to invalidate the paired t-test.

to the t-test). CR final ensemble resulted in higher accuracy with 3 data sets but non of them were statistically significant.

As can be seen from Table 2 SCR initial ensemble results in higher accuracy with 8 data sets than both CM and CR, and the difference between SCR and CR over all data sets is statically significant. The comparisons of individual data sets shows that SCR results in higher accuracy than CM with 3 data sets according to the sign test and with 5 data sets according to the t-test. Since with none of the data sets CM results in higher accuracy (statically significantly), it seems that SCR results in better initial ensembles than CM. Further, the comparison of individual data sets shows that SCR results in higher accuracy than CR with 5 data sets according to both the sign test and the t-test. Since with none of the data sets CR results in higher accuracy, it seems that SCR results in better initial ensembles than CR.

The accuracies of the final ensembles are always as high for SCR than CR and the differences for 5 individual data sets are statistically significant. Thus in general SCR outperforms CR. The accuracies of the final ensembles are higher for SCR than CM for 7 data sets and lower for 2 data sets. With two data sets of the seven the results are statistically significant according to the sign test and with three of the seven data sets according to the t-test.

Over all the data sets, the iterations resulted in 0.6% increase of accuracy for CM (the interval of changes from -1.4% to 1.6%), 0.8% for CR (the interval of changes from -0.5% to 4.4%), and 0.3% for SCR (the interval of changes from -0.1% to 0.9%). The negative values indicate that sometimes the iterations might result overfitting[2] (as with CM-based ensemble on the Lymph data set).

In Table 3 the 2nd and 3rd columns summarize the difference in accuracy before and after iterations for SCR ensembles. The final ensemble resulted in higher accuracy with 6 data sets and lower with 2 data sets. The difference is not statistically significant to any individual data set according to the sign and t-test. Over all the data sets the difference is statistically significant according to the t-test but not according to the sign test. In [11] we presented similar comparison between the initial and final ensembles of CM and CR. In both cases the final ensemble resulted in higher accuracy for 7 data sets but the difference over all the data set was not statistically significant.

Table 3 also includes comparisons of the accuracies of the final ensembles of CM, CR, and SCR with the accuracy achieved using the C4.5 algorithm with no preceding feature selection. In general, SCR resulted in the final ensemble with higher accuracy than C4.5 for all the data sets and this difference is statistically significant. In general, over all the data sets CM and CR did not produce ensembles with higher accuracy than C4.5. When we look situation with respect of the individual data sets SCR produced the final ensemble with higher accuracy than C4.5 for 3 data sets statistically significantly. The accuracies of CR final ensemble and C4.5 are statistically significantly different with 5 data sets: 2 in favour of CR and 3 in favour of C4.5. Of the 6 data sets for which the final

---

[2] The accuracies are calculated using the test set. When the iteration results in overfitting with respect to the validation set then the final accuracy can become smaller.

Table 3. Statistics of the refinement cycle contribution in accuracy for SCR ensembles, and comparisons of the final accuracies of CM, CR, and SCR to the accuracy produced by C4.5 using the sign test and the paired one-tail t-test

Data set	Initial vs final ensemble of SCR		Final ensembles					
			CM/C4.5		CR/C4.5		SCR/C4.5	
	w/t/l	P,1-t.	w/t/l	P,2-t.	w/t/l	P,2-t.	w/t/l	P,2-t.
Car	7/18/5	0.55	11/6/13	1.00	**5/5/20**	**0.00**	13/11/6	0.17
Glass	10/15/5	0.14	**19/6/5**	**0.02**	**21/3/6**	**0.00**	18/3/9	0.12
Iris	2/27/1	0.21	6/21/3	0.35	5/23/2	0.26	4/25/1	0.30
LED_17	8/17/5	0.14	11/5/14	0.49	**7/1/22**	**0.00**	**16/8/4**	**0.00**
Lymph	9/15/6	0.28	16/4/10	0.28	8/11/11	0.50	11/10/9	0.64
Thyroid	5/21/4	0.41	10/12/8	0.74	11/13/6	0.56	10/13/7	0.55
Vehicle	12/11/7	0.08	8/4/18	**0.01**	13/2/15	0.34	**21/2/7**	**0.00**
Wavef.	10/11/9	0.31	**23/2/5**	**0.00**	**23/1/6**	**0.00**	**28/0/2**	**0.00**
Wine	8/11/11	0.50	11/10/9	0.39	13/7/10	0.48	12/10/8	0.29
Zoo	5/22/3	0.38	8/17/5	0.17	**2/6/22**	**0.00**	6/21/3	0.57
Aver	6/2/2	**0.01**	6/1/3	0.15	5/0/5	0.68	**10/0/0**	**0.01**

Table 4. Diversities of the base classifiers for the CM-, CR-, ans SCR-based ensembles

Data set	# of fs	CM-based				CR-based				SCR-based			
		Av. # of sel. fs	Av. diff. in div.	Av. # of cyc-les	Div. of 1st ense-mble	Av. # of sel. fs	Av. diff. in div.	Av. # of cyc-les	Div. of 1st ense-mple	Av. # of sel. fs	Av. diff. in div.	Av. # of cyc-les	Div. of 1st ense-mple
Car	5	4.2	7.3	1.5	14.3	3.2	6.9	1.6	15.3	4.4	2.5	1.4	10.5
Glass	9	4.6	7.5	1.4	27.1	3.3	3.5	3.9	34.5	1.3	3.5	1.4	30.6
Iris	4	1.4	1.8	1.1	2.4	1.7	1.2	1.0	29.7	1.9	0.9	1.0	30.7
LED_17	24	22.4	5.8	1.5	27.9	4.5	1.4	1.6	63.3	9.9	8.3	1.4	20.1
Lymph	18	6.6	7.8	1.6	17.0	5.4	2.5	1.3	26.4	3.1	6.1	1.4	25.8
Thyroid	5	1.8	2.2	1.4	7.3	2.9	1.2	1.2	13.0	2.9	1.2	1.2	13.3
Vehicle	18	12.3	3.9	1.5	24.2	6.5	0.9	1.4	42.8	12.9	1.2	1.3	29.5
Wavef.	21	17.6	3.7	1.3	20.0	8.3	1.3	1.5	39.4	12.2	0.9	1.2	28.7
Wine	12	6.7	1.4	1.3	16.6	3.9	3.1	1.3	17.2	3.8	3.3	1.3	19.2
Zoo	16	13.3	2.6	1.2	13.1	3.9	2.7	1.4	36.6	12.3	8.7	1.2	1.8

CM ensemble has a higher accuracy than C4.5 two situations are statistically significant according to both the sign and t-test. With one of the 3 data sets for which the final CM ensemble has a lower accuracy than C4.5 the difference is statistically significant but only according to the t-test. Because the individual data sets for which the statistically significant higher accuracy was achieved with CM, CR, or SCR than C4.5 were not the same it seems to be that some data set related characteristics have effect.

Table 4 summarizes diversities and the other iteration related characteristics. The second column (# of fs) includes the total number of features of each data set. For each of the three heuristics there are four columns: (1) the average number of features used to construct the base classifiers (Av. # of sel. Fs), (2) the average difference of diversity (multiplied by 100) between the first and last ensemble (Av. diff. in div.), (3) the average number of iterations (refinement cycles)(Av. # of cycles), and (4) diversity (multiplied by 100) of the first ensemble (Div. of 1st ensemble).

When we compare the three heuristics we find that the average number of features selected was biggest using the CM-based heuristic for 6 data sets, and that the only data set for which the CR-heuristic selected the biggest amount of features also the SCR-heuristic selected the same biggest amount of features. Only with two data sets of those 6 ones the CM-based heuristic produced the highest accuracy. Instead, with all the four data sets when SCR-based heuristic selected the biggest number of features it also resulted in the highest accuracy. The minimum average amount of features was selected by CM for 2 data sets, by CR for 5 data sets, and by SCR for 3 data sets. The selection of the smallest number of features by CR never resulted in the highest accuracy as it did in two of the three cases with SCR and one of the two cases with CM. The average percentage of selected features over the ten data sets are: CR 33%, SCR 49%, and CM 69%. An overall glance at the accuracy table (Table 1) with these percentages may result the assumption that CR selects too few features and CM may select unnecessary many features.

From Table 4 is seen that CR produced more diverse initial ensembles than SCR in many cases, and in all cases more diverse than CM, while SCR produced a few more diverse classifiers than CM. On average, after iterations diversity has been increased on for all the heuristics and the data sets. Iterations had the greatest effect for CM, and the smallest effect for CR.

5   Conclusions

In this paper, we have analyzed and experimented with three feature selection heuristics. The correlation-based heuristics and CM-based one rely on slightly different assumptions concerning heterogeneous classification problems. The CM measure increases the class separability encouraging features that are important to distinguish one class from the others. The correlation-based heuristics are intended to select features that are important for one class, and less important for another.

We have evaluated our approach on ten data sets from the UCI machine learning repository. Experiments showed that with both initial and final ensemble the most accurate classifications were produced by SCR, and the less accurate by CR. Iterations had the greatest effect for CM, and the smallest effect for CR. On average, all the three heuristic-based ensembles are competitive with C4.5, but only SCR outperforms C4.5 so that the difference in accuracy is statistically significant.

Acknowledgements

This research is partially supported by the Academy of Finland (project number 52599) and the COMAS Graduate School of the University of Jyväskylä. We would like to thank the UCI machine learning repository, domain theories and data generators for the datasets, and the machine learning library in C++ for the source code used in this study.

References

1. Apte, C., Hong, S.J., Hosking, J.R.M., Lepre, J., Pednault, E.P.D., Rosen, B.K.: Decomposition of Heterogeneous Classification Problems. In X. Liu, P. Cohen, M. Bethold (eds.): Advances in Intelligent Data Analysis, Lecture Notes in Computer Science, Vol. 1280. Springer-Verlag (1997) 17–28
2. Hall, M.: Correlation-based feature selection for discrete and numeric class machine learning. In: Proc. 17th Int. Conf. on Machine learning. Morgan Kaufmann Publishers, CA (2000).
3. Hong, S.J.: Use of contextual information for feature ranking and discretization. IEEE Transactions on knowledge and Data Engineering **9** (1997) 718–730
4. Kohavi, R., Sommerfield, D., Dougherty, J.: Data Mining Using MLC++: A Machine Learning Library in C++. In: Tools with Artificial Intelligence, IEEE CS Press (1996) 234–245
5. Merz, C.J., Murphy, P.M.: UCI Repository of Machine Learning Datasets [http://www.ics.uci.edu/ (mlearn/MLRepository.html]. Dep-t of Information and CS, Un-ty of California, Irvine, CA (1998)
6. Opitz, D., Maclin, R.: Popular Ensemble Methods: An Empirical Study. Artificial Intelligent Research **11** (1999) 169–198
7. Opitz, D.: Feature Selection for Ensembles. In: Proc. of the 16th National Conf. on Artificial Intelligence (AAAI), Orlando (1999) 379–384
8. Oza, N., Tumer, K.: Dimensionality Reduction Through Classifier Ensembles. Tech. Rep. NASA-ARC-IC-1999-126 (1999)
9. Prodromidis, A. L., Stolfo, S. J., Chan P. K.: Pruning Classifiers in a Distributed Meta-Learning System. In: Proc. 1st National Conference on New Information Technologies. (1998) 151–160
10. Puuronen, S., Skrypnyk, I., Tsymbal, A.: Ensemble Feature Selection based on the Contextual Merit. In: Proc. 3rd Int. Conf. on Data Warehousing and Knowledge Discovery (DaWaK'01). September 5-7, 2001 Munich, Germany
11. Puuronen, S., Skrypnyk, I., Tsymbal, A.: Ensemble Feature Selection based on Contextual Merit and Correlation Heuristics. In: Proceedings of the Fifth East-European Conference on Advances in Databases and Information Systems. September 25-28, 2001, Vilnius, Lithuania
12. Quinlan, J.R.: C4.5: programs for machine learning. Morgan Kaufmann, San Mateo, California (1993)

Nonmetric Multidimensional Scaling with Neural Networks

Michiel C. van Wezel, Walter A. Kosters, Peter van der Putten, and
Joost N. Kok

Leiden Institute of Advanced Computer Science (LIACS)
Universiteit Leiden
P.O. Box 9512, 2300 RA Leiden, The Netherlands.
{michiel,kosters,putten,joost}@liacs.nl

Abstract In this paper we present a neural network for nonmetric multidimensional scaling. In our approach, the monotone transformation that is a part of every nonmetric scaling algorithm is performed by a special feedforward neural network with a modified backpropagation algorithm. Contrary to traditional methods, we thus explicitly model the monotone transformation by a special purpose neural network. The architecture of the new network and the derivation of the learning rule are given, as well as some experimental results. The experimental results are positive.

1 Introduction

This paper concerns visualization of multidimensional data using neural nonmetric multidimensional scaling. Generally stated, multidimensional scaling (abbreviated MDS) is a collection of techniques for embedding dissimilarity data in a space with a chosen dimensionality. The embedding is often used for the purpose of data visualization and exploratory data analysis. In this sense, MDS is a competitor for other data visualization techniques, such as a Kohonen neural network and principal component analysis. Traditional MDS techniques are subdivided into metric MDS, where the dissimilarities between objects are assumed to be proportional to Euclidean distances, and nonmetric MDS, where the dissimilarities are only assumed to be monotonically related to Euclidean distances. In the case where the dissimilarities represent, e.g., distances between the capitals of the European countries, the Euclidean distance assumption is realistic. However, in the case of dissimilarities between soda brands that have been reported by a panel of test persons, nonmetric MDS seems more appropriate.

Although traditional multidimensional scaling is a well established pattern recognition technique (see, e.g., [1,2,10,7]), to our knowledge this subject has not received a lot of attention from researchers in the field of neural networks. In particular, all publications concerning neural networks for multidimensional scaling that are known to us concern metric multidimensional scaling. In [9] a simple neural network is given for metric multidimensional scaling. This neural

F. Hoffmann et al. (Eds.): IDA 2001, LNCS 2189, pp. 145–155, 2001.
© Springer-Verlag Berlin Heidelberg 2001

network merely performs a gradient descent on the cost function, which carries the risk of getting stuck in local minima in the error function. To prevent this, in [4,5] Klöck and Buhmann apply annealing methods from statistical mechanics to the metric MDS problem. No neural algorithms have been applied to nonmetric MDS so far.

This paper is organized as follows. First, an exact problem statement is presented for multidimensional scaling in Section 2. Next, the proposed feedforward neural network for nonmetric multidimensional scaling and the corresponding learning rule are given in Section 3. After that, the results of some experiments are given in Section 4 and finally Section 5 gives conclusions and a summary. This paper does not give an in-depth treatment of multidimensional scaling in general. The interested reader is referred to [2,3,10] for more information on the subject.

2   Multidimensional Scaling: Problem Statement

It is impossible to give a definition of the term multidimensional scaling that covers the general use of this term and is yet formal enough to be precise. Some people use the term for a very broad class of data analysis techniques and problems including cluster analysis and factor analysis (see, e.g., [7]), while others use a more narrow definition, and refer to MDS as a class of techniques used to develop a spatial representation of proximity data. The distinction between a broad and a narrow definition is also made in a useful taxonomy introduced in a paper by Carroll and Arabie [1]. This paper is concerned with MDS in the narrow sense: finding a spatial representation of objects based on proximity data.

Using this narrow definition, there are two "kinds" of MDS: metric MDS and nonmetric MDS. Before we are able to describe these variants we need some terminology, which is often used in MDS literature and is also adopted in this paper.

Dissimilarities: In both metric and nonmetric MDS the starting point is a matrix $\boldsymbol{\delta}$ of dissimilarities, of which an element δ_{ij} denotes the dissimilarity between the objects i and j. The number of objects is denoted by n.

Embedding: An embedding of the objects in Euclidean space. The coordinates of an object i in this embedding are denoted by $\mathbf{x}_i$. The dimensionality of the embedding space is denoted by m, so $\mathbf{x}_i = (x_{i1}, \ldots, x_{im})^T$, where ^{T} denotes transpose.

Distances : The (real) Euclidean distance between objects i and j is denoted by $\hat{d}_{ij}$. The distance between the estimates for the spatial representations $\mathbf{x}_i$ and $\mathbf{x}_j$ of objects i and j is denoted by $d_{ij} =\parallel \mathbf{x}_i - \mathbf{x}_j \parallel$, where $\parallel . \parallel$ denotes the Euclidean norm. Collectively the distances are denoted by matrix $\mathbf{d}$.

Disparities: These quantities are used in nonmetric scaling. Disparities $\hat{\boldsymbol{\delta}}$ are as close as possible to distances between the corresponding coordinate estimates d but with the restriction that they are monotonically related to the original dissimilarity data $\boldsymbol{\delta}$.

Using the above concepts, the MDS problem can be accurately described. We differentiate between metric and nonmetric MDS.

Metric multidimensional scaling: As stated earlier, in metric multidimensional scaling, it is assumed that the dissimilarities are *proportional* to Euclidean distances:

$$\delta_{ij} = c\hat{d}_{ij} \tag{1}$$

One way to obtain a spatial representation of dissimilarity data under the above assumption is to minimize the following error function

$$E_{\text{metric\_mds}} = \sum_{ij, i \neq j} (\delta_{ij} - d_{ij})^2. \tag{2}$$

This minimization gives us the correct coordinates up to a scale factor, a translation and a rotation. Torgerson [8] gives a non-iterative procedure for finding these coordinates.

Nonmetric multidimensional scaling: The most fundamental difference between metric MDS and nonmetric MDS is the relaxation of the "distance assumption" (Equation 1) to a "monotonicity assumption". In nonmetric MDS it is assumed that the dissimilarities δ are *monotonically related* to Euclidean distances:

$$\forall i, j, k, \ell : \hat{d}_{ij} < \hat{d}_{k\ell} \Rightarrow \delta_{ij} < \delta_{k\ell} . \tag{3}$$

One can look upon the δ values as being monotonically transformed distance values: $\delta_{ij} = f(\hat{d}_{ij})$ where $f(.)$ is an unknown strict monotonically increasing function. Examples of such functions include some linear, power and logarithmic functions. Nonmetric MDS algorithms estimate a spatial representation for a given dissimilarity matrix in which the rank order of the distances between the embedded objects agrees with the rank order of the original dissimilarities as much as possible.

Traditionally, in nonmetric MDS one attempts to minimize the following cost function, often called stress or stress formula 1 (due to Kruskal, see [6]), in an iterative fashion:

$$E_{\text{nonmetric\_mds}} = \left[\sum_{ij, i \neq j} (\hat{\delta}_{ij} - d_{ij})^2 / \sum_{ij, i \neq j} d_{ij}^2 \right]^{1/2} . \tag{4}$$

Often, the inter-pattern distances are kept normalized ($\sum_{ij, i \neq j} d_{ij}^2 = 1$), in which case the error function reduces to

$$E^{nm} = \sum_{ij, i \neq j} (\hat{\delta}_{ij} - d_{ij})^2, \tag{5}$$

which is computationally simpler. The disparities $\hat{\delta}$ must be re-estimated in each iteration in a so-called nonmetric phase. As stated before, the disparities are chosen to resemble the distances of the current embedding as close as possible subject to a monotonicity constraint:

$$\forall i, j, k, \ell : \delta_{ij} < \delta_{k\ell} \Rightarrow \hat{\delta}_{ij} \leq \hat{\delta}_{k\ell}. \tag{6}$$

This can be achieved by, e.g., setting the disparity $\hat{\delta}_{ij}$ equal to the m-th largest distance $d_{k\ell}$ if the dissimilarity $\delta_{k\ell}$ is the m-th largest dissimilarity. One ends up with an iterative algorithm that is schematically depicted in Figure 1.

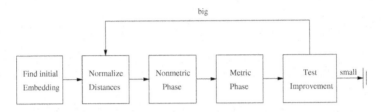

Fig. 1. Schematic representation of nonmetric mds algorithm

By minimizing stress (Equation 4) a spatial representation is found in which the distances between the embedded objects resemble the corresponding disparities as close as possible, while the disparities themselves are monotonically related to the original data. The dissimilarities are thus transformed by the monotone transformation in a way that improves the final fit to the data.

3 A Neural Network Approach to Nonmetric MDS

In this section we present a neural method for nonmetric MDS. The cost function that we minimize in this section is given in Equation 5. This means that we have to normalize the distances $\mathbf{d}$ and the embedding coordinates $\mathbf{x}$ after each iteration.

Metric phase: Recall that nonmetric MDS algorithms usually perform a metric phase and a nonmetric phase in each iteration. In the metric phase, the positions $\mathbf{x}$ of the embedded objects are altered so that the distances $\mathbf{d}$ resemble the disparities $\hat{\boldsymbol{\delta}}$ as well as possible. Thus Equation 5 is minimized as a function of the vectors $\mathbf{x}$. The product rule and the chain rule are needed to compute the partial derivatives (x_{ab} denotes the b-th component of $\mathbf{x}_a$):

$$\frac{\partial E^{nm}}{\partial x_{ab}} = \sum_{ij} \frac{\partial(\hat{\delta}_{ij} - d_{ij})^2}{\partial x_{ab}}. \tag{7}$$

Now

$$\frac{\partial(\hat{\delta}_{ij} - d_{ij})^2}{\partial x_{ab}} = -2(\hat{\delta}_{ij} - d_{ij})\frac{\partial d_{ij}}{\partial x_{ab}}, \tag{8}$$

$$\frac{\partial d_{ij}}{\partial x_{ab}} = \begin{cases} (x_{ab} - x_{jb})/d_{aj} & \text{if } i = a, \\ (x_{ab} - x_{ib})/d_{ia} & \text{if } j = a, \\ 0 & \text{otherwise.} \end{cases} \tag{9}$$

Nonmetric phase: In the nonmetric phase values for the disparity matrix $\hat{\boldsymbol{\delta}}$ are sought that are monotonically related to the original data (the dissimilarities $\boldsymbol{\delta}$), yet as close as possible to the distances $\mathbf{d}$. For accomplishing this task, we designed and implemented a special neural network that is only capable of modeling monotone transformations, and attempted to use this network to model the transformation $f(\boldsymbol{\delta}) = \hat{\boldsymbol{\delta}}$ explicitly. We will refer to this network type as a *mono-nn* in the remainder.

The reason for this explicit modeling using a neural network is threefold. First, explicit modeling gives the user the opportunity to visualize and examine the total monotone transformation, because the neural network performs interpolation. The second reason is that we wanted to investigate the value of neural network techniques in this problem area. The third and most important reason is that the use of interpolation in the nonmetric phase can potentially speed up the nonmetric phase considerably in the case where the dissimilarity matrix is of substantial size. The reason for this is that in a traditional nonmetric phase, one is required to perform a sorting operation of the inter-pattern distances $\mathbf{d}$. Since the fastest sorting algorithm around is the quicksort algorithm and this algorithm has a quadratic worst-case time complexity, the amount of time needed for a traditional nonmetric phase is $\mathcal{O}(n^4)$. When a mono-nn is used, the sorting step is superfluous. The mono-nn can be trained using only a subset of all available data, and it can be used as an interpolator for the remaining data.

This network takes the dissimilarities $\boldsymbol{\delta}$ as inputs and generates the disparities $\hat{\boldsymbol{\delta}}$ as outputs. It uses the distances $\mathbf{d}$ as targets. The network has one hidden layer with non-linear (hyperbolic tangent) transfer functions. The output unit uses the identity as a transfer function. It will be explained shortly how the monotonicity property is ensured. A network of this kind is depicted in Figure 2. Note that there is a bias unit present in the network. This unit is clamped to activation value 1. In the following discussion, a, b, c index units in the input, hidden and output layer respectively; $y_{k|ij}$ denotes the output of unit k (input, hidden or output) when dissimilarity δ_{ij} [1] is offered at the input; $g_k(.)$ denotes the transfer function of unit k. The cost function for this network is (c denotes the only output unit):

$$E^{nn} = \sum_{ij,i\neq j} (y_{c|ij} - d_{ij})^2 = \sum_{ij,i\neq j} E^{nn|ij}. \tag{10}$$

---

[1] We standardized all inputs and targets at the beginning of the nonmetric phase in each iteration.

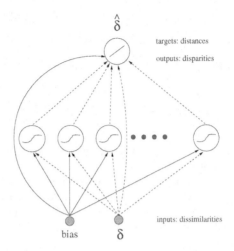

Fig. 2. Monotone neural network for nonmetric phase in MDS.

Furthermore, $w_{k\ell}$ denotes the connection weight of the connection that runs from unit k to unit ℓ and $in_{k|ij}$ denotes the net-input to unit k, so $y_{k|ij} = g_k(in_{k|ij})$. The bias for unit k is denoted by θ_k. Since the individual transfer functions are monotonically increasing, the monotonicity constraint is always satisfied if we use squared network parameters (weights) in the computation for the weighted net-inputs for all weights except the biases:

$$in_{c|ij} = \sum_b w_{bc}^2 y_{b|ij} + w_{ac}^2 y_{a|ij} + \theta_c \qquad in_{b|ij} = \sum_a w_{ab}^2 y_{a|ij} + \theta_b. \qquad (11)$$

The connections for which the *squared* weights are used in the computations are represented by the dashed arrows in Figure 2. Finally, for convenience, we define a matrix of network outputs $\mathbf{Y}$ with $Y_{ij} = y_{c|ij}$.

For a gradient based training procedure, the partial derivatives of the cost function with respect to the network weights have to be computed. Here we use k and ℓ, which can denote any unit, as indexes. We have:

$$\frac{\partial E^{nn|ij}}{\partial w_{k\ell}} = \frac{\partial E^{nn|ij}}{\partial in_{\ell|ij}} \frac{\partial in_{\ell|ij}}{\partial w_{k\ell}} \qquad \frac{\partial E^{nn|ij}}{\partial \theta_\ell} = \frac{\partial E^{nn|ij}}{\partial in_{\ell|ij}} \frac{\partial in_{\ell|ij}}{\partial \theta_\ell}. \qquad (12)$$

If we denote

$$\Delta_{\ell|ij} = \frac{\partial E^{nn|ij}}{\partial in_{\ell|ij}} \qquad (13)$$

and note that

$$\frac{\partial in_{\ell|ij}}{\partial w_{k\ell}} = 2w_{k\ell} y_{k|ij} \qquad \frac{\partial in_{\ell|ij}}{\partial \theta_\ell} = 1 \qquad (14)$$

we obtain

$$\frac{\partial E^{nn|ij}}{\partial w_{k\ell}} = \Delta_{\ell|ij} 2 w_{k\ell} y_{k|ij} \qquad\qquad \frac{\partial E^{nn|ij}}{\partial \theta_\ell} = \Delta_{\ell|ij}. \qquad (15)$$

For the output unit c, with identity activation, the Δ values are easily computed:

$$\Delta_{c|ij} = \frac{\partial E^{nn|ij}}{\partial in_{c|ij}} = 2(a_{c|ij} - d_{ij}). \qquad (16)$$

For the hidden units, with $tanh$ activation, the Δ values are

$$\Delta_{b|ij} = \frac{\partial E^{nn|ij}}{\partial in_{b|ij}} = \Delta_{c|ij} w_{bc}^2 (1 - g_b(a_{b|ij})^2). \qquad (17)$$

Given these formulae, the training algorithm for a monotone neural network is as follows. First, a random object pair i, j is chosen from all possible object pairs. The dissimilarity δ_{ij} for this object pair is used as input for the network. The corresponding output is computed, and the Δ values for all units are computed. Then, the partial derivatives are computed using Equation 15. These partial derivatives are then used to update the network weights, either in batch mode or in pattern mode.

The total nonmetric MDS procedure is greatly accelerated by using a good initial embedding. In the experiments reported in the next section, we used the solution obtained by the metric MDS as the initial embedding. The complete algorithm for nonmetric MDS with a neural network is given in pseudo-code in Figure 3.

---

(0) **initialize** the embedding coordinates $\mathbf{x}$ of all objects randomly.
(1) **initialize** the disparities $\hat{\boldsymbol{\delta}}$ randomly.
(2) (optional) **perform** metric MDS to obtain an initial embedding.
(3) **initialize** weights of monotone neural network randomly.
(4) **for** cycle $= 1$ **to** number of cycles **do**
(5)   {
(6)   **normalize** distances and embedding coordinates.
(7)   /* **nonmetric phase** */
(8)   Train the monotone neural network using M randomly selected
(.)   pattern pairs and gradients via Equation 15.
(9)   Set the disparities $\hat{\boldsymbol{\delta}} = \mathbf{Y}$.
(10)   /* **metric phase** */
(11)   Find new embedding coordinates using $\hat{\boldsymbol{\delta}}$ and Equation 7.
(12)   }

Fig. 3. Algorithm (in pseudo-code) for nonmetric MDS using a gradient based algorithm.

4   Experiments and Results

In this section we describe the experiments we performed with our implementation and the results we obtained. In all experiments we use an embedding dimensionality of 2.

Experiments with the monotonic neural network: We started by testing our special purpose neural network. In order to do this, we created three artificial datasets containing clear non-linearities. The first dataset consisted of 105 points. The x-values were randomly distributed in the $[-2.5 : 2.5]$ interval. The corresponding y values were monotonically related to the x values. This dataset contained a general trend (resembling a tanh(.) function) and some deviations from this trend, but never violating the monotonicity property. The second dataset contained 100 random x values in the $[-2 : 2]$ interval. The y values were uniformly distributed with small standard deviation around the function $0.6 \tanh(5x + 4) + 0.9 \tanh(7x - 1)$. This dataset contains a trend that is monotonically increasing, but the noise makes the total dataset non-monotonically increasing. The third dataset was generated in a similar way around $-0.6 \tanh(5x + 4) + 0.9 \tanh(7x - 1)$. This function is not monotonically increasing. All datasets and the fits obtained by the special networks are shown in Figure 4. Observe that the major nonlinearities are fitted well by our network, but in the case of the first dataset the network was not able to model the subtle nonlinearities. The fit for the third dataset illustrates the monotonicity property of the network.

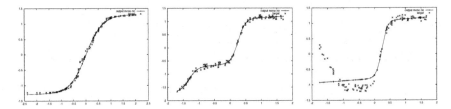

Fig. 4. Curve fitted by mono-nn and original data (scattered crosses) for artificial dataset one (left), two (center) and three (right). See text for more explanation.

Experiment with MDS on artificial data Next, we applied our nonmetric scaling method to an artificial problem. For this problem, we created a dataset, the `square` dataset, consisting of 16 points arranged in a square in $[0 : 1]^2$. The distances between all points were computed and transformed by a nonlinear transformation: $0.5(\tanh(5(x - 0.5)) + 1)$. This yielded the dissimilarities δ that we used as a starting point for our algorithm.

Consider the upper three graphs of Figure 5. The original data that were used to generate the distances (which were in turn transformed to dissimilarities) are shown in the left graph of this figure. The embedding found by a metric

MDS algorithm is shown in the middle graph. The right graph depicts the embedding found by our neural algorithm for nonmetric scaling. It is clearly visible that the latter embedding resembles the original data better. Now consider the lower graphs in Figure 5. The leftmost graph shows the transformation that was applied to the inter-pattern distances to create the dissimilarities. The middle graph shows the function that was learned by the mono-nn together with the target data (the inter-pattern distances) in the final iteration, the right graph shows the inverse of the mono-nn function, which clearly resembles the original transformation.

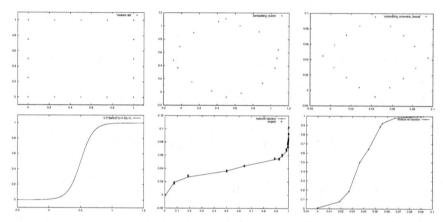

Fig. 5. Upper graphs: Original data, metric MDS embedding and neural nonmetric embedding for the **square** dataset. Lower graphs: Original transformation, mono-nn function and inverse of nn-function for the **square** dataset.

Experiment with MDS on vocational preference data In this experiment we analyzed the small "vocational preferences" dataset. We adopted this dataset from the MDS textbook of Davison [2] who, in turn, adopted it from psychometric literature. The dataset is shown in Table 1. It contains dissimilarities between six 'vocational preference inventory' scales in a sample of 1234 men. In psychological literature these six occupational types are often displayed in a hexagon (see the left side of Figure 6). It is claimed that the vocational interests of persons in occupational types adjacent to each other in the hexagon are more similar than the interests of people with occupational types more distant from each other.

In order to verify this hypothesis, nonmetric scaling was applied to the vocational preferences dataset. We applied our neural nonmetric scaling algorithm to these data and ended up with the embedding shown in the middle of Figure 6. In this embedding, the data points representing the six occupational types have indeed settled themselves in a roughly hexagonal shape.

	R	I	A	S	E	C
Realistic	0.0000	1.0392	1.2961	1.2570	1.1832	1.1314
Investigative	1.0392	0.0000	1.1489	1.1832	1.2961	1.2961
Artistic	1.2961	1.1489	0.0000	1.0770	1.1402	1.3342
Social	1.2570	1.1832	1.0770	0.0000	0.9592	1.1136
Enterprising	1.1832	1.2961	1.1402	0.9592	0.0000	0.8000
Conventional	1.1314	1.2961	1.3342	1.1136	0.8000	0.0000

Table 1. Vocational preference dissimilarity data.

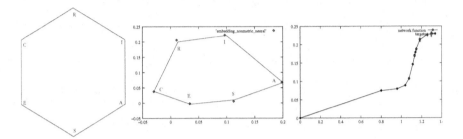

Fig. 6. Theoretical hexagon for vocational preferences (left). Embedding found by neural MDS (middle). Transformation implemented by mono-nn (right).

5   Summary and Conclusions

Multidimensional scaling is a data projection method that uses dissimilarity data as a starting point. This approach makes it different from more often used data projection methods such as principal component analysis. MDS searches for an optimal embedding given a matrix of dissimilarities. There are two types of MDS: metric MDS and nonmetric MDS. This paper focuses on the nonmetric MDS, where the original data are monotonically transformed during the embedding process.

A new neural architecture is proposed for performing the nonmetric phase that is immanent in every nonmetric scaling algorithm. The corresponding learning rule was derived and the network (called mono-nn) was shown to perform well for various test problems. The new scaling method that uses the new network type was tested on a dataset with dissimilarities that were obtained by non-linearly transforming a matrix of distances. The mono-nn was able to correctly model the inverse transformation and the scaling method yielded a good embedding. A real-life dataset from psychometrics was analyzed and the embedding confirmed a hypothesis concerning vocational preferences.

The added value of our scaling method is simple: it explicitly models the nonlinear monotone transformation of the original dissimilarity data. This explicit modeling may give the data analyst more insight into the transformation and it gives the ability to interpolate, contrary to traditional scaling methods. Interpolation gives a potential speed up for large dissimilarity matrices. It must be remarked however, that traditional nonmetric scaling methods are generally faster than our method for small data sets, because of the greater simplicity of the nonmetric phase.

References

1. J.D. Carroll and P. Arabie. Multidimensional scaling. In M.R. Rosenzweig and L.W. Porter, editors, *Annual Review of Psychology*, pages 607–649. 1980.
2. M. L. Davison. *Multidimensional Scaling*. John Wiley and Sons, New York, 1983.
3. J. de Leeuw and W. Heiser. Theory of multidimensional scaling. In P. R. Krishnaiah and L. N. Kanal, editors, *Handbook of Statistics*, volume 2, pages 285–316. North-Holland, 1982.
4. H. Klöck and J.M. Buhmann. Multidimensional scaling by deterministic annealing. In M. Pellilo and E.R. Hancock, editors, *Proceedings of the EMMCVPR'97*, pages 245–260, Venice, 1997.
5. H. Klöck and J.M. Buhmann. Data visualization by multidimensional scaling: A deterministic annealing approach. *Pattern Recognition*, 33(4):651–669, 1999.
6. J. B. Kruskal. Multidimensional scaling by optimizing goodness-of-fit to a nonmetric hypothesis. *Psychometrika*, 29:1–29, 1964.
7. B. D. Ripley. *Pattern Recognition and Neural Networks*. Cambridge University Press, Cambridge, 1996.
8. W. S. Torgerson. *Theory and Methods of Scaling*. John Wiley and Sons, New York, 1958.
9. M. C. van Wezel, J. N. Kok, and W. A. Kosters. Two neural network methods for multidimensional scaling. In *European Symposium on Artificial Neural Networks (ESANN'97)*, pages 97–102, Brussels, 1997. D facto.
10. M. Wish and D. Carroll. multidimensional scaling and its applications. In P. R. Krishnaiah and L. N. Kanal, editors, *Handbook of Statistics*, volume 2, pages 317–345. North-Holland, 1982.

Functional Trees for Regression

João Gama

LIACC, FEP, University of Porto
jgama@ncc.up.pt

Abstract. In this paper we present and evaluate a new algorithm for supervised learning regression problems. The algorithm combines a univariate regression tree with a linear regression function by means of constructive induction. When growing the tree, at each internal node, a linear-regression function creates one new attribute. This new attribute is the instantiation of the regression function for each example that fall at this node. This new instance space is propagated down through the tree. Tests based on those new attributes correspond to an oblique decision surface. Our approach can be seen as a hybrid model that combines a linear regression known to have low variance with a regression tree known to have low bias. Our algorithm was compared against to its components, and two simplified versions, and M5 using 16 benchmark datasets. The experimental evaluation shows that our algorithm has clear advantages with respect to the generalization ability when compared against its components and competes well against the state-of-art in regression trees.

1   Introduction

The generalization capacity of a learning algorithm depends on the appropriateness of its representation language to express a generalization of the examples for the given task. Different learning algorithms employ different representations, search heuristics, evaluation functions, and search spaces. It is now commonly accepted that each algorithm has its own selective superiority [3]; each is best for some but not all tasks. The design of algorithms that explore multiple representation languages and explore different search spaces has an intuitive appeal. This paper presents one such algorithm.

In the context of supervised learning problems it is useful to distinguish between classification problems and regression problems. In the former the target variable takes values in a finite and pre-defined set of un-ordered values, and the usual goal is to minimize a 0-1-loss function. In the latter the target variable is ordered and takes values in a subset of $\Re$. The usual goal is to minimize a squared error loss function. Mainly due to the differences in the type of the target variable successful techniques in one type of problems are not directly applicable to the other type of problems.

Gama [6] has presented a technique to combine classifiers that use distinct representation languages using constructive induction. In this work we study the applicability of a related technique for regression problems. In particular we combine a linear model with a regression tree using constructive induction.

F. Hoffmann et al. (Eds.): IDA 2001, LNCS 2189, pp. 156–166, 2001.

Generalized Linear Models. Generalized Linear Models (GLM) is the most frequently applied statistical technique to set a relationship between several independent variables and a target variable. In the most general terms GLM are of the form $w_0 + \sum w_i \times f_i(x_i)$. GLM estimation is aimed at minimizing the sum of squared deviations of the observed values for the dependent variable from those predicted by the model. One appealing characteristic is that there is an analytical solution for this problem. The coefficients for the polynomial using the least squares error criterion is found by solving the equation: $W = (X^T X)^{-1} X^T Y$ In this paper, we assume that f_i is the identity function leading to the *linear multiple regression*.

Regression Trees. A regression tree uses a divide-and-conquer strategy that decomposes a complex problem into simpler problems and recursively applies the same strategy to the sub-problems. This is the basic idea behind well-known regression tree based algorithms [2,10,14]. The power of this approach comes from the ability to split the space of the attributes into subspaces, whereby each subspace is fitted with different functions. The main drawback of this method is its instability for small variations of the training set [2].

Constructive Induction. Constructive Induction discovers new attributes from the training set and transforms the original instance space into a new high dimensional space by applying attribute constructor operators. The difficulty is how to choose the appropriate operators for the problem in question. In this paper we argue that, in regression domains described at least partially by numerical attributes, techniques based on GLM are a useful tool for constructive induction.

The algorithm that we describe in this work is in the confluence of these three areas. It explores the power of divide-and-conquer from regression trees and the ability of generating hyper-planes from linear-regression. It integrates both using constructive induction. In the next section of the paper we describe our proposal to functional regression trees. In Section 3 we discuss the different variants of regression models. In Section 4 we present related work both in the classification and regression settings. In Section 5 we evaluate our algorithm on 16 benchmark datasets. Last Section concludes the paper.

2   The Algorithm for Regression Trees

The standard algorithm to build regression trees consists of two phases. In the first phase a large decision tree is constructed. In the second phase this tree is pruned back. The algorithm to grow the tree follows the standard divide-and-conquer approach. The most relevant aspects are: the splitting rule, the termination criterion, and the leaf assignment criterion. With respect to the last criterion, the usual rule consists of assignment a constant to a leaf node. This constant is usually the mean of the y values taken from the examples that fall at this node. With respect to the splitting rule, each attribute value defines a possible partition of the dataset. We distinguish between nominal attributes and continuous ones. In the former the number of partitions is equal to the number of values of the attribute, in the latter a binary partition is obtained. To estimate the merit of the partition obtained by a given attribute we use the following heuristic: $\sum_i \frac{(\sum y_i)^2}{n_i}$, where i represents the number of partitions and n_i is the number of examples in partition i. The attribute that maximizes this expression is chosen as test attribute at this node.

The pruning phase consists of traversing the tree in a depth-first fashion. At each non-leaf node two measures should be estimated. An estimate of the error of the subtree below this node, that is computed as a weighted sum of the estimated error for each leaf of the subtree, and the estimated error of the non-leaf node if it was pruned to a leaf. If the latter is lower than the former, the entire subtree is replaced to a leaf. To estimate the error at each leaf we assume a χ^2 distribution of the variance of the cases in it. Following [14] a pessimistic estimate of the MSE at each node t is given by:

$$MSE \times \frac{n-1}{2} \times \left(\frac{1}{\chi^2_{\alpha/2,n-1}} + \frac{1}{\chi^2_{1-\alpha/2,n-1}}\right) \tag{1}$$

where MSE is the mean squared error at this node, n denotes the number of examples at this node, and α is the confidence level.

All of these aspects have several and important variants, see for example [2,14]. Nevertheless all decision nodes contain conditions based in the values of one attribute. The first proposal for a multivariate regression tree has been presented in 1992 by Quinlan [10]. System *M5* builds a tree-based model but can use at the leaves multiple-linear models. The goal of this paper is to study *when* and *where* to use decisions based on a combination of attributes. Instead of considering multivariate models restricted to leaves, we analyze and evaluate multivariate models both at leaves and internal nodes.

2.1   Functional Trees for Regression

In this section we present the general algorithm to construct a functional regression tree. Given a set of examples and an attribute constructor, the main algorithm used to build a decision tree is:

Function Tree(Dataset, Constructor)
1. If Stop_Criterion(DataSet) Return a Leaf Node with a constant value.
2. Construct a model Φ using Constructor
3. For each example $\boldsymbol{x} \in DataSet$
   - Compute $\hat{y} = \Phi(\boldsymbol{x})$
   - Extend $\boldsymbol{x}$ with a new attribute $\hat{y}$.
4. Select the attribute that maximizes some merit-function
5. For each partition i of the DataSet using the selected attribute
   - $\text{Tree}_i = \text{Tree}(\text{Dataset}_i, \text{Constructor})$
6. Return a *Tree*, as a decision node based on the select attribute, containing the Φ model, and descendents Tree_i.

End Function

This algorithm is similar to many others, except in the constructive step (steps 2 and 3). Here a function is built and mapped to a new attribute. In this paper, we restricted the Constructor to the *linear-regression* (LR) function [9]. There are some aspects of this algorithm that should be made explicit. In step 2, a model is built using the Constructor function. This is done using only the examples that fall at this node. Later, in step 3, the model is mapped to one new attribute. The merit of the new attribute is evaluated using the merit-function of the decision tree, and in competition with the original attributes (step 4). The models built by our algorithm have two types of decision nodes: Those based on a test of one of the original attributes, and those based on the values of the constructor function. Once a tree has been constructed, it is pruned back. The general algorithm to prune the tree is:

Function Prune(Tree)
1. Estimate **Leaf_Error** as the error at this node.
2. If Tree is a leaf Return **Leaf_Error**.
3. Estimate **Constructor_Error** as the error of Φ [1].
4. For each descendent i
   - **Backed_Up_Error** += Prune(Tree$_i$)
5. If argmin(Leaf_Error,Constructor_Error,Backed_Up_Error)
   - Is Leaf_Error
     - Tree = Leaf
     - Tree_Error = Leaf_Error
   - Is Model_Error
     - Tree = Constructor Leaf
     - Tree_Error = Constructor_Error
   - Is Backed_Up_Error
     - Tree_Error = Backed_Up_Error
6. Return Tree_Error

End Function

The error estimates needed in step 1 and 3 use equation 1. The pruning algorithm produces two different types of leaves: *ordinary leaves* that predict the mean of the target variable observed in the examples that fall at this node, and *constructor leaves* that is leaves that predict the value of the constructor function learned (in the growing phase) at this node.

By simplifying our algorithm we obtain different conceptual models. Two interesting lesions are described in the following sub-sections.

Bottom-Up Approach We denote as *Bottom-Up Approach* to functional regression trees when the multivariate models are used exclusively at leaves. This is the strategy used for example in M5 [11,16], in Cubist [12], and in RT system [14]. In our tree algorithm this is done restricting the selection of the test attribute (step 4 in the growing algorithm) to the original attributes. Nevertheless we still build, at each node, the linear-regression function. The model built by

---

[1] The Constructor model learned in the growing phase at this node.

the linear-regression function is used later in the pruning phase. In this way, all decision nodes are based in the original attributes. Leaf nodes could contain a constructor model. A leaf node contains a constructor model if and only if in the pruning algorithm the estimated mean-squared error of the constructor model is lower than the *Backed-up-error* and the estimated error of the node has if a leaf replaced it.

Top-Down Approach We denote as *Top-Down Approach* to functional regression trees when the multivariate models are used exclusively at decision nodes (internal nodes). In our algorithm, restricting the pruning algorithm to choose only between the **Backed_Up_Error** and the **Leaf_Error** obtain these kinds of models. In this case all leaves predict a constant value.

Our algorithm can be seen as a hybrid model that performs a tight combination of a decision tree and a linear-regression. The components of the hybrid algorithm use different representation languages and search strategies. While the tree uses a divide-and-conquer method, the linear-regression performs a global minimization approach. While the former performs feature selection, the latter uses all (or almost all) the attributes to build a model. From the point of view of the bias-variance decomposition of the error [1] a decision tree is known to have low bias but high variance, while linear-regression is known to have low variance but high bias. This is the desirable behavior for components of hybrid models.

3 An Illustrative Example

In this section we use the well-known regression dataset *Housing* to illustrate the different variants of regression models. The attribute constructor used is the linear regression model. Figure 1(a) presents a univariate tree for the *Housing* dataset. Decision nodes only contain tests based on the original attributes. Leaf nodes predict the average of y values taken from the examples that fall at the leaf.

In a top-down functional tree (Figure 1(b)) decision nodes could contain (not necessarily) tests based on a linear combination of the original attributes. The tree contains a mixture of built-in attributes, denoted as *LR Node*, and original attributes, *e.g. AGE, DIS*. Any of the linear-regression attributes can be used both at the node where they have been created and at deeper nodes. For example, the *LR Node 19* has been created at the second level of the tree. It is used as test attribute at this node, and also (due to the constructive ability) as test attribute at the third level of the tree. Leaf nodes predict the average of y values of the examples that fall at this leaf. In a bottom-up functional tree (Figure 2(a)) decision nodes only contain tests based on the original attributes. Leaf nodes could predict (not necessarily) values obtained by using a linear-regression function built from the examples that fall at this node. This is the kind of model regression trees that usually appears on the literature. For example, systems M5 [11,16] and RT [14] generate this kind of models.

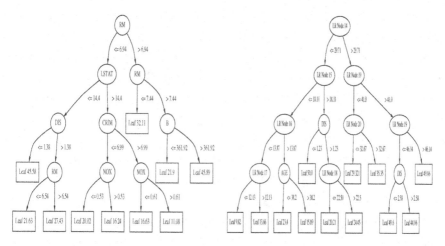

Fig. 1. (a)The Univariate Regression Tree and (b) Top-Down Functional regression tree for the Housing problem.

Figure 2(b) presents the full functional regression tree using both top-down and bottom-up multivariate approaches. In this case, decision nodes could contain (not necessarily) tests based on a linear combination of the original attributes, and leaf nodes could predict (not necessarily) values obtained by using a linear-regression function built from the examples that fall at this node.

4   Related Work

In the context of classification problems, several algorithms have been presented that use at each decision node tests based on linear combination of the attributes [2,8,15,4,5]. Also, Gama [6] has presented *Cascade Generalization*, a method to combine classification algorithms by means of constructive induction. The work presented here near follows this method but in the context of regression domains. Another difference is related to the pruning algorithm. In this work we consider functional leaves.

Breiman *et al.* [2] presents the first extensive and in-depth study of the problem of constructing decision and regression trees. But, while in the case of decision trees they consider internal nodes with a test based on linear combination of attributes, in the case of regression trees internal nodes are always based on a single attribute. Quinlan [10] has presented the system M5. It builds multivariate trees using linear models at the leaves. In the pruning phase for each leaf a linear model is built. Recently, Witten and Eibe [16] have extended M5. A linear model is built at each node of the initial regression tree. All the models along a particular path from the root to a leaf node are then combined into one linear model in a *smoothing* step. Also Karalic [7] has studied the influence of using linear regression in the leaves of a regression tree. As in the work of Quinlan,

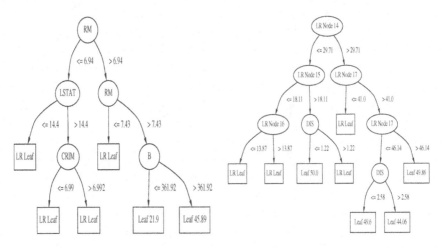

Fig. 2. (a)The Bottom-Up Functional Regression Tree and (b) The Functional Regression Tree for the Housing problem.

Karalic shows that it leads to smaller models with increase of performance. Torgo [13] has presented an experimental study about functional models for regression tree leaves. Later, the same author [14] has presented the system RT. Using RT with linear models at the leaves, RT builds and prunes a regular univariate tree. Then at each leaf a linear model is built using the examples that fall at this leaf.

5   Experimental Evaluation

It is commonly accepted that multivariate regression trees should be competitive against univariate models. In this section we evaluate the proposed algorithm, its lesioned variants, and its components on a set of benchmark datasets. For comparative proposes we evaluate also system M5[2]. The main goal in this experimental evaluation is to study the influence in terms of performance of the position inside a regression tree of the linear models. We evaluate three situations:

- Trees that could use linear combinations at each internal node.
- Trees that could use linear combinations at each leaf.
- Trees that could use linear combinations both at each internal and leaf nodes.

All evaluated models are based on the same tree growing and pruning algorithm. That is, they use exactly the same splitting criteria, stopping criteria, and pruning mechanism. Moreover they share many minor heuristics that individually are too small to mention, but collectively can make difference. Doing so, the differences on the evaluation statistics are due to the differences in the conceptual

---

[2] We have used M5 from the last version of Weka environment. We have used several regression systems. The most competitive was M5.

model. In this work we estimate the performance of a learned model using the *mean squared error* statistic.

5.1   Evaluation Design and Results

We have chosen 16 datasets from the *Repository of Regression problems at LI-ACC*[3]. The choice of datasets was restricted by the criteria that almost all the attributes are ordered with few missing values[4]. To estimate the mean squared error of an algorithm on a given dataset we use 10 fold cross validation. To apply pairwise comparisons we guarantee that, in all runs, all algorithms learn and test on the same partitions of the data. The following table resumes datasets characteristics:

Dataset		#Examples	#Attributes	Dataset		#Examples	#Attributes
Abalone	AB	4177	8 (1 Nom.)	Auto-Mpg	AU	398	6
Cart	Car	40768	10	Computer	CA	8192	21
Cpu	CP	210	7	Diabetes	DI	43	2
Elevators	EL	8752	17	Fried	FR	40768	10
House_16	H16	22784	16	House_8	H8	22784	8
Housing	HO	506	13	Kinematics	KI	8192	8
Machine	MA	209	5	Pole	PO	5000	48
Pyrimidines	PY	74	28	Quake	QU	2178	3

The results in terms of MSE and standard deviation are presented in Table 1. The first two columns refer to the results of the components of the hybrid algorithm. The following three columns refer to the simplified versions of our algorithm and the full model. The last column refers to the M5 system. For each dataset, the algorithms are compared against the full multivariate tree using the *Wilcoxon signed rank-test*. The null hypothesis is that the difference between error rates has median value zero. A $-$ $(+)$ sign indicates that for this dataset the performance of the algorithm was worse (better) than the full model with a p *value* less than 0.01. It is interesting to note that the full model (MT) significantly improves over both components (LR and UT) in 9 datasets out of 16. Table 1 also presents a comparative summary of the results. The first line presents the geometric mean of the MSE statistic across all datasets. The second line shows the average rank of all models, computed for each dataset by assigning rank 1 to the best algorithm, 2 to the second best and so on. The third line shows the average ratio of *MSE*. This is computed for each dataset as the ratio between the *MSE* of one algorithm and the *MSE* of M5. The fourth line shows the number of significant differences using the *signed-rank test* taking the multivariate tree MT as reference. We use the *Wilcoxon Matched-Pairs Signed-Ranks Test* to compare the error rate of pairs of algorithms across datasets. The last line shows the p *values* associated with this test for the *MSE* results on all datasets and taking

---

[3] http://www.ncc.up.pt/~ltorgo/Datasets

[4] The actual implementation ignores missing values at learning time. At application time, if the value of the test attribute is unknown, all descendent branches produce a prediction. The final prediction is a weighted average of the predictions.

Data	L.Regression (LR)	Univ. Tree (UT)	Functional Trees Top	Functional Trees Bottom	MT	M5
AB	$-$ 10.586±1.1	$-$ 6.992±0.5	4.659±0.3	$-$ 4.995±0.3	4.674±0.3	4.524±0.3
AU	11.069±3.3	$-$ 17.407±6.9	8.905±2.5	8.853±2.9	8.914±2.4	7.490±2.8
CA	$-$ 94.277±17.0	$-$ 10.133±0.8	$+$ 6.268±0.7	6.844±0.5	6.343±0.7	$-$ 6.961±1.3
CP	$-$ 3201±3183	2114±2557	1131±2766	1062±2545	1960±2361	1063±1623
Car	$-$ 5.685±0.1	$+$ 0.996±0.0	$-$ 1.012±0.0	0.995±0.0	1.007±0.0	$+$ 0.994±0.0
DI	0.398±0.2	0.469±0.3	0.474±0.2	0.398±0.2	0.398±0.2	$-$ 0.469±0.3
EL^2	$-$ 0.008±0.0	$-$ 0.014±0.0	0.004±0.0	$-$ 0.005±0.0	0.004±0.0	$-$ 0.0048±0.0
FR	$-$ 6.925±0.3	$-$ 3.171±0.1	$-$ 1.783±0.1	$-$ 2.163±0.1	1.772±0.1	$-$ 1.928±0.1
H16	$-$ 2.074e9±2.5e8	$-$ 1.608e9±2.2e8	1.19e9±1.4e8	1.17e9±1.8e8	1.21e9±1.4e8	1.26e9±1.2e8
H8	$-$ 1.73e9±2.1e8	$-$ 1.18e9±1.32e8	1.01e9±1.23e8	1.0e9±1.12e8	1.01e9±1.1e8	9.97e8±9.3e7
HO	23.683±10.5	15.687±4.9	15.785±7.1	11.666±3.3	16.576±8.6	12.875±4.9
KI	$-$ 0.041±0.0	$-$ 0.035±0.0	0.025±0.0	0.027±0.0	0.025±0.0	0.026±0.0
MA	4684±3657	3764±3798	3077±2247	2616±2289	3464±3080	3301±2462
PO	$-$ 939.44±41.9	$+$ 44.93±5.1	86.03±15.2	$+$ 36.57±5.8	86.18±14.8	$+$ 41.95±6.2
PY	0.015±0.0	0.017±0.0	0.012±0.0	0.013±0.0	0.013±0.0	0.012±0.0
QU	0.036±0.0	0.036±0.0	$-$ 0.037±0.0	0.036±0.0	0.036±0.0	0.037±0.0

2)$MSE \times 1000$

Summary of MSE Results

	LR	UT	FT-T	FT-B	FT	M5
Geometric Mean	39.7	22.99	17.15	16.22	17.7	16.4
Average Rank	5.4	4.8	3.0	2.4	2.9	2.3
Average Ratio	4.0	1.47	1.06	0.99	1.1	1
Signi. Wins/Losses	0/10	2/8	1/3	1/3	$-$	2/4
Wilcoxon Test	0.0	0.02	0.21	0.1	$-$	0.23

Table 1. Results in mean squared error (MSE).

MT as reference. All the functional trees have a similar performance. Using the significant test as criteria, FT is the most performing algorithm. It is interesting to note that the bottom-up version is the most competitive algorithm. Nevertheless there is a computational cost associated with the increase in performance verified. To run all the experiments referred here, FT requires almost 1.8 more time than the univariate regression tree.

6   Conclusions

We have presented a hybrid regression tree that combines a linear-regression function with a univariate regression tree by means of constructive induction. At each decision node two inductive steps occurs: the first one consists of building the linear-regression function; the second one consists of applying the regression tree criteria. In the first step the linear-regression function is not used (i.e. not used to produce predictions). All decisions, such as stopping, choosing the splitting attribute, etc, are delayed to the second inductive step. The final decision is made by the regression tree criteria. Using Wolpert's terminology [17] the constructive step performed at each decision node represents a bi-stacked generalization. From this point of view, the proposed methodology can be seen as a general architecture for combining algorithms by means of *constructive induction*, a kind of *local* bi-stacked generalization.

In this paper we have studied *where* to use decisions based on a combination of attributes. Instead of considering multivariate models restricted to leaves, we analyze and evaluate multivariate models both at leaves and internal nodes. Our experimental study suggests that the full model, that is a functional model using linear regression *both* at decision nodes and leaves, improves the performance of the algorithm. A natural extension of this work consists of applying the proposed methodology to classification problems. We have done this work using a discriminant function as attribute constructor. It is subject of another paper. The results on classification are consistent with the conclusions of this paper.

Acknowledgments

Gratitude to the financial support given by the FEDER project, projects Sol Eu-Net, Metal, and ALES, and the Plurianual support of LIACC. We would like to thank Luis Torgo and the referees for their useful comments.

References

1. L. Breiman. Arcing classifiers. *The Annals of Statistics*, 26(3):801–849, 1998.
2. L. Breiman, J. Friedman, R. Olshen, and C. Stone. *Classification and Regression Trees*. Wadsworth International Group., 1984.
3. Carla E. Brodley. Recursive automatic bias selection for classifier construction. *Machine Learning*, 20:63–94, 1995.
4. Carla E. Brodley and Paul E. Utgoff. Multivariate decision trees. *Machine Learning*, 19:45–77, 1995.
5. João Gama. Probabilistic Linear Tree. In D. Fisher, editor, *Machine Learning, Proceedings of the 14th International Conference*. Morgan Kaufmann, 1997.
6. João Gama and P. Brazdil. Cascade Generalization. *Machine Learning*, 41:315–343, 2000.
7. Aram Karalic. Employing linear regression in regression tree leaves. In Bernard Neumann, editor, *European Conference on Artificial Intelligence*, 1992.
8. S. Murthy, S. Kasif, and S. Salzberg. A system for induction of oblique decision trees. *Journal of Artificial Intelligence Research*, 1994.
9. W. Press, S. Teukolsky, W. Vetterling, and B. Flannery. *Numerical Recipes in C: the art of scientific computing 2 Ed.* University of Cambridge, 1992.
10. R. Quinlan. Learning with continuous classes. In Adams and Sterling, editors, *Proceedings of AI'92*. World Scientific, 1992.
11. R. Quinlan. Combining instance-based and model-based learning. In P.Utgoff, editor, *ML93, Machine Learning, Proceedings of the 10th International Conference*. Morgan Kaufmann, 1993.
12. R. Quinlan. Data mining tools See5 and C5.0. Technical report, RuleQuest Research, 1998.
13. Luis Torgo. Functional models for regression tree leaves. In D. Fisher, editor, *Machine Learning, Proceedings of the 14th International Conference*. Morgan Kaufmann, 1997.

14. Luis Torgo. *Inductive Learning of Tree-based Regression Models*. PhD thesis, University of Porto, 2000.
15. P. Utgoff and C. Brodley. Linear machine decision trees. Coins technical report, 91-10, University of Massachusetts, 1991.
16. Ian Witten and Eibe Frank. *Data Mining: Practical Machine Learning Tools and Techniques with Java Implementations*. Morgan Kaufmann Publishers, 2000.
17. D. Wolpert. Stacked generalization. In *Neural Networks*, volume 5, pages 241–260. Pergamon Press, 1992.

Data Mining with Products of Trees

José Tomé A.S. Ferreira, David G.T. Denison, and David J. Hand

Department of Mathematics, Imperial College
180 Queen's Gate, London SW7 2BZ, UK
{jt.ferreira, d.denison, d.j.hand}@ic.ac.uk

Abstract. We propose a new model for supervised classification for data mining applications. This model is based on products of trees. The information given by each predictor variable is separately extracted by means of a recursive partition structure. This information is then combined across predictors using a weighted product model form, an extension of the naive Bayes model. Empirical results are presented comparing this new method with other methods in the machine learning literature, for several data sets. Two typical data mining applications, a chromosome identification problem and a forest cover type identification problem are used to illustrate the ideas. The new approach is fast and surprisingly accurate.

1 Introduction

The aim in supervised classification is to construct a model for predicting the class to which an object belongs on the basis of a vector of measurements of that object. The data from which the rule will be constructed comprises the measurement vectors and class identifiers for a sample of objects taken from the same distribution as the objects to be classified. There now exists a large number of methods for supervised classification - see, for example, [6], [16] and [22]. Many of these methods have been made possible by the advent of powerful computing facilities. Because of this, the emphasis has been on developing highly parameterised flexible models which do well in optimising well-defined performance measures. A criterion which is important in many applications is *simplicity*, as the rule may need to be applied and interpreted by someone who does not understand the underlying statistical and computational methodology. For example, this is the case in medicine, retail banking, marketing and personnel classification and screening.

This paper describes a classification rule which can be applied in such circumstances - Data Mining with Products of Trees (DMPT). It is conceptually straightforward, so that it can be interpreted and explained easily. However, it also exhibits good classification performance. More importantly, it can straightforwardly handle many of the characteristics commonly found in the large datasets found in the emerging field of data mining [7].

In data mining applications we typically find that we are dealing with many thousands of instances and feature vectors that contain both numerical and cate-

F. Hoffmann et al. (Eds.): IDA 2001, LNCS 2189, pp. 167–176, 2001.

gorical predictors. Further, missing values are the norm rather then the exception as in standard statistical analysis. Such data sets are now commonplace, e.g. [9].

In general, models for supervised classification provide estimates of the probabilities $p(Y|\mathbf{x})$, where Y is the class indicator variable taking K discrete values, $C_1, \ldots, C_K$, and $\mathbf{x}$ is a set of P features or attributes. By comparing these probability estimates with each other, or with a threshold, objects can be assigned to classes. Often classification rules shortcut this procedure and estimate class indicator functions, or decision surfaces, directly. However, in this paper we are concerned with rules which estimate $p(Y|\mathbf{x})$ using estimates of the class-conditional distributions $p(\mathbf{x}|Y)$ and the estimated probabilities of belonging to each class $p(Y)$ via Bayes Theorem.

Many different methods exist for estimating the class-conditional distributions. Some assume particular parametric forms, while others are based on non-parametric smoothing methods (e.g. kernel methods). The particular class of models with which we are concerned assumes that

$$p(\mathbf{x}|Y = k) = \prod_{v=1}^{P} p(x_v|Y = k), \qquad k = 1, \ldots, K, \tag{1}$$

where K is the total number of classes. That is, we assume independence between the predictor variables X_v given the class label. This model goes under various names including *independence Bayes* [21], *naive Bayes* [14], and *idiot's Bayes* [18]. The last two names are a reference to the fact that the independence assumption is very seldom true in practice, so that the model has an unrealistic and almost certainly false form. A priori, one would expect this to mean that the model performed poorly in applications. In fact, however, the model often does surprisingly well. Examples of its performance and theoretical reasons why it can do well are presented in [10].

In this paper we are especially concerned with the version of the model in which the marginal distributions $p(x_v|Y)$, $v = 1, \ldots, P$ are approximated by distributions that partition the predictors. An obvious way to partition a predictor is by using a decision-tree approach and this is the general methodology that we propose. However, since the predictors are univariate we have a simpler form for the partition structure than the usual decision tree and that can determine good partitionings quickly.

Such models are not new. They have also been extended in various directions, as described in [4] and [8]. This paper describes a new extension, developing an approach to estimate the partitioning thresholds and the estimated probabilities and then combine the information given by all these.

In terms of simplicity and ease of interpretation, these models are perhaps most usefully compared with tree models (e.g., [2], [19]) and with simple linear regression models. However, whereas tree models recursively refine the complexity of the model, ending with an overcomplex model which is then pruned back, the models described in this paper constrain the complexity by choice of the product form.

2   The Model

2.1   Discovering the Partition Structure

We begin the description of the model by introducing the algorithm for discovering the partition structure. A decision tree separates the feature space into disjoint regions into which it is assumed that the data is homogeneous. The probability of assignment class k is the same for all points in a single region, or terminal node, and equals the empirical relative frequency of that class. In a similar spirit we choose to partition each attribute into disjoint regions. For numerical variables this means splitting up the real line into an unknown number of non-overlapping intervals whose union is the real line (or some subset of it). In order to achieve the same for categorical variables, it is merely necessary to determine which categories can be merged to form an unknown number of cells. Within each cell, the class assignments are assumed to be equal.

With reference to the large body of literature on decision trees, an obvious way to split the feature space in a single variable is to use a stepwise search. Although this quick search method may not find the optimal partitioning it often finds a good approximation to it. To perform the stepwise search we simply start with no partitions for a feature, x_v say, and then search through a list of candidate partition structures and choose the one which most increases the entropy

$$\sum_{i=1}^{2}\sum_{k=1}^{K} p_{kvi} \log p_{kvi},$$

where p_{kvi} is the relative frequency of class k for values of the feature x_v that lie in the ith partition. We repeat such a search strategy recursively on the cells of the partition until we have a large (probably overfitting) model. This ends the forward stepwise selection part of the procedure. Then we perform a backwards deletion search and remove partitions until some stopping criterion is satisfied. Note that we can use essentially the same methodology for both categorical and numeric variables. In this way we can determine a suitable partitioning of each variable in turn.

Other more expensive search techniques could be adopted but we feel that these are unnecessary since it is the setting of the weight function that is of more importance to the overall results.

2.2   Modified Naive Bayes

As we have already noted, the standard naive Bayes assumption is of the form given in (1). We see that this assumes that each variable is of equal importance. One may be able to improve the method by dropping the predictions made by irrelevant variables as suggested before [13], [17], [15]. However, with P predictors to find the optimal set of predictors, for any prespecified criterion, requires the fitting of 2^P models. In this paper we present a different approach where all available variables are used for the classification.

In our approach, we begin by determining the importance of predictor v using a function $w(x_v, v, k)$ - the weight function. The functional formula of w is broad enough to incorporate a large number of beliefs. . Following a standard Bayesian approach, we choose the weights to ensure that the modified likelihood provides as much information as it does for standard naive Bayes methods with P independent variables, so we fix $\sum_{v=1}^{P} w(x_v, v, k) = P$. Adopting this approach, together with the standard naive Bayes assumption, we find that

$$p(Y = k|\mathbf{x}) \propto \left\{ \prod_{v=1}^{P} [p(x_v|Y = k)]^{w(x_v, v, k)} \right\} p(Y = k).$$

Further, taking uniform priors over the class assignments, so that $p(Y = k)$ is the same for $k = 1, \ldots, K$, we find that

$$p(Y = k|\mathbf{x}) \propto \prod_{v=1}^{P} \left\{ \frac{n_{kvj}}{n_{kv}} \right\}^{w(x_v, v, k)} \tag{2}$$

where j is the cell into which variable x_v falls, n_{kvj} the number of class k data set points that fall in cell j and n_{kv} is the number of class k data set points which have a value of predictor v recorded.

2.3   The Weight Function

From the arguments given previously, we can choose the weight function w in any way as long as $\sum_v w(x_v, v, k) = P$. Instead of considering all predictors as equally informative, we propose a more refined approach that aims to take into account the importance of the prediction made for each variable. In this paper, we restrict the weight function to have the form,

$$w_\alpha(x_v, v) \propto \left\{ \sum_{k=1}^{K} |p_{kvi} - \frac{1}{K}|^\alpha \right\}^{\frac{1}{\alpha}}. \tag{3}$$

Although other ideas for the weight function can be found in [5], in what follows, a weight function of the form (3) was used with α being set to two, corresponding to an l_2-form distance.

3   Examples

3.1   Small Data Sets

As stated before, the aim of the DMPT method is not the study of small data sets, for which there are a large number of supervised classification methods (see e.g. [6], [16] and [22]). Despite that, and considering the properties of our model, there are no evident reasons why it should not work equally well in these kind of problems. To illustrate we applied the method to six small data sets form the UCI repository [1] with the results given in Table 1.

The results obtained using DMPT are to be compared with two other methods for supervised classification: the naive Bayes method [14] and the tree-based method C4.5 [19]. The inclusion of the naive Bayes method in the study is due to its well known predictive capabilities and for being the base to many others methods [10] including the one presented here. C4.5 was chosen because it is a well known state-of-the-art method, as well as it is a member of tree-based methods in general.

The three methods were applied to each of the above data sets using two stratified 10-fold cross-validations [2], [11] as in [23] and the mean error rates are presented in Table 1.

Table 1. DMPT, naive Bayes (NB) and C4.5 error rates comparison

Data set	Error rate % DMPT	NB	C4.5	Size	Classes	No.Att.	% Instances with missing values
Breast cancer	2.8	2.7	4.7	699	2	9	2.3
Credit screening	14.8	14.0	14.5	690	2	15	5.4
Hepatitis	15.4	14.2	18.4	155	2	19	48.4
Liver disorders	32.0	35.1	34.1	345	2	6	0.0
Pima Indians	24.0	25.2	25.0	768	2	9	0.0
Zoology	4.2	5.5	7.4	101	7	16	0.0
Average	15.5	16.1	17.4				

As was to be expected, the results are similar for all the classifers with the DMPT method being the one with best results in three out of six occasions. The worst result was on the Hepatitis Prognosis data set where our proposed method performed 1.2% worse (14.2% versus 15.4%) than the naive Bayes method.

3.2   Large Data Sets

This section is divided into two distinct parts: a case study and results on two typical data mining applications. With the inclusion of a case study we pretend to illustrate the interpretability of the proposed method.

The Adult Data Set The Adult data set is also available from the UCI repository [1]. It is made of 48842 instances, for which there are 14 predictors recorded, with some missing values. The classification task is to determine whether a person earns more, or less, than \$50,000 a year (2-class problem). We label the instances with income greater than \$50,000 Class 1 while the remaining are labelled Class 2. The overall frequency of Class 1 is 23.93% and there are six numerical and eight categorical predictors The data set is already divided into training set and testing set, with 32561 and 16281 instances respectively.

The choice of this data set as a case study is easily explained by its features. It is a relatively large data set that can be thought of as a data mining problem as it has both numerical and categorical variables, as well as missing values. These are all features that our methodology can easily cope with.

Table 2 shows the number of resulting partitions in each one of the predictors; for the categorical predictors, the number in parenthesis represents the number of categories. To assess the relevance of the cells, the same Table also shows the maximum (Max) and the minimum (Min) frequency of Class 1 over the partition on the training set. We see that, the number of partitions vary substantially with different predictors and that some predictors have larger discriminatory power than others (as shown by the discrepancy between the maximum and the minimum frequency probability of Class 1 over the set of partitions in that feature).

Table 2. Characteristics of the predictors and resulting partitions for the Adult data set

Predictor	Type	% Missing	No. Partitions	Max	Min
Age	Numerical	0.0	4	0.31	0.01
Workclass	Categorical	5.7	5(8)	0.56	0.00
Fnlwgt	Numerical	0.0	17	0.63	0.04
Education	Categorical	0.0	9(16)	0.74	0.00
Education-num	Numerical	0.0	5	0.74	0.00
Marital-status	Categorical	0.0	4(7)	0.45	0.05
Occupation	Categorical	5.8	8(14)	0.47	0.01
Relationship	Categorical	0.0	5(6)	0.45	0.01
Race	Categorical	0.0	3(5)	0.26	0.09
Sex	Categorical	0.0	2(2)	0.31	0.11
Capital-gain	Numerical	0.0	14	1.00	0.00
Capital-loss	Numerical	0.0	16	1.00	0.00
Hours-per-week	Numerical	0.0	8	0.43	0.00
Native country	Categorical	1.8	10(41)	0.40	0.00

Another interesting fact can be seen from the partitions over numerical predictors. Figure 1 shows the frequency of Class 1 over the partitition for predictors *Education-num* and *Hours-per-week*, which are both numerical. The plots show different patterns, while for predictor *Education-num* the frequency of Class 1 is monotone, for predictor *Hours-per-week* the underlying process seem to be more complex. The relation between *Education-num* and the annual income seems rather intuitive as the more years of education one has, the bigger the probability of earning a greater amount of money. The relation between *Hours-per-week* and the annual income is not that obvious. Here the probability of earning more than \$50,000 is bigger if one works less than 5 hours per week than if one works about 20 hours per week. This is probably due to people collecting very large pensions.

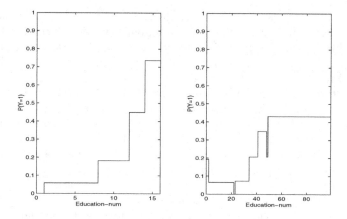

Fig. 1. Frequency of Class 1 over the predictors *Education-num* (left) and *Hours-per-week*

With the partitions established, the classifier is ready to be used. The model, applied to the test set revealed an error rate of 14.05%. The accuracy of the proposed method is very competitive when compared with other state-of-the-art methods as shown in Table 3. Except for the method we propose in this paper, the description of all the classification methods and the results can be found in [12]. It can be seen that the performance of the DMPT method is at the very top. The FSS Naive Bayes method performs as well. The DMPT is also among the least computationally demanding algorithm of those presented.

Table 3. DMPT, naive Bayes (NB) and C4.5 error rates comparison for the Adult data set

Method	Error rate %
DMPT	14.05
FSS Naive Bayes	14.05
NBTree	14.10
IDTM (Decision Table)	14.46
C4.5-auto	14.46
HOODG	14.82
C4.5 rules	14.94
C4.5	15.54
Voted ID3 (0.6)	15.64

Comparative Results Until this moment, we have illustrated the use of the DMPT method with applications that can be said as *conventional* machine learning problems. Now we depart form that and apply DMPT to two data sets that are proper data mining problems.

The first problem is given by the ChromData data set and is a chromosome identification problem. An individual's chromsomes are often displayed in a table called a karyogram, which displays the chromsomes. In order to construct a karyogram, an appropriate dividing cell nucleus is chemically treated to break it apart and to induce a banding pattern on the chromosomes. A high resolution picture is then taken, and various image processing algorithms used to extract the chromosomes, and features of the chromosomes. These features fall into two groups: morphometric (relating to size and shape) and banding (relating to the induced stain patterns). A chromosome includes two pieces of genetic material intertwined, in such a way that there is a central crossover structure - the centromere. Morphometric features include length and width infromation, as well as the features corresponding to the relative position of the centromere. In the data the predictors are ten morphometric features and twenty banding features. There are 26 classes (chromosome types). In the ChromData there are 130,801 instances.

The second problem is given by the Covertype data set [1]. The goal with this classification problem is to be able to identify the forest cover type form cartographic variables only. Each instance is given by ten numerical predictors and two categorical ones (in the original data set, these are given as 44 binary predictors). There are seven kinds of forest cover type. The data set consists in 580,102 instances.

In this section we compare the DMPT method with the naive Bayes classifier, as this is one of the only methods that can be applied, in a reasonable time, for problems of the dimension of those we present here. Both data sets were randomly divided in order to create a test set with 30,000 instances, being the remaining instances used to construct the partition structure.

Both methods were applied using the same partitions, created with the algorithm given in Section 2.1 . No effort was made to optimise the partitions as our only goal is to compare the two methods. The error rate for the ChromData problem was 8.88% with DMPT and 10.02% with naive Bayes; for the Covertype problem, the error rate was 40.24% with DMPT and 44.81% with naive Bayes. As it can be seen, DMPT performs substancially better than naive Bayes. DMPT is 12.8% more accurate than naive Bayes in the ChromData problem and 11.4% in the Covertype problem.

4   Discussion

We proposed a new methodology for the traditional problem of supervised classification - DMPT. It was specially designed to deal with a number of common features to data mining applications: numerical and categorical predictors, missing attribute values, large number of instances to be used as training instances, large number of attributes to be used in as predictors. The DMPT method has shown to be feasible for all these characteristics.

Obviously, the most demanding step, in computational aspects, of our method is the creation of the partitions. In the creation of the partitions, and for numerical predictors, the sorting of the data is the bulk of computation and this is well known to be be an $O(Nlog(N))$ algorithm, where N is the size of the data. This means that the method is feasible even for quite large data sets.

DMPT is a Bayesian-based method that use all information for all the predictors to generate a prediction. Unlike most other methods available for supervised classification, our proposal uses every single predictor for which there is information when generating a prediction. The final output is not a single classification but a probability assessment over the possible classes. This is a common feature of all Bayesian methods. As such misclassification cost can easily be accounted for.

The results presented in this paper were all obtain using a weight function of the form (5). Different applications can benefit from the liberty that DMPT allows when choosing the weight function. Using other forms for the weight function permits the extraction of other caracteristics, as can be seen in [5] and is the subject of future research.

The accuracy results we present in this paper are good but the interpretation we can get form a built model is at least as relevant. The ability of presenting a result to the experts by means of plots like the ones in Fig.1 provides an insightful tool for supporting the results.

Our Matlab implementation of the method can be obtained from the authors.

Acknowldgements The work of J T A S Ferreira was supported by grant SFRH BD 1399 2000 from Fundação para a Ciência e Tecnologia, Ministério para a Ciência e Tecnologia, Portugal.

References

1. Blake, C., Keogh, E. and Merz, C.J. (1998) UCI repository of machine learning databases [http://ics.uci.edu/ mlearn/MLRepository.html]. Department of Information and Computing Science, University of California, Irvine, CA.
2. Breiman, L., Freidman, J.H., Olshen, R.A., and Stone, C.J. (1984) *Classification and Regression Trees*. Belmont, California: Wadsworth.
3. Chipman, H., George, E.I. and McCulloch, R.E. (1998) Bayesian CART model search (with discussion). *J. Am. Statist. Assoc.*, **93**, 935-960.
4. Denison, D.G.T., Adams, N.M., Holmes, C.C. and Hand, D.J. (2000) Bayesian Partition Modelling. *Technical Report*, Imperial College.
5. Ferreira, J.T.A.S., Denison, D.G.T. and Hand, D.J. (2001) Weighted naive Bayes modelling for data mining. *Technical Report*, Imperial College.
6. Hand, D.J. (1997) *Construction and Assessment of Classification Rules*. Chichester, Wiley.
7. Hand, D.J. (1998) Data Mining: Statistics and More?. *The American Statistician*, **52(2)**, 112-118.
8. Hand, D.J. and Adams, N.M. (2000) Defining attributes for scorecard construction in credit scoring. *Journal of Applied Statistics*, **27**, 527-540.
9. Hand, D.J., Blunt, G., Kelly, M.G. and Adams, N.M. (2000) Data Mining for Fun and Profit. *Statistical Science*, **15(2)**, 111-131.

10. Hand, D.J. and Yu, K. (2001) Idiot's Bayes - not so stupid after all? To appear in *International Statistical Review*

11. Kohavi, R. (1995) A study of cross-validation and bootstrap for accuracy estimation and model selection. In *Proceedings of the Fourteenth International Joint Conference on Knowledge Discovery and Artificial Intelligence*, San Mateo,CA, Morgan Kaufmann, 1137-1143.

12. Kohavi, R. (1996) Scaling up the accuracy of Naive-Bayes classifiers: a decision-tree hybrid. In *Proceedings of the Second International Conference on Knowledge Discovery and Data Mining*, 202-207.

13. Kohavi, R. and John, G.H. (1997) Wrappers for feature subset selection. *Artificial Intelligence*, **97(1-2)**, 273-324.

14. Kononenko, I. (1990) Comparision of inductive and naive Bayesian learning approaches to automatic knowledge aquisition. In B. Wielinga *et al.* (eds.) *Current trends in knowledge acquisition*, Amsterdam, IOS Press.

15. Langley, P. (1994) Selection of relevant features in machine learning. In *AAAI Fall Symposium on Relevance*, 140-144.

16. McLachlan, G.J. (1992) *Discriminant Analysis and Statistical Pattern Recognition*. New York: John Wiley and Sons.

17. Moore,. A.W., Hill, D.J. and Johnson, M.P. (1992) An empirical investigation of brute force to choose features, smoothers and function approximators. In Hansin, S. *et al.* (eds.) *Computational Learning Theory and Natural Learning Systems Conference*, Vol. 3, MIT Press.

18. Ohmann, C., Yang, Q., Künneke, M., Stöltzing, H., Thon, K. and Lorenz W. (1988) Bayes theorem and conditional dependence of symptoms: different models applied to data of upper gastrointestinal bleeding. *Methods of information in Medicine*, **27**, 73-83.

19. Quinlan, J.R. (1993) *C4.5: Programs for Machine Learning*. San Mateo, CA: Morgan Kaufmann.

20. Shapire, R.E., Freund, Y., Bartlett, P. and Lee, W.S. (1998) Boosting the margin: A new explanation for the effectiveness of voting methods. *Annals of Statistics*, **26(5)**, 1651-1686.

21. Todd, B.S. amd Stamper, R. (1994) The relative accuracy of a variety of medical diagnostic programmes. *Methods of Information in Medicine*, **33**, 402-416.

22. Webb, A. (1999) *Statistical Pattern Recognition*. London: Arnold.

23. Zheng, Z. and Webb, G.I. (2000) Lazy Learning of Bayesian Rules. *Machine Learning*, **41**, 53-84.

S^3Bagging: Fast Classifier Induction Method with Subsampling and Bagging

Masahiro Terabe[1], Takashi Washio[2], and Hiroshi Motoda[2]

[1] Mitsubishi Research Institute, Inc.
2-3-6 Ohtemachi, Chiyoda, 100-8141 Tokyo, Japan
terabe@mri.co.jp
http://www.ar.sanken.osaka-u.ac.jp/terapreg.html
[2] The Institute of Scientfic and Industrial Research, Osaka University
8-7 Mihogaoka, Ibaraki, 567-0047 Osaka, Japan
{washio, motoda}@sanken.osaka-u.ac.jp
http://www.ar.sanken.osaka-u.ac.jp/{washpreg, motopreg}.html

Abstract. In the data mining process, it is often necessary to induce classifiers iteratively by the human analysts complete to extract valuable knowledge from data. Therefore, the data mining tools need to extract valid knowledge from a large amount of data quickly enough in response to the human demand. One of the approaches to answer this request is to reduce the training data size by subsampling. In many cases, the accuracy of the induced classifier becomes worse when the training data is subsampled. We propose S^3Bagging (**S**mall **S**ub**S**ampled **Bagging**) that adopts both subsampling and a method of committee learning, i.e., Bagging. S^3Bagging can induce classifier efficiently by reducing the training data size by subsampling and parallel processing. Additionally, the accuracy of the classifier is maintained by aggregating the result of each classifier through the Bagging process. The performance of S^3Bagging is investigated by carefully designed experiments.

1 Introduction

In the data mining process, it is often necessary to induce classifiers iteratively by the human analysts complete extracting valuable knowledge from data[1]. Therefore, the data mining tools need to extract valid knowledge from a large amount of data quickly enough in response to the human demand. Many researchers investigated the methods for fast induction of classifiers from large data sets. The main approaches on this issue are classified into the following types[8].

- Design a fast algorithm: This approach includes a wide variety of algorithm design techniques for reducing the asymptotic complexity, for optimizing the search and representation, for finding approximate solutions, and so on.
- Partitioning the data: This approach involves breaking the data set up into subsets, learning from one or more of the subset, and possibly combining the results.

F. Hoffmann et al. (Eds.): IDA 2001, LNCS 2189, pp. 177–186, 2001.
© Springer-Verlag Berlin Heidelberg 2001

The former approach includes developing the heuristics which are effective to restrict the search space of classifier and parallel processing method. The latter approach includes the instance selection and subsampling.

In the field of machine learning, many researchers investigated committee learning method for improving the prediction accuracy of the classifier[17]. In the framework of committee learning, multiple training data sets where the composition of instances is varied from that of the original training data set are prepared by sampling. The committee member classifier (abbreviated here as member classifier) is induced from each training data set. Thus, a committee classifier is constituted by the multiple member classifiers. The classification result by the committee classifier is determined by taking the majority of the classification results of member classifiers.

Boosting[13] and Bagging[2] are the major methods of the committee learning. On the prediction accuracy, Boosting is often superior to Bagging. However, Boosting needs to induce the member classifiers sequentially so that pararell processing is not applicapable. Additionally, it is well known that Boosting tends to be influenced by the noise in training data set comparing with Bagging. On the other hands, Bagging can improve the prediction accuracy generally speaking even if the training data set includes noise. Furthermore, the member classifiers can be induced concurrently so that the committee classifier can be induced in almost the same processing time with the time to induce one member classifier if the parallel processing would be available.

We propose S^3Bagging (**S**mall **S**ub**S**ampled **Bagging**) that adopts both subsampling and a method of committee learning, i.e., Bagging. S^3Bagging is a method which can induce classifier efficiently by reducing the training data size by subsampling and parallel processing. Additionally, the accuracy of the classifiers is maintained by aggregating the result of each classifier through the Bagging process. The performance of S^3Bagging is investigated by carefully designed experiments.

2 S^3 Bagging

In this section, we propose S^3Bagging that adopts both subsampling and a method of committee learning, i.e., Bagging.

First, T training data sets are prepared by subsampling from the original training data set D. The size of each subsampled training data set D_t $(1 \leq t \leq T)$ is $r\%$ of that of the original training data set. The sampling by replacement is adopted as subsampling technique. The subsampling rate r is less than 100% so that the size of subsampled training data set $|D_t|$ is smaller than that of the original training data set $|D|$. Next, a member classifier C_t is induced from each of the training data set D_t. The committee classifier C is constituted by the induced member classifiers C_t $(1 \leq t \leq T)$. The committee classifier C is the classifier which is induced by S^3Bagging.

The committee classifier predicts the class of a test instance based on the class predictions by the member classifiers. Specifically, the committee classifier

Table 1. The specifications of data sets for experiment 1

Data set	# of instances	Attribute Continuous	Attribute Discrete	Class
BREAST-W	699	9	–	2
CREDIT-A	690	6	9	2
CREDIT-G	1,000	7	13	2
HEPATITIS	155	6	13	2
PIMA	768	8	–	2
SICK	3,772	7	22	2
TIC-TAC-TOE	958	–	9	2
VOTE	435	10	16	2

predicts most frequent class predicted by member classifiers. Here, it must be noted that each member classifier can be induced concurrently because each process of subsampling and inducing member classifier is performed independently. Therefore, the processing time to induce committee classifier C, $proc\_time(C)$ is defined as $proc\_time(C) = \max_t proc\_time(C_t)$, where the processing time to induce each member classifier C_t is $proc\_time(C_t)$. The whole processing time to induce committee classifier is almost equivalent to that of inducing each member classifier. Since subsampling is performed, the processing time for induction of committee classifier can also be shortened compared with the case where all original training data are used. Additionally, the prediction accuracy is improved by committee learning.

The main parameter by which S^3Bagging is characterized is subsampling rate r. When the subsampling rate is set large, the prediction accuracy becomes better although the processing time for classifier induction will increase. On the other hands, if the subsampling rate is set small, the prediction accuracy becomes worse although the processing time of the classifier induction will be shortened.

3   Experiments

3.1   Conditions of Experiments

To confirm the effect of the subsampling rate and committee learning on the performance of S^3Bagging, the following experiments are conducted. The performance of S^3Bagging is evaluated from two points of view. One is the prediction accuracy of the induced classifier. The other is the processing time for inducing the classifier. In these experiments, we use a major decision tree generation algorithm, C4.5 Release 8 for classifier induction[10]. All of the functional options of C4.5 are set to defaults. The pruned decision trees are used for evaluation. In Bagging and S^3 Bagging, the number of member classifiers T is set 10 as a general setting[11].

In these experiments, the following five kinds of classifiers are induced for comparison.

- S^3 **Bagging (Committee):** The committee classifier induced by S^3Bagging,
- S^3 **Bagging (Member):** The member classifiers which constitue the committee classifier induced by S^3Bagging,
- **Bagging (Committee):** The committee classifiers induced by Bagging,
- **Bagging (Member)** : The member classifiers which constitute the committee classifier induced by Bagging,
- **All** : The classifier induced from all of the original training data. This is a usual induction method so that the performance of this method is the baseline of the comparison.

Each experimental result of member classifier is the average of that of all member classifiers which constitute the committee classifier.

3.2   Experiment 1: The Subsampling Rate and Prediction Accuracy

To confirm the effect of subsampling rate on prediction accuracy, the experiment with test data sets for machine learning algorithms is conducted. The data sets are selected from UCI Machine Learning Repository[5]. They are selected from the data sets used in the papers of Quinlan[11] and Reinartz[12]. The specifications of the data sets are summarized in Table 1. In order to take into account the influence of the sampling in Bagging and S^3Bagging, the ten fold cross-validation was repeated 10 times and the error rate was evaluated by taking the average over the traials.

The divsrsity of the member classifiers effects the performance of the committee learning. As an index for measuring the diversity of classification results among the member classifiers, entropy $E = -\sum_{i=1}^{|C|} p(c_i) \log_2 p(c_i)$, is calculated, where $|C|$ is the number of classes, $p(c_i)$ is the ratio of member classifier which predicts the class $C = c_i$. The number of classes of all data sets used in this experiments is 2 so that the entropy E takes $0 \leq E \leq 1$. The entropy is 0 when all member classifiers predict the identical class. On the other hands, the entropy is 1 when the member classifiers which classified into each class is the same number.

The error rate of committee classifier is depicted in Figure 1. In all of the data sets, the prediction accuracy becomes better when the subsampling rate is set larger. In many cases, the prediction accuracy is not improved when the subsampling rate is set larger than 30 %. The ratio of error rate of committee classifier to that of member classifier is depicted in Figure 2. This ratio shows that the improvement of prediction accuracy by committee learning. If the ratio is less than 1, the committee learning improves the prediction accuracy. All ratios are smaller than 1. This result shows that the committee learning can improve the prediction accuracy at every subsampling rate. Thus, committee learning has an effect to improve the prediction accuracy not only in Bagging but also in S^3Bagging which adopts subsampling.

We consider the reason why the committee learning improves the prediction accuracy only in the case of TIC-TAC-TOE even when the subsampling rate is set larger than 20%. In order to improve the prediction accuracy of the committee

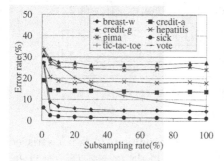

Fig. 1. The error rate of the committee classifier

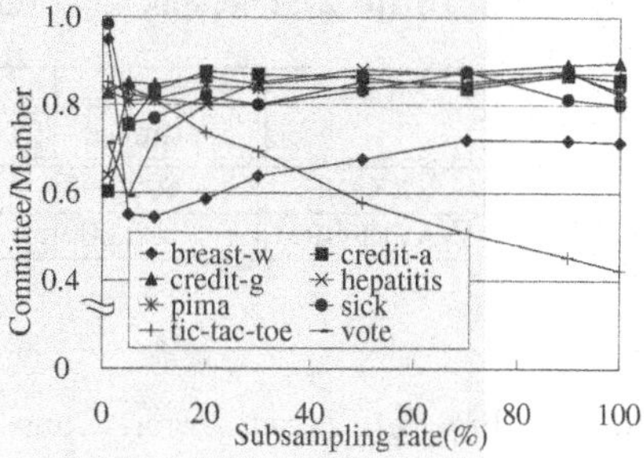

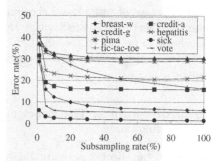

Fig. 2. The ratio of the error rate of committee classifier to that of member classifer

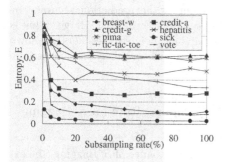

Fig. 3. The entropy of classification by member classifiers composing the committee classifier

Fig. 4. The error rate of the member classifier

classifier compared with that of the member classifiers, it is required to satisfy the following two conditions: (a) the characteristics of the member classifiers of the committee classifier are diverse enough, and (b) the prediction accuracy of each member classifier is high enough. As an index for measuring the diversity of the classification results among member classifiers, the entropy defined previously is calculated. The result is depicted in Figure 3. On the condition (a), the entropy is large[1] in the case of CREDIT-G, PIMA, TIC-TAC-TOE, and HEPATITIS when the subsampling rate is larger than 20%. The large entropy shows that the member classifiers are diverse. Thus, these cases satisfy the condition (a). The error rate of member classifier is depicted in Figure 4. In the case of TIC-TAC-TOE, the error rate is about 27% when the subsampling rate is set 20%, but the error rate is still improved as the subsampling rate becomes large. However, in the case of CREDIT-G, PIMA, and HEPATITIS, the prediction accuracy of member classifier is lager than 22% and the prediction accuracy is not improved even when the

---

[1] Here, it is based on whether entropy is 0.3 or more.

Table 2. The specifications of data sets for experiment 2

Data set	# of instances	Attribute		Class
		Continuous	Discrete	
CENSUS	299,285	7	33	2
WAVEFORM	300,000	21	–	3

subsampling rate is set larger. Consequently, the condition (b) is not satisfied in the case of CREDIT-G, PIMA and HEPATITIS but in the case of TIC-TAC-TOE. By the difference of prediction accuracy of member classifier among the cases, the prediction accuracy of committee classifier is improved only in the case of TIC-TAC-TOE, although the prediction accuracy is not improved on CREDIT-G, PIMA and HEPATITIS.

In the case of BREAST-W, the improvement of the prediction accuracy by committee learning is small when the subsampling rate is set large. This is considered to be due to the small diversity among member classifiers as shown in Figure 3, although the prediction accuracy of each member classifier is high enough.

3.3 Experiment 2: The Effect of Subsampling Rate on S^3Bagging Performance

To confirm the effect of subsampling rate on performance of S^3Bagging, the experiment with large size data is conducted. The processing time (CPU Time) is defined as the sum of processing time for subsampling and induction of member classifiers as mentioned in section 2. In S^3Bagging and Bagging, we assume that each member classifier can be induced in parallel. Therefore, the processing time for inducing committee classifier is the longest processing time among each member classifier induction. A personal computer having the specifications of OS: Linux OS, CPU: PentiumIII 700 MHz, and main memory: 256 M bytes is used in this experiment. For the large size data sets, CENSUS-INCOME (abbreviated here as CENSUS) and WAVEFORM are selected. CENSUS and WAVEFORM are selected from UCI KDD Archive[4] and UCI Machine Learning Repository[5] respectively. These data sets are used in the paper of Provost et al.[9]. The specifications of the data sets are summarized in Table 2.

The experimental results on prediction accuracy are depicted in Figure 5 and 6. The prediction accuracy of committee classifier is always higher than that of member classifier. When the subsampling rate is set 30% in the case of CENSUS and 0.1% in the case of WAVEFORM, the error rate of committee classifier induced by S^3Bagging is almost equivalent to that of the classifier (All) induced from the all of the training data. The experimental results on the processing time are depicted in Figure 7 and 8. Inducing the classifier from all of the training data requires much processing time. The processing time is 300 and 970 seconds in the case of CENSUS and WAVEFORM respectively. On S^3Bagging, the experimental

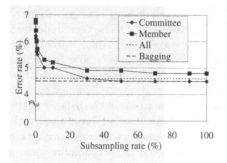

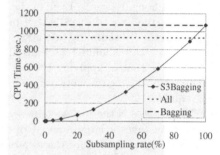

Fig. 5. The subsampling rate vs. error rate: CENSUS

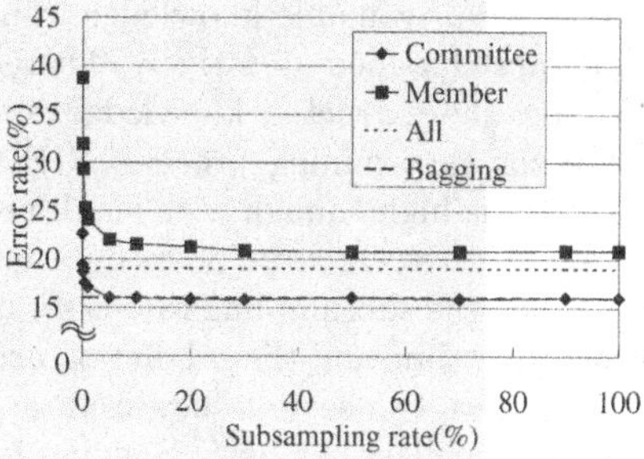

Fig. 6. The subsampling rate vs. error rate: WAVEFORM.

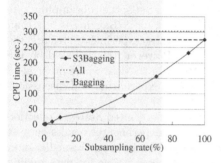

Fig. 7. The subsampling rate vs. processing time: CENSUS

Fig. 8. The subsampling rate vs. processing time: WAVEFORM

result shows that processing time for S^3Bagging is almost proportional to the subsampling rate. Therefore, S^3Bagging can shorten the processing time sharply by setting the subsampling rate small. For example, in the case of CENSUS, the processing time for S^3Bagging is only 22 seconds when the subsampling rate is set 30%. The processing time is less than 8% in the case inducing the classifier from all original training data (All).

4   Discussion

From the experimental results in section 3.2, it is confirmed that the prediction accuracy of the classifier induced by S^3Bagging is not so inferior to that of the classifier induced from all instances of original training data even when the subsampling rate is set 30%. This effect of preventing the deterioration of prediction accuracy by subsampling is derived from the committee learning adopted in S^3Bagging. Next, the experimental result in section 3.3 shows that the processing time for classifier induction is reduced significantly by the effect of subsampling adopted in S^3Bagging. In the data mining process, it is difficult to complete the extraction of valuable knowledge from data at once. Thus, it is often necessary to induce classifiers iteratively. Therefore, it is considered to be very important to show the extracted knowledge in the early stage as a helpful leads to the following data mining process. S^3Bagging extracts classifier whose prediction accuracy is high enough with small processing time even if the data size is large. These features are suitable as data mining tools.

When the subsampling rate is set up, it is necessary to take the balance of the processing time and the prediction accuracy into consideration. From the experimental result in section 3.2, in most of the cases, the degradation of prediction accuracy of classifier by S^3Bagging is about 20% when the subsampling rate is set 30%. From the experimental results with large size data sets in section 3.3, the subsampling rate at which the prediction accuracy of classifier by S^3Bagging is almost equivalent to that of classifier induced from all of the training data is different among data sets. Even in the worst case, CENSUS, the subsampling rate is 30%. In the case of CENSUS, the processing time for S^3Bagging is less than 10% when the subsampling rate is set 30%. From these experimental results, the subsampling rate should be set up about 30%. However, this is based on the result at the time of applying C4.5 as the learner. In the case where other algorithm, such as a neural network, is applied, more experiments and discussions are required.

The CENSUS is the real social research data of the United States. Collection of the data takes a large amount of cost[7]. The experimental results suggest that S^3Bagging can induce the accurate classifier even from much smaller data sets. Therefore, S^3Bagging can save not only processing time for data mining but also the cost of data collection and preparation.

5   Related Work

The effect of the data size reduction by subsampling is investigated by Catlett[6]. In the experiments, the random sampling and the stratified sampling are adopted. They focused on the problem that the C4.5 needs much processing time for inducing decision tree when the data set includes many continuous attributes. Thus, they confirmed that the processing time could be shortened by adopting the subsampling to the training data. In recent years, the parallel processing environment is available so that the novel induction method which can induce the high precision accuracy classifier with small processing time from large size data can be developed.

Harris-Jones et al.[7] investigated the relation between training data size and prediction accuracy of classifier induced by some learning algorithms. Their experimental results shows that the prediction accuracy of classifier reaches a plateau at small subsampling rate in many cases.

Weiss et al.[16] investigated the relation between subsampling rate and the prediction accuracy of classifier induced from the subsampled training data with some kinds of induction algorithms. As the experimental result, the subsampling rates at which the best prediction accuracy classifier is induced are different among each algorithm. Additionally, they conducted the experiments which confirm the change of prediction accuracy of committee classifier when the member classifiers are induced from subsampled training data. In the experiment, they used Arcing[3] as the committee learning method. They show that the deterioration of the prediction accuracy of classifier can be prevented by adopting committee learning. However, they did not discuss from the view point of the processing time for inducing classifier.

Reinartz investigated the effect of the sampling techniques both the prediction accuracy and the processing time[12]. The sizes of data sets used in the experiments are small, and they did not investigate the performance of classifier when the sampling techniques are adopted to the large size data.

6   Conclusion

In this paper, S^3Bagging which adopts both subsampling and Bagging was proposed as the method of inducing a classifier at short processing time. The performance was evaluated through the experiment using test data sets which include large size data set.

S^3Bagging induces the member classifiers in parallel. Additionally, the size of training data is reduced by subsampling so that the processing time for induction of each member classifier becomes shorter. In S^3Bagging, the processing time for inducing classifier is mostly proportional to the subsampling rate. When the subsampling rate is set up to about 30%, the processing time becomes less than 10% compared with the case where all original training data is used. Additionally, by performing committee learning as the way of Bagging, S^3Bagging can prevent the deterioration of the prediction accuracy which is derived from subsampling. From the experimental results, when the subsampling rate is set about 30%, the deterioration of the prediction accuracy could be suppressed to less than 20% in many cases. Thus, the proposed method is effective to induce the classifier whose prediction accuracy is high with smaller processing time under parallel processing environment.

The following issues remain for our future work.

– Experiments on S^3Bagging performance in the parallel processing environment: The execution of S^3Bagging in the parallel processing environment, such as PC cluster, is one of the most important future work.
– Determination of adequate subsampling rate: We proposed the description length of sampled data as a reference index at the time of determining the subsampling rate[14]. However, it is not taking the correlation among attributes into consideration. The new index which reflects the correlation among attributes should be investigated.
– Investigation of the effective presentation method of the extracted information by S^3Bagging: In the process of S^3Bagging, not only committee classifier but also classification of each member classifier and their distribution

are extracted. This information includes useful information for human analysts. The effective method to present this information to the human analysts should be investigated.

References

1. Berry, M.J.A. and Linoff, G.: *Data Mining Techniques: – For Marketing, Sales, and Customer Support*, Weiley (1997).
2. Breiman, L.: Bagging Predictors, *Machine Learning*, **24** (2), pp.123–140 (1996).
3. Breiman, L.: Bias, Variance, and Arcing Classifiers, Technical Report 460 UC-Berkeley, Berkeley, CA. (1996).
4. Bay, S.D.: The UCI KDD Archive, `http://kdd.ics.uci.edu`, Irvine, CA: University of California, Department of Information and Computer Science (1999).
5. Blake, C., Keogh, E. and Merz, C.J.: UCI Repository of Machine Learning Databases, `http://www.ics.uci.edu/~mlearn/MLRepository.html`, Irvine, CA: University of California, Department of Information and Computer Science (1998).
6. Catlett, J.: Megainduction: a Test Flight, In *Proceedings of the Eighth International Workshop on Machine Learning*, pp.596–599 (1991).
7. Harris-Jones, C. and Haines, T.L.: Sample Size and Misclassification: Is More Always Better?, Working Paper AMSCAT-WP-97-118, AMS Center for Advanced Technologies (1997).
8. Provost, F. and Kolluri, V.: A Survey of Methods for Scaling Up Inductive Algorithms, *Knowledge Discovery and Data Mining*, **3** (2), pp.131–169 (1999).
9. Provost, F., Jensen, D. and Oates, T.: Efficient Progressive Sampling, In *Proceedings of the Fifth International Conference on Knowledge Discovery and Data Mining*, pp.23–32 (1999).
10. Quinlan, R.: *C4.5: Programs for Machine Learning*, Morgan Kaufmann (1993).
11. Quinlan, R.: Bagging, Boosting, and C4.5., In *Proceedings of the Thirteenth National Conference on Artificial Intelligence*, pp.725–730 (1996).
12. Reinartz, T.: Similarity-Driven Sampling for Data Mining, Zytkow, J.M. and Quafafou, M. (eds.). *Principles of Data Mining and Knowledge Discovery: Second European Symposium PKDD '98*, pp.423–431 (1998).
13. Shapire, R.E.: A Brief Introduction to Boosting, In *Proceedings of the Sixteenth International Joint Conference on Artificial Intelligence*, pp.1401–1406 (1999).
14. Terabe, M., Washio, T. and Motoda, H.: Fast Classifier Generation by S^3Bagging (in Japanese), *submitted to IPSJ Transaction on Mathematical Modeling and Its Applications*.
15. Weiss, S.M. and Kulikowski, C.A.: *Computer Systems That Learn*, Morgan Kaufmann (1991).
16. Weiss, S.M. and Indurkhya, N.: *Predictive Data Mining – a practical guide –*, Morgan Kaufmann (1998).
17. Zheng, Z.: Generating Classifier Committees by Stochastically Selecting both Attributes and Training Examples, In *Proceedings of the Third Pacific Rim Conference on Artificial Intelligence*, pp.12–33 (1998).

RNA-Sequence-Structure Properties and Selenocysteine Insertion

Rolf Backofen*

Institut für Informatik, LMU München
Oettingenstraße 67, D-80538 München
`backofen@informatik.uni-muenchen.de`

Abstract. Selenocysteine (Sec) is a recently discovered 21st amino acid. Selenocysteine is encoded by the nucleotide triplet UGA, which is usually interpreted as a STOP signal. The insertion of selenocysteine requires an additional signal in form of a SECIS-element (**Sele**nocysteine **I**nsertion **S**equence). This is an mRNA-motif, which is defined by both sequence-related and structure-related properties.

The bioinformatics problem of interest is to design new selenoproteins (i.e., proteins containing selenocysteine), since seleno variants of proteins are very useful in structure analysis. When designing new bacterial selenoproteins, one encounters the problem that the SECIS-element itself is translated (and therefore encodes a subsequence of the complete protein). Hence, changes on the level of mRNA made in order to generate a SECIS-element will also modify the amino acid sequence. Thus, one searches for an mRNA that is maximally similar to a SECIS-element, and for which the encoded amino acid sequence is maximally similar to the original amino acid sequence. In addition, it must satisfy the constraints imposed by the structure.

Though the problem is NP-complete if arbitrary structural constraints are allowed, it can be solved efficiently when we consider the structures as used in SECIS-elements. The remaining problem is to generate a description of the SECIS-element (and its diversity) based on the available data.

1 Selenocysteine Insertion

Selenocysteine (Sec) is a rare amino acid, which was recently discovered as the 21st amino acid [4]. Proteins containing selenocysteine are consequently called *selenoproteins*.

The discovery of selenocysteine was another clue to the complexity and flexibility of the mRNA translation mechanism. Since all 64 codons are either associated with one of the 20 standard amino acids or a STOP-signal, a codon has to be overloaded for encoding this 21st amino acid. Selenocysteine is encoded by the UGA-codon, which is usually a STOP-codon encoding the end of translation

---

* Partially supported by the DFG within the national program SPP 1087 "Selenoprotein – Biochemische Grundlagen und klinische Bedeutung"

F. Hoffmann et al. (Eds.): IDA 2001, LNCS 2189, pp. 187–197, 2001.

by the ribosome (see Figure 1). It has been shown [4] that in the case of selenocysteine, termination of translation is inhibited in the presence of a specific mRNA sequence in the 3'-region after the UGA-codon that forms a hairpin like structure (called Sec insertion sequence (SECIS), see Figure 2). The interesting part is that in bacteria, the SECIS-element itself is translated into an amino acid sequence. Mutagenesis analysis has shown that the sequence, structure, and distance to the UGA-codon is important for selenocysteine insertion. Furthermore, they vary between different organisms [15, 23, 22].

	U	C	A	G	
U	Phe	Ser	Tyr	Cys	U
	Phe	Ser	Tyr	Cys	C
	Leu	Ser	Stop	Stop	A
	Leu	Ser	Stop	Trp	G
C	Leu	Pro	His	Arg	U
	:	:	:	:	

Fig. 1. Translation of RNA. Every amino acid is encoded by a nucleotide triple (called codon). The assignment of codons to amino acid is usually depicted by a three-dimensional array, the genetic code. The selenocysteine codon is encircled.

An mRNA sequence is an element of $\{A, C, G, U\}^*$; A, C, G, U are called *nucleotides*. An important property of the corresponding RNA-molecule is that it can form a distinguished structure. The structure of an RNA-molecule is formed by hydrogen bonds between complementary nucleotide pairs (namely A–U and C–G). Concerning structure, we distinguish three levels of abstractions:

1. primary structure: the mRNA sequence itself.
2. secondary structure: the set of hydrogen bonds that are realized by the corresponding RNA-molecule. Note that any nucleotide can have at most one bond to another nucleotide. The secondary structure of an mRNA-sequence can be determined efficiently by dynamic programming approaches [25, 24, 7, 14].
3. tertiary structure: the real three-dimensional structure of the corresponding RNA-molecule, which is very hard to determine.

Although the real three-dimensional structure of an RNA molecule is biologically important, it is generally agreed that for most purposes, it suffices to consider the secondary structure of the corresponding mRNA-sequence.

In the following, we will consider the problem of designing new selenoproteins. We will give a formal description of this problem. For this formalization, we have presented an efficient algorithm in another paper [2]. This algorithm requires a statistical representation of SECIS-elements. We will then review methods that are available to generate this description.

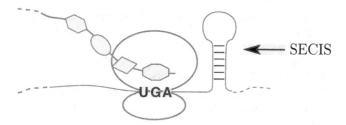

Fig. 2. SECIS-element and translation of mRNA by the ribosome in the case of selenocysteine.

2   Related Computational Problems

Concerning selenoproteins, bioinformatics currently faces the following challenges. An obvious problem is to search for existing but unknown selenoproteins. Since both sequence and structure is important for selenocysteine incorporation, and the different selenoproteins do not have a pronounced consensus sequence (on the amino acid level), the usually BLAST and FASTA search tools will not have enough selectivity. Therefore, new algorithms have to be devised. An example is SECISearch [20], which searches for mammalian selenoproteins by recognizing the corresponding SECIS-element using a hierarchical approach. First, one searches for sequences that satisfy the SECIS-consensus sequence, and allow for the secondary structure required for the SECIS-element. In a second step, the minimal free energy of the sequences is estimated by using the RNAfold program of the Vienna-RNA-Package [14].

However, we consider a different problem, namely modifying existing proteins such that selenocysteine is incorporated instead of another amino acid (e.g., instead of a cysteine). There are mainly two possible applications for newly generated selenoproteins. First, one can generate functionally modified proteins. Experimentally, this has been already done successfully for the bacterium *E. coli* [13], where the mRNA sequence of a protein not containing selenocysteine was engineered (by hand) such that it forms a *E. coli* SECIS (while preserving maximal similarity on the amino acid level). The engineered protein now had a selenocysteine in place of a catalytic cysteine (at position 41), and showed a 4-fold enhanced catalytic activity. Second, and more important, seleno-variants of proteins could be used for the phase determination in an X-ray crystallography (or in the NMR-analysis) of an given protein.

3   Finding New Insertion Sites

We consider the computational properties of substituting the amino acid at a given position of some protein by a selenocysteine. For this purpose, the mRNA-sequence of the protein has to be modified such that an appropriate SECIS-element is generated. Since the SECIS-element is later translated into an amino acid sequence, this modification implies that the amino acid sequence is modified, too. To keep the protein functional, we want to find a SECIS-element which minimizes the changes on the amino acid level.

For this purpose, we first need a description of the SECIS-element, generated from sequences known to be a SECIS-element.[1] For these sequences, the secondary structure usually has been determined. Since the structure of mRNA-motifs are more conserved than the sequence, we use the following two-level approach to describe the SECIS-element. For each possible secondary structure encountered in the set of SECIS-elements (only few), we collect all corresponding sequences into a so-called profile (or consensus sequence). A profile assigns every position of the considered structure a weight. This weight represents the frequence with which a nucleotide is observed at this position. In the following, we will call this *the profile for* a specific secondary structure.

For each combination of structure and corresponding profile, we search for an mRNA-sequence that allows the insertion of Selenocysteine, and satisfies the given structure. This (sub-)problem can be described formally as follows:

Input: An amino acid sequence $A_1 \ldots A_n$, a graph $\Gamma = (V, E)$, $V = \{v_1, \ldots, v_{3n}\}$ (the considered structure), functions $p_1, \ldots, p_{3n}$ (the corresponding profile for this structure, where $p_i : \{A, C, G, U\} \to \mathbb{R}$), and a function sim : $\Sigma_{\text{Amino}} \times \Sigma_{\text{Amino}} \to \mathbb{R}$ measuring the similarity on the level of the amino acid sequence (e.g., given by the PAM250 similarity matrix [6]).

Output: Find mRNA-sequence $N_1, \ldots, N_{3n} \in \{A, C, G, U\}^{3n}$ such that s.t. N_k is assigned to v_k and

1. $\{v_k, v_l\} \in E$ implies that N_k is the complement of N_l,
2. $\sum_{i=1}^{n} f_i(N_{3i-2}, N_{3i-1}, N_{3i})$ is maximized, where

$$f_i(N_{3i-2}, N_{3i-1}, N_{3i}) = \alpha \times \text{sim}\left(A_i, \text{AMINO}(N_{3i-2}, N_{3i-1}, N_{3i})\right)$$

$$+ (1 - \alpha) \times \left[p_{3i-2}(N_{3i-2}) + p_{3i-1}(N_{3i-1}) + p_{3i}(N_{3i})\right].$$

α is a parameter for weighting the similarity on the amino acid level. $\text{AMINO}(N_{3i-2}, N_{3i-1}, N_{3i})$ is the amino acid encoded by the nucleotide codon $N_{3i-2}, N_{3i-1}, N_{3i}$. $\Gamma = (V, E)$ is called the *structure graph*. We denote the above problem by $MRSO(\Gamma, p_1, \ldots, p_{3n}, \text{sim})$. An instance of this problem can be viewed by the following picture:

org. amino seq. $A =$	A_1	$\ldots$	A_i	$\ldots$	A_n	$\leftarrow \text{sim}(A_i, A'_i)$
	$\wr$		$\wr$		$\wr$	
mod. amino seq. $A' =$	A'_1	$\ldots$	A'_i	$\ldots$	A'_n	
	$\uparrow$		$\uparrow$		$\uparrow$	
mod. mRNA seq. $N =$	$N_1 N_2 N_3$	$\ldots N_{3i-2} N_{3i-1} N_{3i}$		$\ldots N_{3n-2} N_{3n-1} N_{3n}$		$\leftarrow p_j(N_j) \ (j \in [1 \ldots 3n])$
	$\wr \ \wr \ \wr$	$\wr \ \ \wr \ \ \wr$		$\wr \ \ \wr \ \ \wr$		
profile positions $=$	$P_1 \ P_2 \ P_3$	$\ldots P_{3i-2} P_{3i-1} P_{3i}$		$\ldots P_{3n-2} P_{3n-1} P_{3n}.$		

---

[1] Of course, the set of known SECIS-elements represent only a portion of set of all possible SECIS-elements.

It is easy to see that the problem cannot be simply solved by determining for each amino acid position $i \in [1..n]$ the codon $N_{3i-2}N_{3i-1}N_{3i}$ which is locally optimal (i.e., maximizes $f_i(N_{3i-2}, N_{3i-1}, N_{3i})$). The reason is that the choice of $N_{3i-2}, N_{3i-1}, N_{3i}$ may fix some other N_k by condition 1 of the problem description (depending on the given secondary structure). But the problem can be solved in linear time if we consider the typical hairpin structures of SECIS-elements. On the other hand, we have shown that the problem becomes NP-complete if we consider more complex structures having crossing edges in the linear embedding of the structure graph. Figure 3 shows two graphs associated two different types of mRNA structures. For the first type, namely the hairpin-like structure as encountered in SECIS-elements, the linear embedding of the structure graph does not contain crossing edges. The second type, namely pseudoknots, is important for programmed frameshifts which allow to encode two different amino acid sequences in one mRNA sequence. This mechanism has been identified in viruses as well as in prokaryotes and eukaryotes [9, 10].

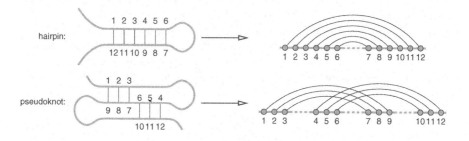

Fig. 3. Linear embedding of hairpin and pseudoknot. We have numbered only the positions that are part of a bond.

The computational properties of this problem are investigated in [2] and summarized in the theorem below. The linear time algorithm is a modified dynamic programming algorithm, and the NP-completeness is shown by reducing $3SAT$ to our problem.

Theorem 1. *Let $(\Gamma, p_1, \ldots, p_{3n}, \text{sim})$ be an instance of MRSO such that the structure graph $\Gamma = (V, E)$ is outer-planar (i.e., there exists a linear embedding that does not contain crossing edges). Then one can find an optimal mRNA-sequence $N_1, \ldots, N_{3n}$ in linear time. Furthermore, the decision version of MRSO is NP-complete in the case of general structure graphs.*

For the general case, we have developed an approximation algorithm [1].

4   Determination of the Profile

There are several sources that provide data suitable for describing SECIS-Elements. The first source comes from so-called *mutagenesis experiments*. In these experiments, original SECIS-elements (called *wild-type*) are mutated, and the mutated variants are experimentally tested for their capability of inserting selenocysteine. On the one hand, this kind of experiments provide a valuable source for the descriptions of SECIS-elements, since the resulting sequences have been verified. On the other hand, usually only a small portion of the sequence space can be explored using this technique.

But there is another form of source that can be used to determine the consensus sequence(s) and structure(s) for SECIS-elements. The SECIS-element enforces incorporation not by their mere existence, but by the fact that there is a special protein which recognizes these elements. This protein is SelB, which is a special elongation factor for selenocysteine. An *elongation factor* is a protein which mediates the elongation of the amino acid sequence by the ribosome (see Figure 4). Thus, the capability of binding to the special elongation factor SelB is an important property that every SECIS-element has to satisfy. Now, there exists a high-throughput method called SELEX (**S**ystematic **E**volution of **L**igands by **Ex**pontential enrichment) [28]. In this method, a huge amount of RNA-sequences is generated and tested in parallel whether they bind to some ligand. Sequences that bind to SelB have been determined using this method [19, 18].

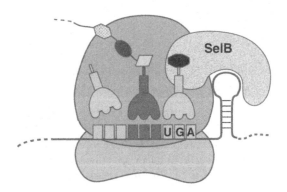

Fig. 4. A more pronounced view of selenocysteine insertion involving special elongation factor SelB.

The remaining problem is to determine the consensus sequence and structure from these data. The usual approach for determining the consensus sequence is to generate a multiple sequence alignment. A *multiple sequence alignment* is a two-dimensional representation of the evolutionary relationship between several sequences. The sequences are listed horizontally, with an additional gap symbols inserted at several places. Thus, every column contains for every sequence either a letter of this sequence, or a gap symbol (see Figure 5). A column in the multiple sequence alignment is called *highly conserved* if the column contains only a few gaps, and the majority of entries in this column are all equal. In Figure 5, columns 1,3,4,7,8,10 denote perfectly conserved columns.

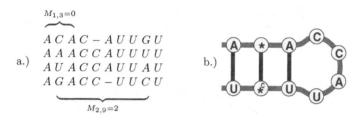

$M_{1,3}=0$

a.)
$$
\begin{array}{l}
A\,C\,A\,C\,-A\,U\,U\,G\,U \\
A\,A\,A\,C\,C\,A\,U\,U\,U\,U \\
A\,U\,A\,C\,C\,A\,U\,U\,A\,U \\
A\,G\,A\,C\,C\,-U\,U\,C\,U
\end{array}
$$

$M_{2,9}=2$

b.)

Fig. 5. Alignment and consensus structure for the sequences $s_1 = ACACAUUGU$, $s_2 = AAACCAUUUU$, $s_3 = AUACCAUUAU$, and $s_4 = AGACCUUCU$. a.) shows an alignment and the mutual information content for selected columns. b.) shows the consensus structure, where $\star$ denotes any nucleotide, and $\star^C$ is the complement of the nucleotide assigned to $\star$.

The problem is that for RNA, the structure is usual more conserved than the sequence. An example are the two columns 2 and 9 in Figure 5, which are not conserved on the sequence level, but perfectly conserved in the corresponding consensus structure. The problem is that usual techniques for multiple sequence alignment will not align columns that are conserved only on the level of structure.

On the other hand, if one has given a correct multiple sequence alignment of RNA sequences, the consensus structure can be deduced from the multiple sequence alignment by evaluating the mutual information content of the columns in the alignment [8]. The mutual information content $M_{i,j}$ of two columns i and j represents the information gained when assuming that i and j depend on each other compared to the situation that they are independent. $M_{i,j}$ is calculated by

$$
M_{i,j} = \sum_{x_i, x_j} f_{x_i x_j} \log \frac{f_{x_i x_j}}{f_{x_i} f_{x_j}},
$$

where x_i (resp. x_j) varies over all possible nucleotides that occur in column i (resp. j), f_{x_i} (resp. f_{x_j}) is the relative frequence of x_i (resp. x_j) in column i (resp. j). $f_{x_i x_j}$ is the relative frequence of occurrences of the pair (x_i, x_j) in columns i and j on the same row. $M_{i,j}$ has its maximal value of 2 if the symbols columns i and j are uniformly distributed (i.e., for all x_i, x_j we have $f_{x_i} = 0.25 = f_{x_j})^2$, but are always forming a complementary base pair, i.e.,

---

2 Recall that we are using the four letter alphabet $\{A, C, G, U\}$.

$$f_{x_i x_j} = \begin{cases} 0.25 & \text{if } x_i \text{ and } x_j \text{ are complemetary base pairs} \\ 0 & \textit{else.} \end{cases}$$

Then one can use a variant of a secondary structure prediction algorithm such as the basic Nussinov/Zuker algorithm [25, 24] (using the mutual information content as a score instead of energy or number of bonds) to calculate a secondary structure S that maximizes

$$\sum_{(i,j) \text{ form a bond in } S} M_{i,j}.$$

Of course, the problem is to generate the initial multiple sequence alignment. If the consensus structure were known, then one could use the apparatus of stochastic context-free grammars to generate good multiple sequence alignments [8, 12, 11, 26, 27, 5]. In this approach, one uses for every bond in the consensus structure a set of rules of the form $r : S \xrightarrow{p_r} xS'x^{\mathtt{C}}$, where S, S' are non-terminals, x ranges over the terminal symbols $\{A, C, G, U\}$ (or over a subset of $\{A, C, G, U\}$), and $x^{\mathtt{C}}$ is the nucleotide complementary to x. p_r is the probability assigned to the rule r. In addition, one needs rules to handle insertion to and deletions from the consensus structure [8]. Figure 6 shows an example of a SCFG for a small hairpin consensus structure. Once we have the SCFG, we can train it by using Expectation-Maximization (via the inside-outside algorithm[21, 3]) to re-estimate the probabilities p_r.

a.)

○ $S \xrightarrow{0.75} a\ S_1\ u$ ○ $S \xrightarrow{0.25} c\ S_1\ g$
○ $S_1 \xrightarrow{0.9} g\ S_2\ c$ ○ $S_1 \xrightarrow{0.1} c\ S_2\ g$
○ $S_2 \xrightarrow{0.5} g\ S_3\ c$ ○ $S_2 \xrightarrow{0.5} c\ S_3\ g$
○ $S_3 \xrightarrow{1} agguсu$

b.)

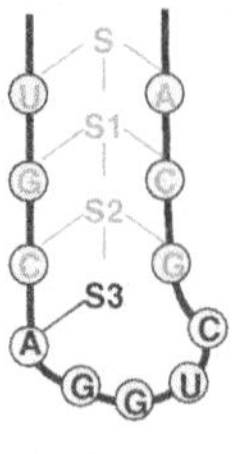

Fig. 6. Stochastic context-free grammar for RNAs. a.) shows a SCFG for a hairpin with 3 bonds, and b.) shows one parse and the corresponding RNA-sequence and structure.

The above described approaches show that we are in a perfect cycle. If we have a good multiple sequence alignment, then we can generate the consensus structure. But good multiple sequence alignments can usually only be generated if the consensus structure is known (via the SCFG-formulation). This is handled in [8] by simply starting with an initial random multiple sequence alignment, then iterating the cycle

$$multiple\ sequence\ alignment$$
$$\downarrow$$
$$consensus\ structure\ +\ SCFG\ +\ parameter\ re\text{-}estimation$$
$$\downarrow$$
$$multiple\ sequence\ alignment$$

until a satisfying result is obtained. A similiar approach using the above described cycle is used in [17, 16], where the multiple sequence alignment is generated by a program that can first, align the sequence locally, and second, explore the structural properties of the aligned sequences.

5 Acknowledgment

I would like to thank Prof. Böck from the Institute of Microbiology, LMU München, for explaining him the biochemical background of selenocysteine, for many suggestion and the fruitful cooperation on the problems discussed in this paper. I would also like to thank N.S. Narayanaswamy, Firas Swidan and Sebastian Will for many helpful and enlightening discussions on related subjects. Furthermore, I would like to thank Andreas Abel, Jan Johannsen, Sven Siebert, and Slim Abdennadher for reading draft versions of this paper.

References

[1] Rolf Backofen, N.S. Narayanaswamy, and Firas Swidan. On the complexity of protein similarity search under mRNA structure constraints. 2001. submitted.

[2] Rolf Backofen, N.S. Narayanaswamy, and Firas Swidan. Protein similarity search under mRNA structural constraints: application to selenocysteine incorporation. 2001. submitted.

[3] J. K. Baker. Trainable grammars for speech recognition. In *Speech Communication Papers for the 97th Meeting of the Acoustical Society of America*, pages 547–550, 1979.

[4] A. Bck, K. Forchhammer, J. Heider, and C. Baron. Selenoprotein synthesis: an expansion of the genetic code. *Trends Biochem Sci*, 16(12):463–467, 1991.

[5] Michael P.S. Brown. Small subunit ribosomal RNA modeling using stochastic context-free grammars. In *ISMB 2000*, pages 57–66, 2000.

[6] M. O. Dayhoff, R. M. Schwartz, and B. C. Orcutt. A model of evolutionary change in proteins. In M. O. Dayhoff, editor, *Atlas of Protein Sequence and Structure*, volume 5, supplement 3, pages 345–352. National Biomedical Research Foundation, Washington, DC, 1978.

[7] Mathews DH, Sabina J, Zuker M, and Turner DH. Expanded sequence dependence of thermodynamic parameters improves prediction of RNA secondary structure. *J Mol Biol*, 288(5):911–40, 1999.

[8] S. R. Eddy and R.Durbin R. RNA sequence analysis using covariance models. *Nucleic Acids Research*, 22(11):2079–2088, 1994.

[9] P. J. Farabaugh. Programmed translational frameshifting. *Microbiology and Molecular Biology Reviews*, 60(1):103–134, 1996.

[10] D. P. Giedroc, C. A. Theimer, and P. L. Nixon. Structure, stability and function of RNA pseudoknots involved in stimulating ribosomal frameshifting. *Journal of Molecular Biology*, 298(2):167–186, 2000.

[11] Leslie Grate. Automatic RNA secondary structure determination with stochastic context-free grammars. In *Proc. of the 3th Int. Conf. on Intelligent Systems for Molecular Biology (ISMB'95)*, pages 136–144, Cambridge, England, 1995. AAAI Press.

[12] Leslie Grate, Mark Herbster, Richard Hughey, David Haussler, Saira Mian, and Harry Noller. RNA modeling using gibbs sampling and stochastic context free grammars. In *Proc. of the 2th Int. Conf. on Intelligent Systems for Molecular Biology (ISMB'94)*, pages 138–146, Stanford, California, 1994. AAAI Press.

[13] Stephane Hazebrouck, Luc Camoin, Zehava Faltin, Arthur Donny Strosberg, and Yuval Eshdat. Substituting selenocysteine for catalytic cysteine 41 enhances enzymatic activity of plant phospholipid hydroperoxide glutathione peroxidase expressed in *escherichia coli. Journal of Biological Chemistry*, 275(37):28715–28721, 2000.

[14] I.L. Hofacker, W. Fontana, P.F. Stadler, S. Bonhoeffer, M. Tacker, and P. Schuster. Fast folding and comparison of RNA secondary structures. *Monatshefte f. Chemie*, 125:167–188, 1994.

[15] Alexander Httenhofer and August Bck. RNA structures involved in selenoprotein synthesis. In R. W. Simons and M. Grunberg-Manago, editors, *RNA Structure and Function*, pages 903–639. Cold Spring Harbour Laboratory Press, Cold Spring Harbour, 1998.

[16] Gorodkin J, Heyer LJ, and Stormo GD. Finding the most significant common sequence and structure motifs in a set of RNA sequences. *Nucleic Acids Res*, 25(18):3724–32, 1997.

[17] Gorodkin J, Stricklin SL, and Stormo GD. Discovering common stem-loop motifs in unaligned RNA sequences. *Nucleic Acids Res*, 29(10):2135–44, 2001.

[18] Stefanie J. Klug, Alexander Httenhofer, and Michael Famulok. In vitro selection of RNA aptamers that bind special elongation factor SelB, a protein with multiple RNA-binding sites, reveals one major interaction domain at the carboxyl terminus. *RNA*, 5:1180–1190, 1999.

[19] Stefanie J. Klug, Alexander Httenhofer, Matthias Kraomayer, and Michael Famulok. In vitro and in vivo characterization of novel mRNA motifs that bind special elongation factor SelB. *Proc. Natl. Acad. Sci. USA*, 94(13):6676–6681, 1997.

[20] G. V. Kryukov, V. M. Kryukov, and V. N. Gladyshev. New mammalian selenocysteine-containing proteins identified with an algorithm that searches for selenocysteine insertion sequence elements. *J Biol Chem*, 274(48):33888–33897, 1999.

[21] K. Lari and S.J. Young. Applications of stochastic context-free grammars using the inside-outside algorithm. *Computer Speech and Language*, 5:237–257, 1991.

[22] Z. Liu, M. Reches, I. Groisman, and H. Engelberg-Kulka. The nature of the minimal 'selenocysteine insertion sequence' (secis) in escherichia coli. *Nucleic Acids Research*, 26(4):896–902, 1998.

[23] Susan C. Low and Marla J. Berry. Knowing when not to stop: selenocysteine incorporation in eukaryotes. *Trends in Biochemical Sciences*, 21(6):203–208, 1996.

[24] Zuker M and Stiegler P. Optimal computer folding of large RNA sequences using thermodynamics and auxiliary information. *Nucleic Acids Res*, 9(1):133–48, 1981.

[25] Ruth Nussinov, George Pieczenik, Jerrold R. Griggs, and Daniel J. Kleitman. Algorithms for loop matchings. *SIAM Journal on Applied Mathematics*, 35(1):68–82, July 1978.

[26] Yasubumi Sakakibara, Michael Brown, Richard Hughey, I. Saira Mian, Kimmen Sjolander, Rebecca C. Underwood, and David Haussler. Recent methods for RNA modeling using stochastic context-free grammars. In *Proc. 5th Symp. Combinatorical Pattern Matching*, 1994.

[27] Yasubumi Sakakibara, Michael Brown, Richard Hughey, I. Saira Mian, Kimmen Sjolander, Rebecca C. Underwood, and David Haussler. Stochastic context-free grammars for tRNA modeling. *Nucleic Acids Research*, 22(23):5112–5120, 1994.

[28] C. Tuerk and L. Gold. Systematic evolution of ligands by exponential enrichment - RNA ligands to bacteriophage-T4 DNA-polymerase. *Science*, 249(4968):505–510, 3 1990.

An Algorithm for Segmenting Categorical Time Series into Meaningful Episodes

Paul Cohen[1] and Niall Adams[2]

[1] Department of Computer Science. University of Massachusetts
cohen@cs.umass.edu
[2] Department of Mathematics. Imperial College, London
n.adams@ic.ac.uk

Abstract. This paper describes an unsupervised algorithm for segmenting categorical time series. The algorithm first collects statistics about the frequency and boundary entropy of ngrams, then passes a window over the series and has two "expert methods" decide where in the window boundaries should be drawn. The algorithm segments text into words successfully in three languages. We claim that the algorithm finds meaningful episodes in categorical time series, because it exploits two statistical characteristics of meaningful episodes.

1 Introduction

Most English speakers will segment the 29 characters in "itwasabrightcoldday-inapriland" into nine words. We draw segment boundaries in eight of the 28 possible locations so that the sequences of characters between the boundaries are meaningful. In general, there is an exponential number of ways to draw segment boundaries in ways that produce meaningless segments (e.g., "itw" "asab" "rig"...) but we somehow manage to find the "right" segmentation, the one that corresponds to meaningful segments. It seems likely that we do it by recognizing words in the sequence: The task is more difficult if the characters constitute words in an unknown language, or if they are simply coded. For example, "juxbtbcsjhiudpmeebzjobqsjmboe" is formally (statistically) identical with "itwasabrightcolddayinapriland": one sequence is obtained from the other by replacing each letter with the adjacent one in the alphabet. Yet one is easily segmented, the other is not.

This paper asks whether there is a way to find meaningful units in time series other than recognizing them. It proposes two statistical characteristics of meaningful units, and reports experiments with an unsupervised segmentation algorithm based on these characteristics. We offer the conjecture that these characteristics are domain-independent, and we illustrate the point by segmenting text in three languages.

F. Hoffmann et al. (Eds.): IDA 2001, LNCS 2189, pp. 198–207, 2001.

2   The Segmentation Problem

Suppose we remove all the spaces and punctuation from a text, can an algorithm figure out where the word boundaries should go? Here is the result of running the Voting Experts algorithm on the first 500 characters of *1984*. The * symbols are induced boundaries:

Itwas * a * bright * cold * day * in * April * andthe * clockswere * st * ri * king * thi * rteen * Winston * Smith * his * chin * nuzzl * edinto * his * brea * st * in * aneffort * to * escape * the * vilewind * slipped * quickly * through * the * glass * door * sof * Victory * Mansions * though * not * quickly * en * ought * oprevent * aswirl * ofgrit * tydust * from * ent * er * inga * long * with * himThe * hall * ways * meltof * boiled * cabbage * and * old * ragmatsA * tone * endof * it * acoloured * poster * too * large * for * indoor * dis * play * hadbeen * tack * ed * tothe * wall * It * depicted * simplya * n * enormous * face * more * than * ametre * widethe * faceof * aman * of * about * fortyfive * witha * heavy * black * moustache * and * rugged * ly * handsome * featur

The segmentation clearly is not perfect: Words are run together (Itwas, aneffort) and broken apart (st * ri * king). More seriously, some words are split between segments ("to" in en * ought * oprevent), although the algorithm loses only a small fraction of words this way.

To define the segmentation problem, we must first distinguish patterns from episodes. An episode is a pattern that means something in a domain. This definition holds for any notion of pattern and any domain. Thus, "black" and "un" are patterns and also episodes (both are morphemes), whereas "bla" is a pattern but not an episode. In computer vision, induced lines are patterns, and some of them are also episodes by merit of corresponding to edges in the scene. In chess, sixteen pawns arranged neatly in a square in the center of the board doubtless constitute a pattern, but the arrangement is meaningless in the domain, so it is not an episode. The segmentation problem is: *Given a time series of categorical data that contains episodes but no episode boundary markers (e.g., spaces, punctuation) insert boundary markers so that the subsequences between the markers are episodes, i.e., meaningful.*

It is not difficult to write algorithms to find patterns in time series of categorical data, but the majority of these patterns will not be episodes. For example, the following subsequences are the 100 most frequently-occurring patterns in the first 10,000 characters of Orwell's text, but most are not morphemes, that is, meaningful units:

th in the re an en as ed to ou it er of at ing was or st on ar and es ic el al om ad ac is wh le ow ld ly ere he wi ab im ver be for had ent itwas with ir win gh po se id ch ot ton ap str his ro li all et fr andthe ould min il ay un ut ur ve whic dow which si pl am ul res that were ethe wins not winston sh oo up ack ter ough from ce ag pos bl by tel ain

3   Related Work

Many methods have been developed for segmenting time series. Of these, many deal with continuous time series, and so are not directly comparable to the problem we are considering here. Some methods for categorical series are based on compression (e.g., [5,8]), but as we just saw, compression finds common, not necessarily meaningful subsequences. Some methods are trained to find instances of patterns or templates (e.g., [2,4]) but we wanted an unsupervised method. There is some work on segmentation in the natural language and information retrieval literature, for instance, techniques for segmenting Chinese, which has no word boundaries in its orthography. The method in [8], is similar to ours, though it requires supervised training on very large corpora. Magerman and Marcus' [3] approach to parsing based on mutual information statistics is similar to our notion of boundary entropy (see below). We know of no related research on characteristics of meaningful subsequences, that is, statistical markers of boundaries of meaning-carrying subsequences.

4   Characteristics of Episodes

Although we are far from a theory of episodes—a theory to tell us which subsequences of a series are meaningful—we have observed four empirical characteristics of episodes. Two of them are not implemented in the current algorithm: one of these relies on the fact that random coincidences are rare, so coincidences often mark episode boundaries [1]; the other exploits the idea that adjacent episodes are often generated by processes that have different underlying probability distributions, so one can infer episode boundaries by looking for points at which the sequences to the left and right of the boundary have different estimated distributions [6,7]. The two characteristics of episodes that we have implemented here are called boundary entropy and frequency:

Boundary entropy Every unique subsequence is characterized by the distribution of subsequences that follow it; for example, the subsequence "en" in this sentence repeats five times and is followed by tokens c and ". This distribution has an entropy value (0.72, as it happens). In general, every subsequence of length n has a boundary entropy, which is the entropy of the distribution of subsequences of length m that follow it. If a subsequence S is an episode, then the boundary entropies of subsequences of S will have an interesting profile: They will start relatively high, then sometimes drop, then peak at the last element of S. The reasons for this are, first, that the predictability of elements within an episode increases as the episode extends over time; and, second, that the element that immediately follows an episode is relatively uncertain. Said differently, within episodes, we know roughly what will happen, but at episode boundaries we become uncertain.

Frequency Episodes, recall, are meaningful sequences, they are patterns in a domain that we call out as special, important, valuable, worth committing to memory, worth naming, etc. One reason to consider a pattern meaningful is that one can use it for something, like prediction. (Predictiveness is another characteristic of episodes nicely summarized by entropy.) Rare patterns are less useful than common ones simply because they arise infrequently, so all human and animal learning places a premium on frequency. In general, episodes are common patterns, but not all common patterns are episodes, as we saw earlier.

5   The Voting Experts Algorithm

The voting experts algorithm includes experts that attend to boundary entropy and frequency, and is easily extensible to include experts that attend to other characteristics of episodes. The algorithm simply moves a window across a time series and asks, for each location in the window, whether to "cut" the series at that location. Each expert casts a vote. Each location takes n steps to traverse a window of size n, and is seen by the experts in n different contexts, and may accrue up to n votes from each expert. Given the results of voting, it is a simple matter to cut the series at locations with high vote counts. Here are the steps of the algorithm:

Build an ngram tree of depth $n + 1$. Nodes at level i of an ngram tree represent ngrams of length i. The children of a node are the extensions of the ngram represented by the node. For example, a b c a b d produces the following ngram tree of depth 2:

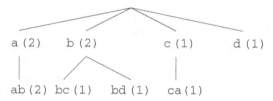

Every ngram of length 2 or less in the sequence a b c a b d is represented by a node in this tree. The numbers in parentheses represent the frequencies of the subsequences. For example, the subsequence ab occurs twice, and every occurrence of a is followed by b.

For the first 10,000 characters in Orwell's text, an ngram tree of depth 7 includes 33774 nodes, of which 9109 are leaf nodes. That is, there are over nine thousand unique subsequences of length 7 in this sample of text, although the average frequency of these subsequences is 1.1—most occur exactly once. The average frequencies of subsequences of length 1 to 7 are 384.4, 23.1, 3.9, 1.8, 1.3, 1.2, and 1.1.

Calculate boundary entropy. The boundary entropy of an ngram is the entropy of the distribution of tokens that can extend the ngram. The entropy of a distribution of a random variable x is just $- \sum Pr(x) \log Pr(x)$. Boundary entropy is easily calculated from the ngram tree. For example, the node a has entropy equal to zero because it has only one child, ab, whereas the entropy of node b is 1.0 because it has two equiprobable children, bc and bd. Clearly, only the first n levels of the ngram tree of depth $n + 1$ can have node entropy scores.

Standardize frequencies and boundary entropies. In most domains, there is a systematic relationship between the length and frequency of patterns; in general, short patterns are more common than long ones (e.g., on average, for subsets of 10,000 characters from Orwell's text, 64 of the 100 most frequent patterns are of length 2; 23 are of length 3, and so on). Our algorithm will compare the frequencies and boundary entropies of ngrams of different lengths, but in all cases we will be comparing how *unusual* these frequencies and entropies are, relative to other ngrams of the same length. To illustrate, consider the words "a" and "an". In the first 10000 characters of Orwell's text, "a" occurs 743 times, "an" 124 times, but "a" occurs only a little more frequently than other one-letter ngrams, whereas "an" occurs much more often than other two-letter ngrams. In this sense, "a" is ordinary, "an" is unusual. Although "a" is much more common than "an" it is much less unusual relative to other ngrams of the same length. To capture this notion, we standardize the frequencies and boundary entropies of the ngrams. (To standardize a value in a sample, subtract the sample mean from the value and divide by the sample standard deviation. This has the effect of expressing the value as the number of standard deviations it is away from the sample mean.) Standardized, the frequency of "a" is 1.1, whereas the frequency of "an" is 20.4. In other words, the frequency of "an" is 20.4 standard deviations above the mean frequency for sequences of the same length. We standardize boundary entropies in the same way, and for the same reason.

Score potential segment boundaries. In a sequence of length k there are $k - 1$ places to draw boundaries between segments, and, thus, there are 2^{k-1} ways to divide the sequence into segments. Our algorithm is greedy in the sense that it considers just $k - 1$, not 2^{k-1}, ways to divide the sequence. It considers each possible boundary in order, starting at the beginning of the sequence. The algorithm passes a window of length n over the sequence, halting at each possible boundary. All of the locations within the window are considered, and each garners zero or one vote from each expert. Because we have two experts, for boundary-entropy and frequency, respectively, each possible boundary may garner up to $2n$ votes. This is illustrated below. A window of length 3 is passed along the sequence itwasacold.

Initially, the window covers itw. The entropy and frequency experts each decide where they could best insert a boundary within the window (more on this, below). The entropy expert favors the boundary between t and w, while the frequency expert favors the boundary between w and whatever comes next. Then the window moves one location to the right and the process repeats. This time, both experts decide to place the boundary between t and w. The window moves again and both experts decide to place the boundary after s, the last token in the window. Note that each potential boundary location (e.g., between t and w) is seen n times for a window of size n, but it is considered in a slightly different context each time the window moves. The first time the experts consider the boundary between w and a, they are looking at the window itw, and the last time, they are looking at was.

```
entropy    | i  t | w | a  s  a  c  o  l  d  .  .  .
frequency  | i  t  w| a  s  a  c  o  l  d  .  .  .

entropy    i | t | w  a | s  a  c  o  l  d  .  .  .
frequency  i | t | w  a | s  a  c  o  l  d  .  .  .

entropy    i  t | w  a  s| a  c  o  l  d  .  .  .
frequency  i  t | w  a  s| a  c  o  l  d  .  .  .

           | i | t | w  a| s | a  c  o  l  d  .  .  .
             0  0  3  1  0  2
```

In this way, each boundary gets up to $2n$ votes, or $n = 3$ votes from each of two experts. The wa boundary gets one vote, the tw boundary, three votes, and the sa boundary, two votes.

The experts use slightly different methods to evaluate boundaries and assign votes. Consider the window itw from the viewpoint of the boundary entropy expert. Each location in the window bounds an ngram to the left of the location; the ngrams are i, it, and itw, respectively. Each ngram has a standardized boundary entropy. The boundary entropy expert votes for the location that produces the ngram with the highest standardized boundary entropy. As it happens, for the ngram tree produced from Orwell's text, the standardized boundary entropies for i, it, and itw are 0.2, 1.39 and 0.02, so the boundary entropy expert opts to put a boundary after the ngram it.

The frequency expert places a boundary so as to maximize the sum of the standardized frequencies of the ngrams to the left and the right of the boundary. Consider the window itw again. If the boundary is placed after i, then (for Orwell's text) the standardized frequencies of i and tw sum to 1.73; if the boundary is placed after it, then the standardized frequencies of it and w sum to 2.9; finally, if it is placed after itw, the algorithm has only the standardized frequency of itw to work with; it is 4.0. Thus, the frequency expert opts to put a boundary after itw.

Segment the sequence. Each potential boundary in a sequence accrues votes, as described above, and now we must evaluate the boundaries in terms of the votes and decide where to segment the sequence. Our method is a familiar "zero crossing" rule: If a potential boundary has a locally maximum number of votes, split the sequence at that boundary. In the example above, this rule causes the sequence itwasacold to be split after it and was. We confess to one embellishment on the rule: The number of votes for a boundary must exceed a threshold, as well as be a local maximum. We found that the algorithm splits too often without this qualification. In the experiments reported below, the threshold was always set to n, the window size. This means that a location must garner half the available votes (for two voting experts) and be a local maximum to qualify for splitting the sequence.

Let us review the design of the experts and the segmentation rule, to see how they test the characteristics of episodes described earlier. The boundary entropy expert assigns votes to locations where the boundary entropy peaks, locally, implementing the idea that entropy increases at episode boundaries. The frequency expert tries to find a "maximum likelihood tiling" of the sequence, a placement of boundaries that makes the ngrams to the left and right of the boundary as likely as possible. When both experts vote for a boundary, and especially when they vote repeatedly for the same boundary, it is likely to get a locally-maximum number of votes, and the algorithm is apt to split the sequence at that location.

6   Evaluation

We removed spaces and punctuation from text and assessed how well the voting experts algorithm could induce word boundaries. In these experiments, boundaries stand in six relationships to episodes.

1. The boundaries coincide with the beginning and end of the episode;
2. The episode falls entirely within the boundaries and begins or ends at one boundary.
3. The episode falls entirely within the boundaries but neither the beginning nor the end of the episode correspond to a boundary.
4. One or more boundaries splits an episode, but the beginning and end of the episode coincide with boundaries.
5. Like case 4, in that boundaries split an episode, but only one end of the episode coincides with a boundary.
6. The episode is split by one or more boundaries and neither end of the episode coincides with a boundary.

These relationships are illustrated graphically, following the convention that a horizontal line denotes an episode, and vertical lines denote induced boundaries:

The cases can be divided into three groups. In cases 1 and 4, boundaries correspond to both ends of the episode; in cases 2 and 5, they correspond to one end of the episode; and in cases 3 and 6, they correspond to neither end. We call these cases *exact, dangling,* and *lost* to evoke the idea of episodes located exactly, dangling from a single boundary, or lost in the region between boundaries.

We ran the voting experts algorithm on the first 50,000 characters in Orwell's *1984*, spaces and punctuation removed. The window length was 6. The algorithm induced 11210 boundaries, for a mean episode length of 4.46. The mean word length in the text was 4.49. The algorithm induced boundaries at 74.9% of the true word boundaries (the hit rate) missing 25.1% of the word boundaries. 25.4% of the induced boundaries did not correspond to word boundaries (the false positive rate). Exact cases, described above, constitute 55.2% of all cases; that is, 55.2% of the words were bounded at both ends by induced boundaries. Dangling and lost cases constitute 39.5% and 5.3% of all cases, respectively. Said differently, only 5.3% of all words in the text got lost between episode boundaries. These tend to be short words, in fact, 58% of the lost words are of length 3 or shorter. In contrast, all of the words for which the algorithm found exact boundaries are of length 3 or longer.

It is easy to ensure that all word boundaries are found, and no word is lost: Induce a boundary between each letter. However, this strategy would induce a mean episode length of 1.0, much shorter than the mean word length, and the false-positive rate would approach 100%. In contrast, the voting experts algorithm finds roughly the same number of episodes as there are words in the text, and loses very few words between boundaries.

The effects of the corpus size and the window length are shown in the following graph. The proportion of "lost" words (cases 3 and 6, above) is plotted on the vertical axis, and the corpus length is plotted on the horizontal axis. Each curve in the graph corresponds to a window length, n. The proportion of lost words becomes roughly constant for corpora of length 10,000 and higher.

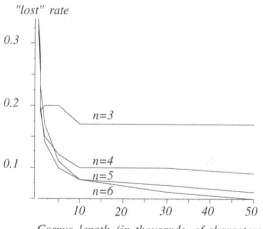

Corpus length (in thousands of characters)

Said differently, corpora of this length seem to be required for the algorithm to estimate boundary entropies and frequencies accurately. As to window length, recall that a window of length n means each potential boundary is considered n times by each expert, in n different contexts. Clearly, it helps to increase the window size, but the benefit diminishes.

The appropriate control conditions for this experiment were run and yielded the expected results: The algorithm performs very poorly given texts of random words, that is, subsequences of random letters. The algorithm performs marginally less well when it is required to segment text it has not seen. For example, if the first 10,000 of Orwell's text are used to build the ngram tree, and then the algorithm is required to segment the next 10,000 characters, there is a very slight decrement in performance.

As a test of the generality of the algorithm, we ran it on corpora of Roma-ji text and a segment of Franz Kafka's *The Castle* in the original German. Roma-ji is a transliteration of Japanese into roman characters. The corpus was a set of Anime lyrics, comprising 19163 roman characters[1]. For comparison purposes we selected the first 19163 characters of Kafka's text and the same number of characters from Orwell's text. As always, we stripped away spaces and punctuation, and the algorithm induced word boundaries. Here are the results:

	Hit rate	F. P. rate	Exact	Dangling	Lost
English	.71	.28	.49	.44	.07
German	.79	.31	.61	.35	.04
Roma-ji	.64	.34	.37	.53	.10

Clearly, the algorithm is not biased to do well on English, in particular, as it actually performs best on Kafka's text, losing only 4% of the words and identifying 61% exactly. The algorithm performs less well with the Roma-ji text; it identifies fewer boundaries accurately (i.e., places 34% of its boundaries within words) and identifies fewer words exactly. The explanation for these results has to do with the lengths of words in the corpora. We know that the algorithm loses disproportionately many short words. Words of length 2 make up 32% of the Roma-ji corpus, 17% of the Orwell corpus, and 10% of the Kafka corpus, so it is not surprising that the algorithm performs worst on the Roma-ji corpus and best on the Kafka corpus.

7   Conclusion

The voting experts algorithm segments characters into words with some accuracy, given that it is a greedy, unsupervised algorithm that requires relatively little data. In future work, the algorithm will be augmented with other experts besides the two described here. In particular, we have identified two other features of episodes— meaning-carrying subsequences—and we are building experts to

---

[1] We are grateful to Ms. Sara Nishi for collecting this corpus.

detect these features. The idea that meaningful subsequences differ from meaningless ones in some formal characteristics—that syntactic criteria might help us identify semantic units—has practical as well as philosophical implications.

Acknowledgments

Prof. Cohen's work was supported by Visiting Fellowship Research Grant number GR/N24193 from the Engineering and Physical Sciences Research Council (UK). Additional support for Prof. Cohen and for Dr. Adams was provided by DARPA under contract(s) No.s DARPA/USASMDCDASG60-99-C-0074 and DARPA/AFRLF30602-00-1-0529. The U.S. Government is authorized to reproduce and distribute reprints for governmental purposes notwithstanding any copyright notation hereon. The views and conclusions contained herein are those of the authors and should not be interpreted as necessarily representing the official policies or endorsements either expressed or implied, of DARPA or the U.S. Government.

References

1. Cohen, Paul. Fluent learning: Elucidating the structure of episodes. This volume.
2. M. Garofalakis, R. Rastogi, and K. Shim. Spirit: sequential pattern mining with regular expression constraints. In Proc. of the VLDB Conference, Edinburgh, Scotland, September 1999.
3. Magerman D. and Marcus, M. 1990. Parsing a natural language using mutual information statistics. In Proceedings of AAAI-90, Eighth National Conference on Artificial Intelligence, 984989
4. H. Mannila, H. Toivonen, and A. I. Verkamo. Discovery of frequent episodes in event sequences. Data Mining and Knowledge Discovery, 1(3), 1997.
5. Nevill-Manning, C.G. and Witten, I.H. (1997) Identifying Hierarchical Structure in Sequences: A linear-time algorithm, Volume 7, pages 67-82.
6. Tim Oates, Laura Firoiu, Paul Cohen. Using Dynamic Time Warping to Bootstrap HMM-Based Clustering of Time Series. In Sequence Learning: Paradigms, Algorithms and Applications. Ron Sun and C. L. Giles (Eds.) Springer-Verlag: LNAI 1828. 2001
7. Paola Sebastiani, Marco Ramoni, Paul Cohen. Sequence Learning via Bayesian Clustering by Dynamics. In Sequence Learning: Paradigms, Algorithms and Applications. Ron Sun and C. L. Giles (Eds.) Springer-Verlag: LNAI 1828. 2001
8. Teahan, W.J., Y. Wen, R. McNab and I.H. Witten. A compression-based algorithm for Chinese word segmentation. Computational Linguistics, v 26, no 3, September, 2000, p 375-393.
9. Weiss, G. M., and Hirsh, H. 1998. Learning to Predict Rare Events in Categorical Time-Series Data, Proceedings of the 1998 AAAI/ICML Workshop on Time-Series Analysis, Madison, Wisconsin.

An Empirical Comparison of Pruning Methods for Ensemble Classifiers

Terry Windeatt and Gholamreza Ardeshir

Centre for Vision, Speech and Signal Processing
School of Electronics Engineering, Information Technology and Mathematics,
Guildford, Surrey, Gu2 7XH, UK
T.Windeatt, g.ardeshir@eim.surrey.ac.uk

Abstract. Many researchers have shown that ensemble methods such as Boosting and Bagging improve the accuracy of classification. Boosting and Bagging perform well with unstable learning algorithms such as neural networks or decision trees. Pruning decision tree classifiers is intended to make trees simpler and more comprehensible and avoid over-fitting. However it is known that pruning individual classifiers of an ensemble does not necessarily lead to improved generalisation. Examples of individual tree pruning methods are Minimum Error Pruning (MEP), Error-based Pruning (EBP), Reduced-Error Pruning(REP), Critical Value Pruning (CVP) and Cost-Complexity Pruning (CCP). In this paper, we report the results of applying Boosting and Bagging with these five pruning methods to eleven datasets.

1 Introduction

The idea of ensemble classifiers (also known as multiple classifiers) is based on the observation that achieving optimal performance in combination is not necessarily consistent with obtaining the best performance for a single classifier. However certain conditions need to be satisfied to realise the performance improvement, in particular that the constituent (base) classifiers be not too highly correlated [24]. Various techniques have been devised to reduce correlation between classifiers before combining including: (i) reducing dimension of training set to give different feature sets, (ii) incorporating different types of base classifier, (iii) designing base classifiers with different parameters for same type of classifier, (iv) resampling training set so each classifier is specialised on different subset, and (v) coding multi-class binary outputs to create complementary two-class problems. In this paper, we consider two popular examples belonging to category (iv), that rely on perturbing training sets.

Training on subsets appears to work well for unstable classifiers, such as neural networks and decision trees, in which a small perturbation in the training set may lead to a significant change in constructed classifier. Effective methods for improving unstable predictors based on perturbing the training set prior to combining, include Bagging [2] and Boosting [11]. Bagging (from Bootstrap Aggregating) forms replicate training sets by sampling with replacement, and

F. Hoffmann et al. (Eds.): IDA 2001, LNCS 2189, pp. 208–217, 2001.

combines the resultant classifications with a simple majority vote. Boosting, which combines with a fixed weighted vote is more complex than Bagging in that the distribution of the training set is adaptively changed based upon the performance of sequentially constructed classifiers. Each new classifier is used to adaptively filter and re-weight the training set, so that the next classifier in the sequence has increased probability of selecting patterns that have been previously misclassified. In [3], the Bagging family is said to perform better with large tree base classifiers while the Boosting family is is said to perform better with small trees. Several studies have compared Boosting and Bagging [1] [16] [9].

Size of tree can be changed by pruning, although as we explain in Sect. 2, the pruning process is not always simple. It might seem desirable to characterise performance as a function of degree of pruning. However pruning methods in general were not designed to operate with an independent parameter to change tree complexity in a smooth fashion and results are usually presented with or without pruning applied. There have been two recent large-scale studies of Bagging and Boosting with tree classifiers using MC4 [1] and C4.5 [9]. In [9], the test data was used to determine if pruning was required, on the assumption that internal cross-validation on the training set would come to the same conclusion. No pattern was discerned about whether to prune over thirty-three datasets, except that with Adaboost, pruning versus no-pruning did not register a difference on significance tests in any of the thirty-three domains. In [1], all experiments were carried out with pruning applied, but Bagging was repeated without pruning to help understand why Bagging produces larger trees. It appeared that trees generated from bootstrap samples were initially smaller than MC4, but ended up larger after pruning. The authors concluded that pruning reduces bias but increases variance. and that bootstrap replicates inhibit Reduced Error Pruning.

Ensemble classifiers have been empirically shown to give improved generalisation for a variety of classification data. However there is discussion as to why they appear to work well. Two concepts have been proposed for analysing behaviour, bias/variance from regression theory and the margin concept [12]. However they appear to provide useful perspectives for the way ensemble classifiers operate rather than an understanding of how ensemble classifiers reduce generalisation error. In absence of a complete theory, empirical results continue to provide useful insight to their operation.

In the past much effort has been directed toward developing effective tree pruning methods (for a review see [10]) in the context of a single tree. For a tree ensemble, besides individual tree (base classifier) pruning, it is also possible to consider ensemble pruning. In [7], five ensemble pruning methods used with Adaboost are proposed, but the emphasis there is on efficiency, i.e. finding a minimal number of base classifiers without significantly degrading performance. The goal of single tree pruning in an ensemble, is to produce a simpler tree base classifier which gives improved ensemble performance.

Decision trees can be divided into two categories according to type of test: multivariate decision trees and univariate trees. Multivariate decision trees are those in which at each internal node several attributes are tested together while in univariate decision trees one attribute will be tested at each internal node. ID3 [17],C4.5 [18], CART [4] are examples of univariate decision tree and LMDT is a multivariate decision tree[22]. In this paper we have used C4.5, (which uses gain ratio in construction mode to determine which attribute to test), with several individual decision tree pruning methods.

In this paper, we briefly review five methods of decision tree pruning in section 2. These are Minimum Error Pruning (MEP), Error-based Pruning(EBP), Reduced-Error Pruning (REP), Critical Value Pruning (CVP) and Cost-Complexity Pruning (CCP). In section 3, we will explain the method used in the experiments and the results of Boosting and Bagging over eleven datasets when trees are pruned with different pruning algorithms.

2   Pruning Decision Trees

Decision tree pruning is a process in which one or more subtrees of a decision tree are removed. The need for pruning arises because the generated tree can be large and complex, so it may not be accurate or comprehensible. Complexity of a univariate decision tree is measured as the number of nodes, and the reasons for complexity are mismatch of representational biases and noise [5]. It means that the induction algorithm is unable to model some target concepts, and also that in some algorithms, (e.g. C4.5), subtree replication causes the tree to be too large and to overfit [17].

According to [10] there are different types of pruning methods but post-pruning is more usual [5]. The disadvantage of pre-pruning methods is that tree growth can be prematurely stopped, since the procedure estimates when to stop constructing the tree. A stopping criterion that estimates the performance gain expected from further tree expansion is applied, and tree expansion terminates when the expected gain is not accessible [10], [18]. A way around this problem is to use a post-pruning method which grows the full tree and retro-spectively prunes, starting at the leaves of the tree. Post-pruning methods remove one or more subtrees and replace them by a leaf or one branch of that subtree. One class of these algorithms divides the training set into a growing set and pruning set. The growing set is used to generate the tree as well as prune, while the pruning set is used to select the best tree [4]. In the case of shortage of training set, the cross-validation method is used i.e. the training set is divided into several equal-sized blocks and then on each iteration one block is used as pruning set and the remaining blocks used as a growing set. Another class of post-pruning algorithm uses all the training set for both growing and pruning [19]. However it is then necessary to define an estimate of the true error rate using the training set alone.

All the pruning methods considered here use post-pruning, and therefore construct the full tree before applying the pruning criteria.

2.1   Error-Based Pruning (EBP)

EBP was developed by Quinlan for use in C4.5. It does not need a separate pruning set, but uses an estimate of expected error rate. A set of examples covered by the leaf of a tree is considered to be a statistical sample from which it is possible to calculate confidence for the posterior probability of mis-classification. The assumption is made that the error in this sample follows a binomial distribution, from which the upper limit of confidence [15] is the solution for p of

$$CF = \sum_{x=0}^{E} \binom{N}{x} p^x (1-p)^{N-x} \tag{1}$$

where N is number of cases covered by a node and E is number of cases which is covered by that node erroneously (As C4.5 we have used an approximate solution for equation 1).

A default confidence level of 25% is suggested, and the upper limit of confidence is multiplied by the number of cases which are covered by a leaf to determine the number of predicted errors for that leaf. Further the number of predicted errors of a subtree is the sum of the predicted errors of its branches. If the number of predicted errors for a leaf is less than the number of predicted errors for the subtree in which that leaf is, then the subtree is replaced with the leaf. We tried changing the confidence level to vary the degree of pruning but found that this is not a reliable way of varying tree complexity [23].

2.2   Minimum Error Pruning (MEP)

MEP was introduced by Niblett and Bratko, and uses Laplace probability estimates to improve the performance of ID3 in noisy domains [14]. Cestnik and Bratko have changed this algorithm by using more general Bayesian approach to estimating probabilities which they called *m-probability estimation* [6]. In this algorithm, the parameter m is changed to vary degree of tree pruning. Their suggestion is that perhaps the parameter can be adjusted to match properties of learning domain such as noise. To prune a tree at a node, the first step is to calculate the expected error rates of its children. The expected error rate of a node is the minimum of $1 - p_i(t)$ where $p_i(t)$ is the probability of ith class of examples reaching that node and is determined by

$$p_i(t) = \frac{n_i(t) + p_{ai} \cdot m}{n(t) + m} \tag{2}$$

where $n_i(t)$ in the number of examples reaching the node and belong to the ith class, $n(t)$ is the total number of example reaching the node t and p_{ai} is the a priori probability of the ith class.

They called the expected error rate the *static error*. In the second step, *dynamic error* of the node is calculated, where *dynamic error* is defined as the weighted sum of the static errors of its children [6]. The node will be pruned if its static error is greater than its dynamic error and will be replaced by a leaf.

2.3   Reduced Error Pruning (REP)

REP requires a separate pruning set, and was proposed by Quinlan [19]. It simply replaces each internal node (non-leaf node) by the best possible branch with respect to error rate over the pruning set. Branch pruning is repeated until there is an increase in the pruning set error rate. The procedure is guaranteed to find the smallest, most accurate subtree with respect to the pruning set.

2.4   Critical Value Pruning (CVP)

CVP was proposed in [13], and operates with a variety of node selection measures. The idea is to set a threshold, the *critical value* which defines the level at which pruning takes place. An internal node is only pruned if the associated selection measures for the node and all its children do not exceed the *critical value*. The full tree is pruned for increasing critical values giving a sequence of trees, and then the best tree is selected on the basis of predictive ability. A number of suggestions were made for finding the best tree in the sequence, the obvious one being to use a separate pruning set as in REP.

2.5   Cost-Complexity Pruning (CCP)

CCP was developed for the CART system, and produces a sequence of trees by pruning those branches that give lowest increase in error rate per leaf over the training set [4]. Let the sequence of subtrees be denoted by $T_1 > T_2 > ... > t$ in which T_1 is the original tree and T_2 has been generated by pruning T_1 and finally t is the root. Leaves of the subtree s are examined and assigned a measure α representing the increase in error rate per leaf

$$\alpha = \frac{M}{N(L(s) - 1)} \tag{3}$$

where N is the number of training examples, M is the additional number of misclassified examples when the leaf is removed, $L(s)$ is number of leaves in the subtree s [19]. To produce T_{i+1} from T_i, all nodes in T_i with the lowest α are pruned.

In order to select the best tree in the sequence, either cross-validation on the training set or a separate pruning set is employed. The selected tree is either

1. the smallest tree with error rate less than minimum observed error rate (0SE rule)
2. the smallest tree with error rate less than minimum observed error rate plus one standard error (1SE rule).

However in Breiman's Bagging [2], since each tree is built on a bootstrap replicate he uses the full training set to prune.

3   Experiments

To generate the decision tree that has been used as base classifier in the ensemble we have used C4.5 algorithm. After building the tree, we have used five different pruning algorithms MEP,EBP, REP, CVP, CCP where EBP is the pruning method used in the C4.5 system.

The datasets which have been used in the experiments can be found on UCI web site [21]. However, we have used the datasets which have been downloaded from Quinlan web site at the University of New South Wales [20], which have been split into training and test sets. Table 1 gives the description of the datasets, showing that, with the exception of Waveform-21, they have been divided into 70/30 training/testing split according to class distribution. All experiments have been carried out with number of base classifiers set to ten as in [16]. Each experiment is repeated ten times, and where a separate pruning set is employed this includes a random 70/30 growing/pruning split of the training set.

Table 1. Specification of Datasets

| Name | Training set | Test set | Class | Attributes | |
				Cont.	Disc.
Breast	466	233	2	10	–
BreastCancer	191	95	2	–	9
Crx	490	200	2	6	9
Glass	142	72	6	9	–
Heart	180	90	2	13	–
Hypothyroid	2108	1055	2	7	18
Iris	100	50	3	4	–
Labor-neg	40	17	2	8	8
Soybean-large	455	228	19	–	35
Vote	300	135	2	–	16
soybean-small	31	16	4	35	–
waveform-21	300	4700	3	21	–

In order to test whether degree of pruning could be varied to optimise generalisation performance we selected the MEP method, and varied parameter m as follows: $m = 0, .01, .5, 1, 2, 3, 4, 8, 12, 16, 32, 64, 128, 999, 9999$. For each value of m, resampling is started from the same random state. For values of $m > 4$ size of tree indicated that varying m was not a reliable way of varying degree of pruning, which agrees with the assessment in [10]. Table 2 shows for C4.5, Bagged C4.5 and Boosted C4.5 the ratio of unpruned error rate to minimum (over m) pruned error rate.

To have a comparison with cross-validation, we have applied 10-fold cross validation to the datasets, with results shown in table 4. We also applied Mc Nemar's test with 5% confidence level [8] to determine whether difference between

Table 2. Ratio of unpruned error rate to minimum pruned error rate using MEP

Name	C4.5	Boosted C4.5	Bagged C4.5
Breast	1.00	1.00	1.00
Breast-Cancer	1.72	1.07	1.23
Crx	1.18	1.00	1.00
Glass	1.04	1.09	1.00
Heart	1.00	1.15	1.06
Hypothyroid	1.00	1.00	1.00
Iris	1.00	1.00	1.00
Labor-neg	1.00	1.00	1.00
Soybean-large	1.01	1.01	1.00
Vote	1.75	1.25	1.25
Waveform-21	1.00	1.10	1.00
Average	1.15	1.06	1.05

Boosting and Bagging was significant, and table 4 shows the number of folds that showed a significant difference. Only two datasets appeared significantly different , Waveform21 and Soybean.

Table 3. Mean Test Error of Boosting and Bagging and Ratio Boosting-Bagging using EBP

Name	C4.5	Boosted C4.5	Bagged C4.5	Ratio
Breast	5.58	3.86	3.86	1.00
Breast-Cancer	40.00	33.68	29.47	1.14
Crx	21.00	17.00	18.00	0.94
Glass	37.50	29.17	30.56	0.95
Heart	21.11	20.00	18.89	1.06
Hypothyroid	1.70	1.23	0.95	1.29
Iris	4.00	6.00	4.00	1.50
Labor-neg	29.41	23.53	23.53	1.00
Soybean-large	34.64	26.75	32.89	0.81
Vote	7.41	2.96	2.96	1.00
Waveform-21	30.74	20.09	22.09	0.91
Average				1.05

The results shown in table 3 are the average test errors of C4.5, Boosting and Bagging with EBP. The ratio of Boosting/Bagging error rates is shown for each dataset and averaged over all datasets. Similar tables were produced for the other pruning methods with average ratios: REP:0.98, CCP(0SE):1.03, CCP(1SE):1.03, CVP:1.02.

Table 4. Mean test error for 10-fold Cross Validation with Boosting and Bagging, and ratio Boosting-Bagging using EBP

Name	C4.5	Boosted C4.5	Bagged C4.5	Ratio	McNemar's Test
Breast	5.29	4.43	4.29	1.03	0
Breast Cancer	31.60	28.50	29.53	0.97	0
Crx	16.23	14.49	14.78	0.98	0
Glass	37.43	27.69	28.41	0.97	0
Heart	22.22	19.26	20.00	0.96	0
Hypothyroid	1.30	1.36	1.23	1.11	0
Iris	3.33	4.67	4.67	1.00	0
Labor-neg	41.67	28.33	31.67	0.89	0
Soybean-Large	42.42	35.27	24.71	1.43	2
vote	5.06	4.60	4.83	0.95	0
Waveform-21	25.14	17.32	18.40	0.94	1
Average				1.02	

To compare the different pruning methods tables 5 and 6 show, for Boosting and Bagging respectively, the relative performance of the five pruning methods (for MEP $m = 2$). In tables 5 and 6 the pruning method with the minimum error rate is set to 1.00.

Table 5. Relative Test Error of Boosting with different Puning Methods

Name	MEP	EBP	REP	CVP(0-SE)	CVP(1-SE)	CCP
breast	1.00	1.50	1.73	1.63	1.57	1.58
breastCancer	1.02	1.25	1.00	1.01	1.28	1.20
crx	1.02	1.02	1.02	1.00	1.07	1.06
glass	1.19	1.00	1.45	1.24	1.40	1.24
heart	1.00	1.38	1.35	1.37	1.45	1.26
hypothyroid	4.33	1.02	1.16	1.13	1.08	1.00
iris	1.07	1.07	1.00	1.04	1.04	1.00
labor-neg	1.00	1.00	1.02	1.12	1.18	1.18
soybean-large	2.11	1.16	1.02	1.00	1.05	1.00
vote	1.00	1.00	1.05	1.15	1.20	1.20
waveform-21	1.04	1.00	1.22	1.14	1.18	1.17
Average	1.50	1.13	1.18	1.17	1.23	1.17

4   Discussion

Table 2 shows that, on average over these datasets, Boosting and Bagging can both benefit from varying parameter m in the MEP method. However, as noted

Table 6. Relative Test Error of Bagging with different Puning Methods

Name	MEP	EBP	REP	CVP(0-SE)	CVP(1-SE)	CCP
breast	1.22	1.00	1.18	1.01	1.07	1.02
breastCancer	1.12	1.12	1.00	1.00	1.17	1.14
crx	1.00	1.06	1.07	1.00	1.03	1.03
glass	1.14	1.00	1.49	1.23	1.20	1.23
heart	1.00	1.00	1.09	1.09	1.08	1.11
hypothyroid	5.50	1.00	1.06	1.03	1.03	1.08
iris	1.00	1.00	1.50	1.40	1.60	1.50
labor-neg	1.00	1.00	1.20	1.25	1.43	1.02
soybean-large	2.73	1.30	1.15	1.00	1.02	1.04
vote	1.00	1.00	1.05	1.07	1.05	1.10
waveform-21	1.29	1.00	1.09	1.07	1.09	1.07
Average	1.64	1.04	1.17	1.11	1.16	1.12

previously as m increases for some datasets there was not a monotonic decrease with tree complexity. It should also be noted from table 4 that the average ratio for Boosting/Bagging for these datasets is 1.02, compared with a value of 0.93 over twenty-two datasets in [16].

The comparison of pruning methods in tables 5 and 6 shows that EBP performs best on average for Bagging and Boosting, and MEP worst. However the individual results indicate that there is no pattern corresponding to which pruning method performs best. For example, MEP performs best on three datasets. This suggests that if the type or level of pruning could be suitably chosen, performance would improve.

5   Conclusion

The fact that decision trees have a growing phase and pruning phase that are both data-dependent makes it difficult to match level or type of pruning to an ensemble of tree classifiers. Our results indicate that if a single pruning method needs to be selected then overall the popular EBP makes a good choice.

Acknowledgement

Gholamreza Ardeshir is grateful to the Ministry of Culture and Higher Education of Iran for its financial support during his Ph.D studies.

References

1. E. Bauer and R. Kohavi. An empirical comparison of voting classification algorithms: Bagging, boosting, and variants. *Maching Learning*, 36(1):105–142, 1999.
2. Leo Breiman. Bagging predictors. *Machine Learning*, 26(2):123–140, 1996.
3. Leo Breiman. Some infinity theory for predictor ensembles. Technical report, TR 577, Department of Statistics, University of California, Berkeley, 2000.

4. Leo Breiman, J. H. Freidman, R. A. Olshen, and C.J. Stone. Classification and regression trees. *Wadsworth International Group*, 1984. ftp://ftp.stat.berkeley.edu/pub/users/breiman.

5. L. A. Breslow and D. W. Aha. Simplifying decision trees: A survey. *Knowledge Engineering Review*, pages 1–40, 1997.

6. B. Cestnik and I. Bratko. On estimating probabilities in tree pruning. In Kodratoff Y., editor, *Machine Learning - EWSL-91.European Working Session on Learning Proceedings*, pages 138–50. Springer-Verlag, 1991.

7. T. G. Dietterich D. Margineantu. Pruning adaptive boosting. In *International Conference on Machine Learning*, pages 211–218. Morgan Kaufmann, 1997.

8. Thomas G. Detterich. Approximate statistical tests for comparing supervised classification learning algorithms. *Neural Computation*, 10:1895–1923, 1998.

9. T. G. Dietterich. An experimental comparison of three methods for constructing ensembles of decision trees: Bagging, boosting, and randomization. *Machine Learning*, 40(2):139–158, 2000.

10. F. Esposito, D. Malerba, and G. Semeraro. A comparative analysis of methods for pruning decision trees. *IEEE Transactions on Pattern Analysis and Machine Intelligence*, 19(5):476–491, May 1997.

11. Y. Freund and R. Schapire. Experiments with a new boosting algorithm. *Machine Learning:Proceedings of the Thirteenth International Conference*, pages 148–156, 1996.

12. G. James. *Majority Vote Classifiers: Theory and Applications*. PhD thesis, Dept. of Statistics, Univ. of Stanford, May 1998. http://www-stat.stanford.edu/ gareth/.

13. J. Mingers. Expert systems-rule induction with statistical data. *Operational Research Society*, 38:39–47, 1987.

14. T. Niblett and I. Bratko. Learning decision rules in noisy domains. In *Expert System 86, Cambridge*. Cambridge University Press, 1986.

15. J. Ross Quinlan. Personal communication from Quinlan.

16. J. Ross Quinlan. Bagging, boosting, and c4.5. In *Fourteenth National Conference on Artificial Intelligence*, 1996.

17. R. Quinlan. Induction of decision tree. *Machine Learning*, 1:81–106, 1986.

18. R. Quinlan. *C4.5: Programs for Machine Learning*. Morgan Kaufmann, San Mateo,California, 1993.

19. R.J. Quinlan. Simplyfying decision trees. *International Journal of Man-Machine Studies*, 27:221–234, 1987.

20. http://www.cse.unsw.edu.au/~ quinlan/.

21. These datasets can be found on: www.ics.uci.edu/ mlearn/ MLSummary.html.

22. P. E. Utgoff and C. E. Brodley. Linear machine decision trees. Technical report, Department of Computer Science, University of Massachusetts, Amhers, 1991.

23. T. Windeatt and G. Ardeshir. Boosting unpruned and pruned decision trees. In *Applied Informatics, Preceedings of the IASTED International Symposia*, pages 66–71, 2001.

24. T. Windeatt and R. Ghaderi. Binary labelling and decision level fusion. *Information fusion*, 2(2):103–112, 2001.

A Framework for Modelling Short, High-Dimensional Multivariate Time Series: Preliminary Results in Virus Gene Expression Data Analysis

Paul Kellam[1], Xiaohui Liu[2]*, Nigel Martin[3], Christine Orengo[4],
Stephen Swift[2], and Allan Tucker[2]

[1] Department of Immunology and Molecular Pathology, University College London,
Gower Street, London, WC1E 6BT, UK
[2] Department of Information Systems and Computing, Brunel University, Uxbridge,
Middlesex, UB8 3PH, UK
[3] School of Computer Science and Information Systems, Birkbeck College, Malet
Street, London, WC1E 7HX, UK
[4] Department of Biochemistry and Molecular Biology, University College London,
Gower Street, London, WC1E 6BT, UK

Abstract. Short, high-dimensional Multivariate Time Series (MTS) data
are common in many fields such as medicine, finance and science, and any
advance in modelling this kind of data would be beneficial. Nowhere is
this more true than functional genomics where effective ways of analysing
gene expression data are urgently needed. Progress in this area could help
obtain a "global" view of biological processes, and ultimately lead to a
great improvement in the quality of human life. We present a computational framework for modelling this type of data, and report preliminary
experimental results of applying this framework to the analysis of gene
expression data in the virology domain. The framework contains a three-
step modelling strategy: correlation search, variable grouping, and short
MTS modelling. Novel research is involved in each step which has been
individually tested on different real-world datasets in engineering and
medicine. This is the first attempt to integrate all these components into
a coherent computational framework, and test the framework on a very
challenging application area, which has produced promising results.

1 Introduction

Short, high-dimensional MTS data are common in many fields such as medicine,
finance and science, and any advance in modelling this kind of data would be
beneficial. One of the most exciting applications of this type of research is the
analysis of gene expression data, often a short and high-dimensional MTS. For
the first time, DNA microarray technology has enabled biologists to study all the
genes within an entire organism to obtain a global view of gene interaction and

---

* Contact e-mail: `Xiaohui.Liu@brunel.ac.uk`

F. Hoffmann et al. (Eds.): IDA 2001, LNCS 2189, pp. 218–227, 2001.

regulation. This technology has great potential in providing a deep understanding of biological processes, and ultimately, could lead to a great improvement in the quality of human life. However, this potential will not be fully realised until effective ways of analysing gene expression data collected by DNA array technology have been found.

As gene expression data is essentially a high-dimensional but very short MTS, it makes sense to look into the possibility of using time series modelling techniques to analyse these data. There are many statistical MTS modelling methods such as the Vector Auto-Regressive (VAR) process, Vector Auto-Regressive Moving Average [10], non-linear and Bayesian systems [2], but few of these methods have been applied to the modelling of gene expression data. This is perhaps not surprising since the very nature of these time series data means that a direct and fruitful application of these methods or indeed their counterparts in dynamic systems and signal processing literature will be extremely hard. For example, some methods such as the VAR process place constraints on the minimum number of time series observations in the dataset; also many methods require distribution assumptions to be made regarding the observed time series, e.g. the maximum likelihood method for parameter estimation.

Our recent work, based on the modelling of MTS data in medical and engineering domains, has demonstrated that it is possible to learn useful models from short MTS using a combination of statistical time series modelling, relevant domain knowledge and intelligent search techniques. In particular, we have developed a fast Evolutionary Program (EP) to locate variables that are highly correlated within high-dimensional MTS [13], strategies for decomposing high-dimensional MTS into a number of lower-dimensional MTS [16], a modelling method based on Genetic Algorithms (GA) and the VAR process which bypasses the size restrictions of the statistical methods [14], and a method for learning Dynamic Bayesian Networks (DBN) structure and parameters for explanation from MTS [17]. In Sec. 2 we propose a computational framework for modelling short high-dimensional MTS which has been applied to the analysis of virus gene expression data with promising results as discussed in Sec. 3.

2 Framework

We have examined the application of our framework to learning MTS models from gene expression data. A MTS is defined as a series of nT observations, x_{it}, $i = 1, \ldots, n$, $t = 1, \ldots, T$, made sequentially through time where t indexes the different measurements made at each time point, and i indexes the number of variables in the time series.

Figure 1 shows the methodology adopted, which reduces the dimensionality of the MTS before model building, and explicitly manages the temporal relationships between variables. The methodology involves three main processes.

Firstly, a **Correlation Search** is performed on the data, producing a set of strong correlations between variables over differing time lags.

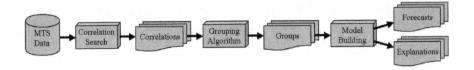

Fig. 1. Methodology

The correlations produced from the MTS data are then fed into a **Grouping algorithm** producing a set of groups of variables where there are a high number of correlations between members of the same group, but a low number of correlations between members of different groups.

These groups are then used as a basis for the **Model Building** process. In this paper the models we build are DBNs, but our methodology can employ a variety of other techniques such as the VAR process. Forecasts and explanations are then produced using the discovered models.

2.1 Correlation Search

The first stage of the methodology constructs a list, Q, of correlations which contains pairs of highly correlated variables over all possible integer time lags from zero to some positive maximum. Each element of Q is a *triple* made up of two variables and a time lag, *lag*. For example, the triple $(x_1, x_2, 2)$ represents the correlation between x_1 and x_2 with a time lag of 2. The search can be either exhaustive or approximate in nature depending upon the application in question. For example, in [13] evolutionary algorithms have been used for performing this search efficiently for large and time-limited domains. The construction of Q is constrained so that Q only contains one of the triples (x_i, x_j, lag) and (x_j, x_i, lag), namely the one with the highest correlation, and no triples of the form (x_i, x_i, lag). Q is then used in conjunction with the grouping strategy described in Sec. 2.2.

In these experiments we use Pearson's Correlation Coefficient (PCC) [12], defined in (1), but other metrics have been successfully employed in our methodology [16]. We can calculate the PCC between two variables over differing time lags by shifting one variable in time.

$$PCC(x_1, x_2, lag) = \frac{\sum\limits_{i=1}^{T-lag} (x_{1i} - \mu_{x_1})(x_{2(i+lag)} - \mu_{x_2})}{\sqrt{\sum\limits_{i=1}^{T-lag} (x_{1i} - \mu_{x_1})^2 \sum\limits_{i=1}^{T-lag} (x_{2(i+lag)} - \mu_{x_2})^2}}, \qquad (1)$$

where μ_x denotes the mean of the MTS variable x. PCC measures linear relationships between two variables, either discrete or continuous.

2.2   Grouping

The grouping stage makes use of Falkenauer's Grouping Genetic Algorithm (GGA) [6]. This uses a search procedure in conjunction with the results of the correlation search and a metric, *Partition Metric*, to score possible groupings. Note that as a result of this, the size of Q will substantially affect the final groupings discovered. Groupings are scored according to *Partition Metric* which groups variables together that have strong mutual dependency and separates them into different groups where the dependency is low.

Let $G = \{g_1, \ldots, g_m\}$ be a partition of $\{x_1, \ldots, x_n\}$. That is,

$$\bigcup_{i=1}^{m} g_i = \{x_1, \ldots, x_n\} \text{ and } g_i \cap g_j = \emptyset, \ 1 \leqslant i < j \leqslant m.$$

Given a list of correlations, Q, and $g \in G$, let

$$g/Q = \{(x_i, x_j, lag) \in Q : x_i, x_j \in g\}, \ 1 \leqslant i \neq j \leqslant n.$$

For any $g \in G$, $|g/Q| \leqslant \frac{1}{2}|g|(|g| - 1)$ since there are no autocorrelations in Q. Hence, the number of non-correlations in g is $\frac{1}{2}|g|(|g| - 1) - |g/Q|$. We define the *Partition Metric* of G, denoted $f(G)$, to be:

$$f(G) = \sum_{i=1}^{m} \left(|g_i/Q| - \left(\frac{1}{2}|g_i|(|g_i| - 1) - |g_i/Q| \right) \right) = \sum_{i=1}^{m} \left(2|g_i/Q| - \frac{1}{2}|g_i|(|g_i| - 1) \right).$$

Intuitively, we sum the difference between the number of correlations and non-correlations over each group in G. Clearly we wish to maximise *Partition Metric*, which is discussed in detail in [16].

2.3   Modelling

Within this framework, a modelling paradigm has been developed to explain and forecast MTS. This makes use of DBNs which offer an ideal way to perform both explanation and forecasting. A DBN allows the testing of the effect of changing a biological process by manipulating key components of the system. This is of particular interest to the pharmaceutical industry in trying to identify the best target in a modelled disease process for therapeutic intervention.

A well-known model for performing probabilistic inference about a discrete system is the Bayesian Network (BN) [11], and its dynamic counterpart can model a system over time [4]. The N nodes in a Dynamic Bayesian Network (DBN) represent the n variables at differing time slices (where $N \leqslant nT$). Therefore links occur between nodes over different time lags (non-contemporaneous links) and within the same time lag (contemporaneous links). Essentially a DBN can be considered as a Markov Model where conditional independence assumptions are made. Figure 2(a) shows a DBN with five variables over six time lags where each node represents a variable at a certain time slice and each link represents a conditional dependency between two nodes. Figure 2(b) shows the same

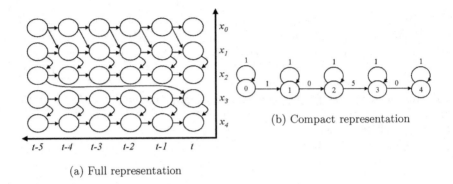

(a) Full representation

(b) Compact representation

Fig. 2. A DBN with five variables over six time lags

network in a more compact representation where numbers within nodes represent the variable and numbers with links represent time lag.

Given some evidence about a set of variables at time t, algorithms exist that can infer the most probable explanations for the current observations. When learning DBNs from high dimensional MTS, the search spaces can be huge. EP-Seeded GA is an algorithm for learning DBN models efficiently [17] which does not suffer from local maxima due to the global nature of its search. It only searches for non-contemporaneous links assuming that all dependencies span at least one time slice.

Figure 3 illustrates the EP-Seeded GA process. It involves applying an EP in order to find a list of highly-correlated triples, $List$. This is similar to the evolutionary correlation search [13] as discussed in Sec. 2.1 but uses log likelihood [3] on discretised data rather than a correlation coefficient and auto-correlations may be included. $List$ is then used to form the initial population of a GA [8] which consists of subsets of $List$, denoted T_i in Fig. 3. A form of uniform crossover [15] is used to generate new candidate networks where the fitness function is the log likelihood of each network.

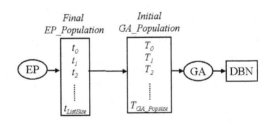

Fig. 3. EP-Seeded GA

3   Application to Viral Gene Expression Data

Viruses have been studied extensively in molecular biology in order to understand cellular processes. Viruses also permit the study of complex multigene expression in the context of the human cell. Herpesviruses maintain an episomal genome of over 100 genes in the nucleus of infected cells. Under appropriate cellular stimulation a highly controlled cascade of viral transcription occurs with the clearly defined endpoint of new virus production. This therefore provides an excellent system for bioinformatic analysis of a transcriptome in the context of human cells. We have produced a DNA array of all known and putative open reading frames of human herpesvirus 8, consisting of 106 genes ($n = 106$) expressed at eight time points ($T = 8$).

The correlation search and grouping stages were applied to the expression data after it had been pre-processed. This involved taking log ratios and then mean centring, both standard techniques for normalising expression data. For the model building experiments in this paper, the gene expression data is discretised into 3 states based on whether a gene is under expressed (state 0), not expressed (state 1) or over expressed (state 2), taken from the work in [9]. Correlation search and grouping are carried out on the continuous data.

Previous work has concentrated mostly on the clustering and classification of gene expression data. For example, in [5,9] methods such as hierarchical clustering, K-means clustering and self-organising maps have been used to arrange genes into similar groups based on their gene expression profile. Some initial work has been performed on modelling gene expression profiles using BNs [7]. However this work has not taken into account the time aspect of the expression data as we do here.

3.1   Correlation Search and Grouping Results

The correlation search used is an exhaustive method with maximum time lag of 3. The GGA was designed to look for between one and eight groups that maximised the number of correlations within a group. The number of correlations needed is 649, based on the calculations described in [16]. The resulting number of correlations, using PCC, has an absolute average of 0.975 and an absolute standard deviation of 0.010. The groups formed are shown in Table 1. Biologically, these groups made sense to a varying degree. Some contained genes with phased inductions around similar time points (e.g. groups 0 and 1) and another contained mostly *host cell* (*housekeeping*) genes (e.g. group 7). However, some of these housekeeping genes were found in unlikely groups (e.g. group 5). A possible reason for this is that taking log ratios exaggerates the effects of negative expression values. Taking this into account, models that explain a state 1 to state 2 change are likely to have the more reliable biological meaning.

3.2   EP-Seeded GA Results

The application of the EP-Seeded GA algorithm to learning DBNs from the eight grouped gene expression datasets has resulted in some highly connected

Table 1. GGA results

Group (Size)	Gene Identifier
0 (10)	A1 F7 G2 G11 H3 H4 J5 K8 E3 I9
1 (15)	A2 A5 A6 A9 B1 G9 H5 C5 C9 C10 C12 J9 J12 D3 F1
2 (22)	A3 A4 A7 A8 A12 B5 B6 B8 B9 H8 H11 B12 K10 K11 C6 C7 J1 D11 D12 I5 E6 I6
3 (15)	K1 B2 B10 B11 G1 G5 G12 H10 C2 C3 C4 D1 D2 J6 D8
4 (12)	A10 A11 F11 K12 H2 H9 K5 C8 J7 J8 D9 E5
5 (11)	K4 F6 F9 F10 G4 G6 J10 D4 D10 K9 I2
6 (11)	F5 F8 F12 G3 H6 H7 J4 K6 I4 E7 I7
7 (10)	G8 J2 E1 I1 E2 I3 E4 E8 I8 E9

networks. The fitness of these networks (their log likelihood) are documented in Table 2 along with the results of applying a well-known greedy algorithm, K2, introduced in [3] for learning static networks, that has been adapted to learn DBNs in [17]. The fitness is clearly greater in the DBNs resulting from the EP-Seeded GA algorithm for all groups. Figure 4 shows the compact representation of two of the networks discovered by EP-Seeded GA (for groups 0 and 2).

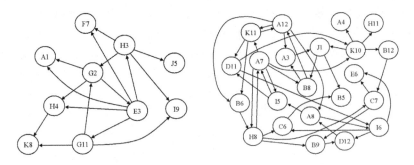

Fig. 4. Sample DBNs learnt using EP-Seeded GA on the expression data from group 0 and group 2 respectively (time lags and auto-regressive links have not been included)

Table 2 shows the *Weighted-Kappa* (WK) value of the forecasts for each group given the first 3 values of each gene. WK is used to rate agreement between the classification decisions made by two observers [1]. In this case observer 1 is the real data and observer 2 is the forecast data. For each group, the number of correct and incorrect forecasts is divided up into a 3 by 3 contingency table (for each state of the discrete variable). Once this table has been completed, WK can be computed using the procedure detailed in [1]. On the 5 step forecast, only the first 3 values of each gene were entered into the DBN. Based on this information, the DBN was used to forecast the next 5 values of each gene. On the 1 step ahead forecast the last 3 values of the genes were entered and the DBN was used to predict the next value for each gene. This was repeated for the next 5 time slices. Any WK value above 0.6 is considered *Good* and anything over 0.8 is considered *Very Good* [1]. It can be seen that all of the forecasts generated *Good* or *Very Good* WK values which implies a good model of the data has been learnt. As expected, the 1 step ahead forecast performed better in most cases than the 5 step forecast although the differences are small in each case.

One advantage of DBNs over other MTS models is that they are able to generate *transparent* explanations given certain observations based on the data. In this context transparent means that the model and explanations are easily interpretable by users who are not experts in statistical and artificial intelligence models. Figure 5 illustrates two sample explanations generated from the networks in Fig. 4. The explanations in Fig. 5 show some of the nodes in the group 0 and group 2 networks relevant to the observed variables (in grey). For example, Fig. 5(a) contains the observations that gene J5 is in state 1 (not expressed) currently and was in state 0 (under expressed) one time slice previously. Some of the other related variables include H4, G11 and E3 which influence the observed variables over differing time lags and with differing probabilities.

One of the best tests for model building using this type of data is whether the model makes biological sense. This can be investigated by decoding the model and applying virology domain knowledge. This was performed for the explanation in Fig. 5(b). The model implies B12 (K-bZIP) and H8 (ORF 45) work in a network to affect C7 (ORF 57). The induction of HHV8 genes as the virus begins its lytic replication cycle is controlled by the viral transactivator ORF 50. K-bZIP and ORF 57 are both up-regulated by ORF 50 but despite this the predictive model suggests subtle differential temporal regulation. Such differential gene expression by ORF 50 is known for other viral genes. Kb-ZIP is a basic leucine zipper protein suggesting a function as a transcription factor and/or DNA binding protein. Recent experimental data shows Kb-ZIP co-localises with the viral genome replication complex in specialised structures in the cell nucleus

Table 2. Final fitness (log likelihhood) for K2 and EP-Seeded GA and Weighted-Kappa scores (WK) for the discrete data forecasts using the EP-Seeded GA DBNs.

Group (Size)	K2 Fitness	EP-Seeded GA Fitness	WK Forecast (1 Step Ahead)	WK Forecast (5 Step Ahead)
0 (10)	68.564	73.471	0.861	0.920
1 (15)	79.934	92.273	0.796	0.753
2 (22)	128.670	136.101	0.865	0.794
3 (15)	95.184	104.434	0.848	0.761
4 (12)	62.765	68.902	0.855	0.754
5 (11)	74.857	81.276	0.824	0.882
6 (11)	77.779	82.461	0.900	0.900
7 (10)	68.904	71.746	0.946	0.918

and may be involved with linking lytic genome replication with the cell cycle.
ORF 45 has no known function but has an amino acid sequence reminiscent of
a nuclear protein with transcription factor and/or DNA binding activity. This
suggests both Kb-ZIP and ORF 45 are involved in similar processes and indicates
obvious validating experiments. ORF 57 is a *keep mediator* of virus lytic gene
expression and specifically transports certain viral RNAs from the nucleus to the
cytoplasm. Therefore, the fine temporal control of the HHV8 lytic cycle suggests
that initial lytic genome replication events precede but are directly linked to a
network involving ORF 57 mediated gene expression.

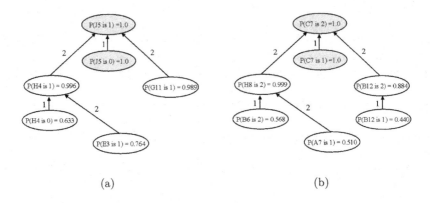

(a)   (b)

Fig. 5. Sample explanations generated using the DBNs in Fig. 4

4   Concluding Remarks

Modelling short MTS data is a challenging task, particularly when the dataset is
high-dimensional. Little work appears to have used time series methods to model
this kind of data. Here we suggest a computational framework for modelling such
data that explicitly manages the temporal relationships between variables within
each step. We have obtained preliminary results when applying this method to
the analysis of virus gene expression data. The groups found and their corresponding networks have varying degrees of biological support. However, we have
identified ways to overcome some of the problems encountered by concentrating
on the data pre-processing stage. The forecasts produce very good WK statistics and some of the explanations have shown interesting biological connotations,
indicating that DBNs can model expression data with a high degree of accuracy.

Ongoing and future work will investigate the use of VAR processes and continuous BNs to model non-discrete MTS data, and the application of these methods to other datasets such as yeast expression datasets and visual field data [14].
We also intend to refine our framework in order to improve scalability and efficiency.

Acknowledgements. This work is supported in part by the BBSRC, the MRC and the EPSRC in the UK. The authors would like to thank Jason Crampton for his constructive remarks on the *Partition Metric* and the manuscript in general, and Richard Jenner for preparing the gene expression data.

References

1. Altman, D.G.: Practical Statistics for Medical Research. Chapman and Hall (1997)
2. Casdagli, M., Eubank, S.: Nonlinear Modeling and Forecasting. Addison Wesley (1992)
3. Cooper, G.F., Herskovitz, E.: A Bayesian Method for the Induction of Probabilistic Networks from Data. Machine Learning **9** (1992) 309–347
4. Dagum, P., Galper, A., Horvitz, E., Seiver, A.: Uncertain Reasoning and Forecasting. International Journal of Forecasting **11** (1995) 73–87
5. Eisen, M.B., Spellman, P.T., Brown, P.O., Botstein, D.: Cluster Analysis and Display of Genome-Wide Expression Patterns. Proceedings of the National Academy of Science **95** (1998) 14863-14868
6. Falkenauer, E.: Genetic Algorithms and Grouping Problems. Wiley (1998)
7. Friedman, N., Linial, M., Nachman, I., Pe'er, D.: Using Bayesian Networks to Analyze Expression Data. Journal of Computational Biology **7** (2000) 601-620
8. Holland, J.H.: Adaptation in Natural and Artificial Systems. University of Michigan Press (1995)
9. Jenner, R.G., Alba, M.M., Boshoff, C., Kellam, P.: Kaposi's Sarcoma-Associated Herpesvirus Latent and Lytic Gene Expression as Revealed by DNA Arrays. Journal of Virology **75** No. 2 (2001) 891-902
10. Lütkepohl, H.: Introduction to Multivariate Time Series Analysis. Springer-Verlag (1993)
11. Pearl, J.: Probabilistic Reasoning in Intelligent Systems, Networks of Plausible Inference. Morgan Kaufmann (1988)
12. Snedecor, G., Cochran, W.: Statistical Methods. Iowa State University Press, (1967)
13. Swift, S., Tucker, A., Liu, X.: Evolutionary Computation to Search for Strongly Correlated Variables in High-Dimensional Time-Series. In: Hand, D.J., Kok, J.N., Berthold, M.R. (eds.): Advances in Intelligent Data Analysis 99, Springer-Verlag (1999) 51–62
14. Swift, S., Liu, X.: Predicting Glaucomatous Visual Field Deterioration Through Short Multivariate Time Series Modelling. Artificial Intelligence in Medicine, Elsevier (to appear)
15. Syswerda, G.: Uniform Crossover in Genetic Algorithms. Proceedings of the 3rd International Conference on Genetic Algorithms, Morgan Kaufmann (1989) 10–19
16. Tucker, A., Swift, S., Liu, X.: Grouping Multivariate Time Series via Correlation. IEEE Transactions on Systems, Man, and Cybernetics, Part B: Cybernetics **31** (2001) 235–245
17. Tucker, A., Liu, X., Ogden-Swift, A.: Evolutionary Learning of Dynamic Probabilistic Models with Large Time Lags. International Journal of Intelligent Systems **16**, Wiley, (2001) 621–645

Using Multiattribute Prediction Suffix Graphs for Spanish Part-of-Speech Tagging

José L. Triviño-Rodriguez and Rafael Morales-Bueno*

University of Málaga, Málaga, Spain,
{trivino,morales}@lcc.uma.es,
http://www.lcc.uma.es

Abstract. An implementation of a Spanish POS tagger is described in this paper. This implementation combines three basic approaches: a single word tagger based on decision trees; a POS tagger based on a new learning model called the *Multiattribute Prediction Suffix Graph*; and a feature structure set of tags. Using decision trees for single word tagging allows the tagger to work without a lexicon that enumerates possible tags only. Moreover, it decreases the error rate because there are no unknown words. The feature structure set of tags is advantageous when the available training corpus is small and the tag set large, which can be the case with morphologically rich languages such as Spanish. Finally, the multiattribute prediction suffix graph model training is more efficient than traditional full-order Markov models and achieves better accuracy.

1 Introduction

Many words in Spanish function as different parts of speech (POS). Part-of-speech tagging is the problem of determining the syntactic part of speech of an occurrence of a word in context. Because most of the high-frequency Spanish words function as several parts of speech, an automatic system for POS tagging is very important for most other high-level natural language text processing.

There are many approaches to the performance of this task, but they could be classified into two main ones:

- The linguistic approach, in which the language model is written by a linguist, generally in the form of rules or constraints. An example of this approach is the tagger developed by Voutilainen [20].
- The automatic approach, in which the language model is automatically obtained from corpora. These corpora can be either raw or annotated. In this way, the automatic taggers are classified into supervised and unsupervised learning:
  - Supervised learning is applied when the model is obtained from annotated corpora.

---

* This work has been partially supported by the FACA project, number PB98-0937-C04-01, of the CICYT, Spain. FACA is a part of the FRESCO project

F. Hoffmann et al. (Eds.): IDA 2001, LNCS 2189, pp. 228–237, 2001.
© Springer-Verlag Berlin Heidelberg 2001

- Unsupervised learning is applied when the model is obtained from raw corpora training.

The most noticeable examples of automatic tagging approaches are Markov chains [3], [2], [7], [9], [6], [4], prediction suffix trees [17], neural networks [15], decision trees [5], [10], and transformation-based error driven learning [1].

The most widely used methods for POS tagging are stochastic methods based on fixed-order Markov chains models and hidden Markov models. However, the complexity of Markov chains grows exponentially with its order L, and hence only low-order Markov chains could be considered in practical applications. Moreover, it has been observed that, in many natural sequences, memory length depends on the context and is not fixed. In order to solve this problem, in 1996, Dana Ron [13], [14] described a learning model of a subclass of PFAs (*Probabilistic Finite Automatas*) called PSAs (*Probabilistic Suffix Automatas*). A PSA is like a Markov chain with variable memory length. Thus, this model avoids the exponential complexity relation with the order L and so can be applied to improve Markov chains.

However, there are real problems, such as part-of-speech tagging where analyzing the part of speech of the last words is not enough. In order to study these kinds of problems several values must be considered, e.g., gender, number, etc. Nevertheless, both PSAs and Markov chains do not model sequences conditioned by parallel sequences.

The only way to apply Markov chains to these kinds of problems is by using the Cartesian product of alphabets of attributes of the problem as the alphabet of the Markov chain. However, this makes the length of the Markov chain grow exponentially with the number of attributes. Moreover, the cardinal of the alphabet grows significantly. This makes the number of states grow. A great number of states decreases the number of samples used to compute every probability in the model, and so it decreases confidence in the model.

Using PSAs in order to model these kinds of problems has exponential complexity, too. On the one hand, PSAs have all the problems of Markov chains. On the other hand, PSAs cannot handle different memory lengths for every attribute. This is because all attributes do not need the same memory length; but if the Cartesian product of alphabets of attributes is used, all attributes must have the same memory length.

To avoid these problems, we have developed a new model called the Multi-attribute Prediction Suffix Graph (MPSG) [18]. The MPSG model is based on the Prediction Suffix Tree (PST) model [13] (The PST model was introduced by Rissanen [12] and has been used for universal data compression [21]).

This paper describes how the MPSG model could be applied to part-of-speech tagging. This model allows us to compute the next symbol probability function of a sequence of data conditioned by previous values of several parallel sequences. The model can learn sequences generated by one attribute if the sequences generated by the rest of the attributes are known. Thus, this model allows us to analyze independently the memory length in every attribute needed to know the next symbol probability distribution of a target attribute. Hence, this tagger uses a feature structure set of tags like the model described by Kempe [8] for POS tagging. Due to the fact that this model computes every attribute in an independent way, the number of samples used to compute the next symbol probability distribution does not decrease very much.

Finally, most taggers deal with the problem of determining the syntactic part of speech of an occurrence of a word in context, but they cannot determine the set of correct tags for a word without any such context. Instead, most taggers use a lexicon that lists possible tags. In morphologically rich languages, this lexicon must be very large in order to represent the language correctly. In order to solve this problem, a single word tagger has been added to the system described. This single word tagger [16] is based on a modified version of the ID3 algorithm described by Quinlan [11] and has been used successfully in [17].

This paper is organised as follows: first, the MPSG model is described. Next, its application to POS tagging is shown. Finally, the results obtained from this model are analysed. The central aim of this paper is the application of MPSGs to POS tagging. Therefore, the MPSG model will only be described informally. A formal description can be found in [18].

2   The MPSG Model

2.1   Previous Definitions

Informally, an MPSG is similar to a multiattribute Markov chain. In this sense, a state in an MPSG is like a state in a Markov chain. However, states in an MPSG are composed of several attributes. An attribute A^i is a feature of an object and is constituted by a nonempty and finite set of values. If every allowed value is represented by a symbol then the set of values compounds an alphabet over A^i. Thus, we can express it in terms of strings over the alphabet A^{i^*}. Therefore, a *state* s over a set of attributes $A = \{A^1, A^2, \ldots, A^n\}$ in an MPSG is a set of n strings. Each string is compounded by symbols of a different attribute of A. The set of all states s over A is denoted by S^A. Thus, a state could be described as a vector with n components. From this point on in this paper, each component will be identified by the superindex of its attribute. Different values of an attribute will be identified by subindexes.

Next, several informal definitions will be given in order to describe the hypothesis model given later.

Definition 1. *Given the state s over a set of attributes A ($s \in S^A$), the expanded attribute of a state represents the last expanded attribute of a state during the construction of an MPSG. In other words, if a state underwent its last expansion in the attribute Y, the learning algorithm of the MPSGs assumes that the previous attributes of Y have been sufficiently expanded, and so only attributes greater than or equal to Y could be expanded.*

Example 1 (Expanded attribute). Let $A = \{X, Y, Z\}$ be a set of attributes and let the state $s = < x_1 x_1, y_2, \mathbf{e} >$ be a state over S^A. The expanded attribute of this state is attribute 2 because attribute 1 has been sufficiently expanded in order to determine the next symbol probability distribution; otherwise, attribute 1 must have been expanded before expanding attribute 2.

Definition 2. *The direct parent of a state s is the only state s' that could be expanded directly (through the expansion attribute of s' or greater) in order to get s. Therefore, s' is the only state that can be the parent of s in an MPSG.*

Example 2 (Direct parent). Next, several states and their direct parents are shown.

state	direct parent
$< x_1, y_2 y_1, \mathbf{e} >$	$< x_1, y_1, \mathbf{e} >$
$< x_1 x_1 x_2, y_2, \mathbf{e} >$	$< x_1 x_1 x_2, \mathbf{e}, \mathbf{e} >$
$< x_1 x_2, \mathbf{e}, z_1 >$	$< x_1 x_2, \mathbf{e}, \mathbf{e} >$
$< x_1, \mathbf{e}, \mathbf{e} >$	$< \mathbf{e}, \mathbf{e}, \mathbf{e} >$

Definition 3. *Now, we consider the concept of the expansion symbol of a state s. A state is the context in which the next symbol probability distribution is computed. If a state is not sufficiently large to correctly determine this probability distribution then the state must be expanded, that is, the context information stored in the state must be increased. The expansion symbol of a state is the information added to a state when it is expanded from its direct parent.*

Example 3 (Expansion symbol). Several states and their expansion symbols are now shown:

state	direct parent	expansion symbol
$< x_1, y_2 y_1, \mathbf{e} >$	$< x_1, y_1, \mathbf{e} >$	y_2
$< x_1 x_1 x_2, y_2, \mathbf{e} >$	$< x_1 x_1 x_2, \mathbf{e}, \mathbf{e} >$	y_2
$< x_1 x_2, \mathbf{e}, z_1 >$	$< x_1 x_2, \mathbf{e}, \mathbf{e} >$	z_1
$< x_1, \mathbf{e}, \mathbf{e} >$	$< \mathbf{e}, \mathbf{e}, \mathbf{e} >$	x_1

Definition 4. *If a state is not large enough to determine this probability distribution correctly, then the state must be expanded; that is, the context information stored in the state must be increased. Expansion means increasing the memory length by going back further into the past.*

The MPSG learning algorithm increases the memory length of a state progressively (one attribute at a time) in an ordered way from the first attribute. Thus, given a state s, there exists only one state that can be expanded in an ordered way to obtain s. However, if the attributes are not considered in an ordered way, then there exists a set of states that can be expanded to obtain s. This is the set of *indirect parents of s.*

2.2   Multiattribute Prediction Suffix Graph (MPSG)

A state in an MPSG is composed of the last symbols in the data sequence, that is, the last values of each attribute. The state $\mathbf{E} =< \mathbf{e}^1, \mathbf{e}^2, \ldots, \mathbf{e}^n >$ compounded by n empty strings is always in the MPSG. This state expresses the next symbol probability distribution when the context of the symbols is not considered. Thus, the function γ associated with this state is the stationary probability distribution of an MPSG. Clearly, the state $\mathbf{E}$ has no direct parent because there is no state that can be expanded to obtain a state compounded by n empty strings.

A multiattribute prediction suffix graph (MPSG) G is a directed graph where each node in the graph is labeled with a state of G. These states represent the memory that determines the next symbol probability function. The node labeled with n empty strings is always in the graph. This node expresses the next symbol probability distribution independently of any context. Nodes are joined by edges labeled by pairs (σ, i) where $\sigma \in A$ and $i \in \mathbf{N}$. These pairs allow us to find every state in the graph efficiently. There are three kinds of edges:

Expansive edges. Let s, s' be two states. Let us consider that s is the direct parent of s'. Let us consider that s' gives us more information than s in order to determine the next symbol probability distribution. Two probability distributions differ substantially if the ratio between distributions is greater than a lower bound. Then, an edge from s to s' is added to the graph. This edge is called an expansive edge because the edge increases the memory length of an attribute.

Compression edges. Let us consider that s and s' are the same states that were mentioned in the case of expansive edges, with the same features. Let us now assume that there exists a state t that holds: t is a indirect parent of s'; t determines the same next symbol probability distribution as s'. Then, an edge from s to t is added to the graph. This edge is called a compression edge because the same next symbol probability distribution is determined with shorter memory length.

Backward edges. Let us assume that both s and s' determine the same next symbol probability distribution; that s and s' have the same memory length; and that s' has more attributes than s with shorter memory length. Then, an edge from the largest direct suffix of s in the graph to s' is added to the graph. This edge is called a backward edge because s' is to the left of s in the layout of the graph.

Definition 5. *Let $A = \{A^1, \ldots, A^n\}$ be a set of attributes, each one compounded by a set of symbols $\forall_{i=1}^n A^i = \{a_1^i, \ldots, a_{m_i}^i\}$, then an MPSG G^x over A for the attribute A^x is a directed graph that satisfies:*

1. *Every node is labeled with a pair (s, γ_s) where:*
   - *$s \in S^A$ is a state of G^x.*
   - *$\gamma_s : A^x \to [0, 1]$ is the next symbol probability function associated with s. This function must satisfy: $\forall s \in G^x, \sum_{a^x \in A^x} \gamma_s(a^x) = 1$.*
2. *The node $\mathbf{E}$ labeled with n empty string is always in G^x.*

3. *If a node s is in G^x then its direct parent is in G^x.*
4. *Every edge is labeled by a pair (b, r) where:*
   - *$b \in A^i, 1 \leq i \leq n$ is a symbol of an attribute of A.*
   - *$r \in \mathbf{N}, 1 \leq r \leq L$ is a number in $\mathbf{N}$. L is the maximum memory length of the MPSG.*
5. *From every node there is at most an edge for each symbol and attribute A.*
6. *If a node s and its direct parent s' are in G^x, then there is an expansive edge from s' to s labeled with the expansion symbol of s and the length of the last non-empty attribute of s.*
7. *Let s and s' be two nodes in G^x, a compression edge labeled with a pair (b, r) could join the node s' to s iff there exists a state t that is not in G^x and the following conditions are satisfied:*
   (a) s' is the direct parent of t.
   (b) s is a state to the right of s' in the layout of the MPSG.
   (c) There does not exist a state t'' in the MPSG greater than s that can be joined with s' with a compression edge.
   (d) r is the length of the last non-empty attribute of t.
   (e) b is the expansion symbol of t.
8. *Let s and s' be two nodes in G^x, then a backward edge labeled with a pair (b, r) could join the node s' to s iff there exists a state t that is not in G^x and the following conditions are satisfied:*
   (a) s can be expanded directly to obtain t.
   (b) s' is a state to the right of s in the layout of the MPSG.
   (c) There does not exist a state t' that is greater than s' and the backward edge can start from t'.
   (d) r is the length of the last non-empty attribute of t.
   (e) b is the expansion symbol of t.

Example 4 (MPSG). In figure 1 an example of an MPSG is shown. This MPSG is compounded by the next nodes and edges:

nodes =
$$\{< \mathbf{e}, \mathbf{e}, \mathbf{e} >, < x_1, \mathbf{e}, \mathbf{e} >, < x_2, \mathbf{e}, \mathbf{e} >, < x_3, \mathbf{e}, \mathbf{e} >,$$
$$< \mathbf{e}, y_1, \mathbf{e} >, < x_1 x_2, \mathbf{e}, \mathbf{e} >, < x_2, y_1, \mathbf{e} >, < \mathbf{e}, y_1 y_1, \mathbf{e} >,$$
$$< x_2, y_1, z_1 >, < \mathbf{e}, y_2, y_1, \mathbf{e} >, < \mathbf{e}, y_2, y_1, z_2 >\}$$

expansive edges =
$$\{(< \mathbf{e}, \mathbf{e}, \mathbf{e} >, < x_1, \mathbf{e}, \mathbf{e} >, x_1, 1), (< \mathbf{e}, \mathbf{e}, \mathbf{e} >, < x_2, \mathbf{e}, \mathbf{e} >, x_2, 1),$$
$$(< \mathbf{e}, \mathbf{e}, \mathbf{e} >, < x_3, \mathbf{e}, \mathbf{e} >, x_3, 1), (< \mathbf{e}, \mathbf{e}, \mathbf{e} >, < \mathbf{e}, y_1, \mathbf{e} >, y_1, 1),$$
$$(< x_2, \mathbf{e}, \mathbf{e} >, < x_1 x_2, \mathbf{e}, \mathbf{e} >, x_1, 2), (< x_2, \mathbf{e}, \mathbf{e} >, < x_2, y_1, \mathbf{e} >, y_1, 1),$$
$$(< \mathbf{e}, y_1, \mathbf{e} >, < \mathbf{e}, y_1 y_1, \mathbf{e} >, y_1, 2), (< x_2, y_1, \mathbf{e} >, < x_2, y_1, z_1 >, z_1, 1),$$
$$(< \mathbf{e}, y_1, \mathbf{e} >, < \mathbf{e}, y_2 y_1, \mathbf{e} >, y_2, 2), (< \mathbf{e}, y_2 y_1, \mathbf{e} >, < \mathbf{e}, y_2 y_1, z_2 >, z_2, 1)\}$$

compression edges =
$$\{(< x_2, y_1, \mathbf{e} >, < \mathbf{e}, y_2 y_1, \mathbf{e} >, y_2, 2)\}$$

backward edges =
$$\{(< \mathbf{e}, y_2 y_1, \mathbf{e} >, < x_2, y_1, z_1 >, z_1, 1)\}$$

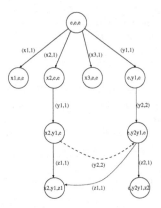

Fig. 1. This figure shows an MPSG. Dashed edges denote compression edges. Dotted edges denote backward edges.

2.3   Tagging with MPSGs

Part-of-speech tagging is the problem of determining the syntactic part of speech of an occurrence of a word in context. Therefore, given a sequence of words $w_1, \ldots, w_o$, POS tagging is the problem of determining the tag sequence $t_1, \ldots, t_o$ that is most likely for $w_1, \ldots, w_o$. This could be computed by maximizing the joint probability of $w_1, \ldots, w_o$ and $t_1, \ldots, t_o$:

$$
\begin{aligned}
\tau(w_1, \ldots, w_o) &= \underset{t_1, \ldots, t_o}{argmax} \, Pr(t_1, \ldots, t_o | w_1, \ldots, w_o) \\
&= \underset{t_1, \ldots, t_o}{argmax} \, \frac{Pr(t_1, \ldots, t_o, w_1, \ldots, w_o)}{Pr(w_1, \ldots, w_o)} \\
&= \underset{t_1, \ldots, t_o}{argmax} \, Pr(t_1, \ldots, t_o, w_1, \ldots, w_o)
\end{aligned}
\tag{1}
$$

The equation 1 could be laid out as follows:

$$
Pr(t_1, \ldots, t_o, w_1, \ldots, w_o) =
$$
$$
\prod_{i=1}^{o} Pr(t_i | t_1, \ldots, t_{i-1}) * Pr(w_i | t_1, \ldots, t_i, w_1, \ldots, w_{i-1})
\tag{2}
$$

Up to this point, no assumptions about the probabilities have been made, but the probabilities required by equation 2 are not empirically collectable. In order to solve this problem, it is necessary to make the following assumptions about these probabilities [1]:

---

[1] These assumptions were described by Charniak [2]. That is, that the probability of the tag given the past depends only on the last two tags, and the probability of the word given the past depends only on its tag.

$$Pr(t_i|t_1,\ldots,t_{i-1}) = Pr(t_i|t_{i-L},\ldots,t_{i-1}) \tag{3}$$

$$Pr(w_i|t_1,\ldots,t_i,w_1,\ldots,w_{i-1}) = Pr(w_i|t_i) \tag{4}$$

From equations 3 and 4, equation 2 can be expressed as follows:

$$Pr(t_1,\ldots,t_o,w_1,\ldots,w_o) =$$

$$\prod_{i=1}^{o} Pr(t_i|t_{i-L},\ldots,t_{i-1}) * Pr(w_i|t_i) \tag{5}$$

where L is the maximum length of the model.

This tagger is based on a feature structure set of tags. Therefore, the term t_i in equation 5 is a vector of n symbols[2] (a symbol for each attribute). However, the Viterbi algorithm [19] can only compute the probability of a sequence of a single attribute. In order to solve this problem, a set of n MPSGs (an MPSG for each attribute) has been computed independently. The equation 5 is computed from this set of trees as follows:

$$Pr(t_1,\ldots,t_o,w_1,\ldots,w_o) =$$

$$= \prod_{j=1}^{o} Pr(t_1^j,\ldots,t_o^j,w_1,\ldots,w_o)$$

$$= \prod_{i=1}^{o}(Pr(w_i|t_i) * \prod_{j=1}^{o} Pr(t_i^j|t_{i-L}^j,\ldots,t_{i-1}^j)) \tag{6}$$

Given a set of MPSGs M, $Pr(t_i^j|t_{i-L},\ldots,t_{i-1})$ is estimated by $\gamma_s^j(t_i^j)$ where s is the longest suffix of $t_{i-L},\ldots,t_{i-1}$ labeling a state in M^j, and γ_s^j is the next symbol probability function related with the state s in M^j.

The static probability $Pr(w_i|t_i)$ is estimated indirectly from $Pr(t_i|w_i)$ using Bayes' Theorem:

$$Pr(w_i|t_i) = \frac{Pr(w_i) * Pr(t_i|w_i)}{Pr(t_i)}$$

The terms $Pr(w_i)$ are constant for a given sequence w_i and can therefore be omitted from the maximization. The values of $Pr(t_i)$ can be computed as follows:

$$Pr(t_i) = \prod_{j=1}^{n} \gamma_{\mathbf{e}}^j(t_i^j) \tag{7}$$

The static parameters $Pr(t_i|w_i)$ are computed by the single word tagging.

3 Experimental Results

The order L of the MPSG tagger has been set to 4. The tagger has been trained on a Spanish 45000-word tagged corpus. An on-line demo of this tagger can be found in *http://www.lcc.uma.es/~trivino*.

---

[2] We take into account 18 attributes for each word, thus $n = 18$ in our tagger

The accuracy obtained is 97.32%. This is not the best accuracy achieved by a Spanish tagger [17]. However, it must be taken into account that the tagger has been trained with a small sample. This tagger is more complex than the VMM tagger, so it needs a larger number of words to perform the training task. Moreover, many of the mistakes are due to incorrect tags in the training sample. In Table 1 a comparison between the accuracy of several taggers is shown.

Tagger	Language	Corpus (words)	Accuracy (%)
Triviño MPSG tagger	Spanish	45000	97.32
Triviño VMM tagger	Spanish	45000	98.58
Padró	Spanish/English	10^6	97.45
Charniak	English	10^6	96.45
Kempe	French	$2*10^6$	96.16
Brill	English	350000	96.0
Singer	English	10^6	95.81
Kempe	French	10000	88.89

Table 1. Comparison between several taggers

Our results (97.32%) are better than Singer's tagger (95.81%) based on VMM because the single word tagger has added lexical information to our tagger. This can decrease the error rate when errors due to bad tags for rare words are avoided by the single word tagger. However, it is difficult to compare these results with other works, since the accuracy varies greatly depending on the corpus, tag set, etc. Performance could be measure more precisely by training with a larger sample. Thus, the performance could decrease or increase if different corpus styles are used.

4 Conclusions and Future Work

In this paper, a high-accuracy Spanish tagger has been presented. This tagger has three main features: a single word tagger based on decision trees, a feature structure set of tags, and a multiattribute variable memory Markov model.

The results obtained show that joining an accurate single word tagger with a feature structure set of tags and an MPSG produces an improvement in the performance of the tagger.

On the other hand, most of the errors arise because there are incorrect tags in the training corpus. Thus, a further line of research is to analyze how noise in the sample affects the accuracy of the model.

References

1. E. Brill. Some advances in transformation-based part of speech tagging. In *Proceedings of AAAI94*, page 6, 1994.

2. Eugene Charniak, Curtis Hendrickson, Neil Jacobson, and Mike Perkowitz. Equations for part-ofspeech tagging. In *Proceedings of the Eleventh National Conference on Artificial Intelligence*, pages 784–789, 1993.

3. K.W. Church. A stochastic parts program and noun phrase parser for unrestricted text. In *Proceedings of ICASSP*, 1989.

4. D. Cutting, J. Kupiec, J. Pederson, and P. Sibun. A practical part-of-speech tagger. In *Proceedings of the Third Conference on Applied Natural Language Processing*. ACL, 1992.

5. W. Daelemans, J. Zavrel, P. Beck, and S.Gillis. MTB: A memory-based part-of-speech tagger generator. In *Proceedings of 4th Workshop on Very Large Corpora*, Copenhagen, Denmark, 1996.

6. R. Garside, G. Leech, and G. Sampson. *The Computational Analysis of English*. London and New York: Longman, 1987.

7. Fred Jelinek. Robust part-of-speech tagging using a hidden Markov model. Technical report, IBM, 1985.

8. André Kempe. Probabilistic tagging with feature structures. In *Coling-94*, volume 1, pages 161–165, August 1994.

9. B. Merialdo. Tagging english text with a probabilistic model. *Computational Linguistics*, 2(20):155–171, 1994.

10. L. Márquez and H. Rodríguez. Towards learning a constraint grammar from annotated corpora using decision trees. ESPRIT BRA-7315 Acquilez II, Working Paper, 1995.

11. J.R. Quinlan. Induction of decision trees. *Machine Learning*, (1):81–106, 1986.

12. J. Rissanen. A universal data compression system. *IEEE Trans. Inform. Theory*, 29(5):656–664, 1983.

13. D. Ron. *Automata Learning and its Applications*. PhD thesis, MIT, 1996.

14. Dana Ron, Yoram Singer, and Naftali Tishby. The power of amnesia: Learning probabilistic automata with variable memory length. *Machine Learning*, (25):117–149, 1996.

15. H. Schmid. Part-of-speech tagging with neural networks. In *Proceedings of 15th International Conference on Computational Linguistics*, Kyoto, Japan, 1994.

16. J.L. Triviño. SEAM. Sistema experto para anlisis morfológico. Master's thesis, Universidad de Málaga, 1995.

17. J.L. Triviño and R. Morales. A spanish POS tagger with variable memory. In *IWPT 2000, Sixth International Workshop on Parsing Technologies*, February 2000.

18. J.L. Triviño and R. Morales. MPSGs (multiattribute prediction suffix graphs). Technical report, Dept. Languages and Computer Sciences. University of Málaga, 2001.

19. A.J. Viterbi. Error bounds for convolutional codes and an asymptotical optimal decoding algorithm. In *Proceedings of IEEE*, volume 61, pages 268–278, 1967.

20. A. Voutilainen and T. Järvinen. Specifying a shallow grammatical representation for parsing purposes. In *Proceedings of the 7th meeting of the European Association for Computational Linguistics*, pages 210–214, 1995.

21. M. Weinberger, A. Lempel, and K. Ziv. A sequential algorithm for the universal coding of finite-memory sources. *IEEE Trans. Inform. Theory*, 38:1002–1014, 1982.

Self-Supervised Chinese Word Segmentation

Fuchun Peng and Dale Schuurmans

Department of Computer Science
University of Waterloo
200 University Avenue West
Waterloo, Ontario, Canada, N2L 3G1
{f3peng,dale}@ai.uwaterloo.ca

Abstract. We propose a new unsupervised training method for acquiring probability models that accurately segment Chinese character sequences into words. By constructing a core lexicon to guide unsupervised word learning, self-supervised segmentation overcomes the local maxima problems that hamper standard EM training. Our procedure uses successive EM phases to learn a good probability model over character strings, and then prunes this model with a mutual information selection criterion to obtain a more accurate word lexicon. The segmentations produced by these models are more accurate than those produced by training with EM alone.

1 Introduction

Unlike English and other western languages, many Asian language such as Chinese, Japanese, and Thai, do not delimited words by white-space. Word segmentation is therefore a key subproblem to language processing tasks (such as information retrieval) in these languages. Unfortunately, segmenting an input sentence into words is a nontrivial task in such cases. For Chinese, there has been a significant amount of research on techniques for discovering word segmentations; see for example [14]. The main idea behind most of these techniques is to start with a lexicon that contains the set of possible Chinese words and then segment a concatenated Chinese character string by optimizing a heuristic objective such as maximum length match, mutual information, or maximum likelihood. This approach implies, however, that one of the main problems in Chinese word segmentation is constructing the original lexicon.

Methods for constructing lexicons can be classified as either supervised or unsupervised. In supervised construction, one has to segment a raw unsegmented corpus by hand and then collect all the words from the segmented corpus to build a lexicon. Unfortunately, since there are over 20,000 Chinese characters, among which 6763 are most commonly used, building a complete lexicon by hand is impractical. Therefore a number of unsupervised segmentation methods have been proposed recently to segment Chinese and Japanese text [1,3,8,12,9]. Most of these approaches use some form of EM to learn a probabilistic model of character sequences and then employ Viterbi-decoding-like procedures to segment new text into words. One reason that EM algorithm is widely adopted for

F. Hoffmann et al. (Eds.): IDA 2001, LNCS 2189, pp. 238–247, 2001.
© Springer-Verlag Berlin Heidelberg 2001

unsupervised training is that it is guaranteed to converge to a good probability model that locally maximizes the likelihood or posterior probability of the training data [6]. For Chinese segmentation, EM is usually applied by first extracting a lexicon which contains the candidate multi-grams from a given training corpus, initializing a probability distribution over lexicon, and then using the standard iteration to adjust the probabilities of the multi-grams to increase the posterior probability of the training data.

One advantage of unsupervised lexicon construction is that it can automatically discover new words once other words have acquired high probability [4]. For example, if one knows the word "*computer*" then upon seeing "*computerscience*" it is natural to segment "*science*" as a new word. Based on this observation, we propose a new word discovery method that is a variant of standard EM training, but avoids getting trapped in local maxima by keeping two lexicons: a *core* lexicon which contains words that are judged to be familiar, and a *candidate* lexicon which contains all other words that are not in the core lexicon. We use EM to maximize the likelihood of the training corpus given the two lexicons; which automatically suggests new words as candidates for the core. However, once new words have been added to the core, EM is reinitialized by giving half of the total probability mass to the core lexicon, thus allowing core words to guide the segmentation and pull EM out of poor local maxima.

A problem with maximum likelihood training, however, is that it tends to put probability mass on conjunctions of shorter words. Note that since likelihood is defined by a product of individual chunk probabilities (making the standard assumption that segments are independent), the more chunks a segmentation has, the smaller its likelihood will tend to be. For example, in English, given a character sequence *sizeofthecity* and a uniform distribution over multi-grams, the segmentation *sizeof|thecity* will have higher likelihood than segmentation *size|of|the|city*. Therefore, maximum likelihood will prefer fewer chunks in its segmentation and tend to put large probability on long non-word sequences like *sizeof* and *thecity*. If one can break such sequences into shorter legal words, *size, of, the, city*, the lexicon will be much smaller and the training and segmentation performance should be improved. To this end, we employ a mutual information criterion to eliminate longer agglomerations in favor of shorter primitives (once the EM optimization has stabilized). Not only does this have the advantage of producing a smaller core lexicon, it also has the side effect of driving EM out of poor local maxima [6,2] and yielding better segmentation performance. The remainder of the paper describes the self-supervised training procedure in detail, followed by the mutual information lexicon pruning criterion, experiments, error analysis, and discussion.

2   Self-Supervised Segmentation

We first develop a technique to help EM avoid poor local maxima when optimizing the probability model. This is done by dynamically repartitioning the vocabulary and reinitializing EM with successively better starting points.

Assume we have a sequence of characters $C = c_1 c_2 ... c_T$ that we wish to segment into chunks $S = s_1 s_2 ... s_M$, where T is the number of characters in the sequence and M is the number of words in the segmentation. Here chunks s_i will be chosen from the core lexicon $V_1 = \{s_i, i = 1, ..., |V_1|\}$ or the candidate lexicon $V_2 = \{s_j, j = 1, ..., |V_2|\}$. If we already have the probability distributions $\theta = \{\theta_i | \theta_i = p(s_i), i = 1, ..., |V_1|\}$ defined over the core lexicon and $\phi = \{\phi_j | \phi_j = p(s_j), j = 1, ..., |V_2|\}$ over the candidate lexicon, then we can recover the most likely segmentation of the sequence $C = c_1 c_2 ... c_T$ into chunks $S = s_1 s_2 ... s_M$ as follows. First, for any given segmentation S of C, we can calculate the joint likelihood of S and C by

$$prob(S, C | \theta, \phi) = \prod_{i=1}^{M_1} p(s_i) \prod_{j=1}^{M_2} p(s_j) = \prod_{k=1}^{M} p(s_k)$$

where M_1 is the number of chunks occurring in the core lexicon, M_2 is the number of chunks occurring in the candidate lexicon, and s_k can come from either lexicon. (Note that each chunk s_k must come from exactly one of the core or candidate lexicons.) Our task is to find the segmentation S^* that achieves the maximum likelihood:

$$S^* = \underset{S}{argmax}\{prob(S|C; \theta, \phi)\} = \underset{S}{argmax}\{prob(S, C|\theta, \phi)\} \qquad (1)$$

Given a probability distribution defined by θ and ϕ over the lexicon, the Viterbi algorithm can be used to efficiently compute the best segmentation S of character string C. However, *learning* which probabilities to use given a training corpus is the job of the EM algorithm.

2.1   Parameter Re-estimation

Following [6], the update Q function that we use in the EM update is given by

$$Q(k, k+1) = \sum_S prob(S|C; \theta^k, \phi^k) log(prob(C, S|\theta^{k+1}, \phi^{k+1})) \qquad (2)$$

Maximizing (2) under the constraints that $\sum_i \theta_i^{k+1} = 1$ and $\sum_j \phi_j^{k+1} = 1$ yields the parameter re-estimation formulas

$$\theta_i^{k+1} = \frac{\sum_S \#(s_i, S) \times prob(S, C|\theta^k, \phi^k)}{\sum_{s_i} \sum_S \#(s_i, S) \times prob(S, C|\theta^k, \phi^k)} \qquad (3)$$

$$\phi_j^{k+1} = \frac{\sum_S \#(s_j, S) \times prob(S, C|\theta^k, \phi^k)}{\sum_{s_j} \sum_S \#(s_j, S) \times prob(S, C|\theta^k, \phi^k)} \qquad (4)$$

where $\#(s_i, S)$ is the number of times s_i occurring the segmentation S. These are the standard re-estimation formulas, and are the same for θ and ϕ except that each will be reinitialized differently in successive optimizations (see below).

In both cases the denominator is a weighted sum of the number of words in all possible segmentations, the numerator is a normalization constant, and (3) and (4) therefore are weighted frequency counts. Thus, the updates can be efficiently calculated using the forward and backward algorithm, or efficiently approximated using the Viterbi algorithm; see [13] and [5] for detailed algorithms.

2.2   Self-Supervised Training

The main difficulty with applying EM to this problem is that the probability distributions are complex and typically cause EM to get trapped in poor local maxima. To help guide EM to better probability distributions, we partition the lexicon into a core and candidate lexicon, V_1 and V_2; where V_1 is initialized to be empty and V_2 contains all words. In a first pass, starting from the uniform distribution, EM is used to increase the likelihood of the training corpus C_1. When the training process stabilizes, the N words with highest probability are selected from V_2 and moved to V_1, after which all the probabilities are rescaled so that V_1 and V_2 each contain half the total probability mass. EM is then run again. The rationale for shifting half of the probability mass to V_1 is that this increases the influence of core words in determining segmentations and allows them to act as more effective guides in processing the training sequence. We call this procedure of successively moving the top N words to V_1 *forward selection.*

Forward selection is repeated until the segmentation performance of Viterbi on the validation corpus C_2 leads to a decrease in F-measure (which means we must have included some erroneous core words). After forward selection terminates, N is decremented by 5, and we carry out a process of *backward deletion,* where the N words with the lowest probability in V_1 are moved back to V_2, and EM training is successively repeated until F-measure again decreases on the validation corpus C_2 (which means we must have deleted some correct core words). The two procedures of forward selection and backward deletion are alternated, decrementing N by 5 at each alternation, until $N \leq 0$; as shown in Fig. 1. As with EM, the outcome of this self-supervised training procedure is a probability distribution over the lexicon that can be used by Viterbi to segment test sequences.

3   Mutual Information Lexicon Pruning

Both EM and self-supervised training acquire probability distributions over the entire lexicon. However, as pointed out by [12], the lexicon is the most important factor in the word segmentation, and therefore a better lexicon is more critical than a better model. Unfortunately, by maximizing likelihood, either through EM or self-supervised training, erroneous agglomerations tend to get naturally introduced in the lexicon. This means that after a high-likelihood model has been learned, we are still motivated to prune the lexicon to remove erroneous non-word entries. We do this by invoking a mutual information based criterion.

```
0. Input
      Completely unsegmented training corpus C1
      Validation corpus C2

1. Initialize
      V1 = empty;
      V2 contains all potential words;
      OldFmeasure = infinite small;
      bForwardSelection=true;
      set N to a fix number;

2. Iterate
      while (N > 0){
          EM Learning based on current V1 and V2 until converge;

          Calculate NewFmeasure on validation corpus C2;

          if(NewFmeasure < OldFmeasure){
              // change selection direction
              bForwardSelection = ¬bForwardSelection;
              N = N-5;
          }

          OldFmeasure = NewFmeasure;

          //SelectCoreWords(true) does forward selectoin
          //SelectCoreWords(false) does backward selectoin
          SelectCoreWords(bForwardSelection);
      }

3. Test
      Test on test corpus C3
```

Fig. 1. Self-supervised Learning

Recall that the mutual information between two random variables is given by

$$MI(X,Y) = \sum_{x,y} p(x,y) log \frac{p(x,y)}{p(x) \times p(y)} \tag{5}$$

where a large value indicates a strong dependence between X and Y, and zero indicated independence. In our case, we want to test the dependence between two chunks s_1 and s_2. Given a long word, say $s=$ "abcdefghijk", we consider splitting it into its most likely two-chunk segmentation, say $s_1 =$ "abcd" and $s_2 =$ "efghijk". Let the probabilities of the original string and the two chunks be $p(s)$, $p(s_1)$, and $p(s_2)$ respectively. The *pointwise mutual information* [10] between s_1 and s_2 is

$$MI(s_1, s_2) = log \frac{p(s)}{p(s_1) \times p(s_2)}. \tag{6}$$

To apply this measure to pruning, we set two thresholds $\gamma_1 > \gamma_2$. If the mutual information is higher than the threshold γ_1, we say s_1 and s_2 are strongly correlated, and do not split s. (That is, we do not remove s from the lexicon.) If mutual information is lower than the lower threshold γ_2, we say s_1 and s_2 are independent, so we remove s from the lexicon and redistribute its probability to s_1 and s_2. If mutual information is between the two thresholds, we say s_1 and s_2 are weakly correlated, and therefore shift some of the probability from s to s_1 and s_2, by keeping a portion of s's probability for itself (1/3 in our experiments) and distributing the rest of its probability to the smaller chunks, proportional to their probabilities. The idea is to shift the weight of the probability distribution toward shorter words. This splitting process is carried out recursively for s_1 and s_2. The pseudo code is illustrated in Fig. 2.

```
1: (s₁, s₂) = mostlikely_split(s);

2: MI = log   p(s)
            ─────────────
            p(s₁)×p(s₂)

3: if(MI > γ₁){//strongly dependent
        return -1;
   }
   else if(MI < γ₂){//independent
        probSum = p(s₁) + p(s₂);
        p(s₁)+ = p(s) × p(s₁)/probSum;
        p(s₂)+ = p(s) × p(s₂)/probSum;
        p(s) = 0;
        return 1;
   }
   else{//weakly dependent
        probDistribute = p(s)/3;
        probSum = p(s₁) + p(s₂);
        p(s) = probDistribute;
        p(s₁)+ = 2 × probDistribute × p(s₁)/probSum;
        p(s₂)+ = 2 × probDistribute × p(s₂)/probSum;
        return 0;
   }
```

Fig. 2. Mutual information probabilistic lexicon pruning

4   Experiments

To compare our technique to previous results, we follow [8,12] and measure performance by precision, recall, and F-measure on detecting word boundaries. Here, a word is considered to be correctly recovered iff [11]:

1. a boundary is correctly placed in front of the first character of the word,
2. a boundary is correctly placed at the end of the last character of the word,
3. and there is no boundary between the first and last character of the word.

Let N_1 denote the number of words in the test corpus C_3, let N_2 denote the number of words in the recovered corpus C_3', and let N_3 denote the number of words correctly recovered. Then the precision, recall and F measures are defined

$$\text{precision: } p = \frac{N_3}{N_2}$$
$$\text{recall: } r = \frac{N_3}{N_1}$$
$$\text{F-measure: } F = \frac{2 \times p \times r}{p+r}$$

We use a training corpus, C_1, that consists of 90M (5,032,168 sentences) of unsegmented Chinese characters supplied by the author of [8], which contains one year of the "People's Daily" news service stories (www.snweb.com). The test corpus C_3 is ChineseTreebank from LDC[1] (1M, 10,730 sentences), which contains 325 articles from "Xinhua newswire" between 1994 and 1998 that have been correctly segmented by hand. We remove all white-space from C_3 and create an unsegmented corpus C_3''. We then use the algorithm described in Section 2 to recover C_3' from C_3''. The validation corpus, C_2, consists of 2000 sentences randomly selected from the test corpus.

According to the *1980 Frequency Dictionary of Modern Chinese* (see [7]), the top 9000 most frequent words in Chinese consist of 26.7% unigrams, 69.8%

---

[1] http://www.ldc.upenn.edu/ctb/

bigrams, 2.7% trigrams, 0.007% 4-grams, and 0.002% 5-grams. So in our model, we limit the length of Chinese words up to 4 characters, which is sufficient for most situations.

The experimental results are shown in Table 1—Results 1 are obtained by standard EM training, Results 2 are obtained by self-supervised training, Results 3 are obtained by repeatedly applying lexicon pruning to standard EM training, and Results 4 are obtained by repeatedly applying lexicon pruning to self-supervised training. The row labeled Perfect lexicon is obtained by using maximum length match with the complete lexicon of the test corpus, which contains exactly all the words occurring in the test corpus. Soft-counting is the result of [8][2], which is also a EM-based unsupervised learning algorithm. The Word-based results are from [12] which uses a suffix tree model. Finally, the Perfect lexicon[12] results are obtained using a lexicon from another test corpus.

	p	r	F	final lex size
Results 1	44.6%	37.3%	40.0%	19044012
Results 2	55.7%	53.9%	54.1%	19044012
Results 3	73.2%	71.7%	72.1%	1670317
Results 4	75.1%	74.0%	74.2%	1670317
Perfect lexicon	92.2%	91.8%	91.9%	10363
Soft-counting[8]	71.9%	65.7%	68.7%	
Word-based[12]	84.4%	87.8%	86.0%	
Perfect lexicon[12]	95.9%	93.6%	94.7%	

Table 1. Experimental results

There are several observations to make from these results. First, self-supervised training improves the performance of standard EM training from 40% to 54.1%. Second, mutual information pruning gives even greater improvement (from 40% to 72.1%), verifying the claim of [12] that the lexicon is more influential than the model itself. The lexicon pruning reduces the lexicon size from 19044012 to 1670317, which makes the lexicon much smaller. By combining the two strategies, we obtain further improvement (from 40% to 74.2%), which is promising performance considering that we are using a purely unsupervised approach. By comparison, the result given by a perfect lexicon is 91.9%. Finally, the two improvement strategies of self-supervised training and lexicon pruning are not entirely complementary and therefore the resulting performance does not increase additively when both are combined (72.1% using lexicon pruning alone to 74.2% using both).

A direct comparison can be made to [8] because it also investigates a purely unsupervised training approach without exploiting any additional context information and uses the same training set we have considered. When we compare the results, we find that self-supervised training plus lexicon pruning achieves

---

[2] They did not supply F-measure, we calculate it for comparison.

both a higher precision and recall than [8], and obtains a **5.5%** (from 68.7% to 74.2%) improvement in F-measure. One problem with this comparison, however, is that [8] does not report precisely how the testing was performed. It is also possible to compare to [12], which uses a suffix tree model and employs context information (high entropy surroundings) to achieve an 86% F-measure score. This result suggests the context information is very important. However, because of a different test set (our test set is the 1M ChineseTreebank from LDC, whereas their test data is 61K pre-segmented by NMSU segmenter [9] and corrected by hand), the comparison is not fully calibrated. In the perfect lexicon experiments, [12] achieves higher performance (94.7% F-measure), whereas only 91.9% is achieved in our experiments. This suggests that we may obtain better performance when testing on the data used in [12]. Nevertheless, the result of [12] appears to be quite strong, and demonstrates the utility of using local context rather than assuming independence between words, as we have done.

5   Error Analysis

Fig. 3 shows two categories of errors that are typically committed by our model. In each case, the left string shows the correct word and the right bracketed string shows the recovered word segmentation. The first error category is *date and number*. In Chinese, dates and numbers are represented by 10 characters. Unlike Arab numbers, these 10 Chinese number characters are not different from other Chinese characters. Most dates and numbers are not correctly recognized because they do not have sufficient joint statistics in the training corpus to prevent them from being broken down into some smaller numbers. For example, *"1937 year"* is broken into *"19", "3", "7 year"*; *"2 wan 3 qian 1 bai 1 shi 4"* is broken into *"2 wan", "3 qian", "1", "bai 1 shi 4"* (*wan* denotes 10-thousand, *qian* denotes thousand, *bai* denotes hundred, and *shi* denotes ten). It turns out that one can easily use heuristic methods to correct errors in these special cases. For example, if a string of concatenated characters are all number characters, then the string is very likely to be a single date or number.

The second error category is the recognition of compound nouns. For example, the compound *"total marks"* is recovered as two words *"total"* and *"marks"*; *"team Australia"* is recovered to *"team"* and *"Australia"*. One reason for the failure to correctly recover compounds is that we are limiting the maximum length of a word to 4 characters, whereas many compounds are longer than this limit. However, simply relaxing the length limit creates a significant sparse data problem. The recognition of noun compounds appears to be difficult, and a general approach may have to consider language constraints.

6   Related Work

Our work is related to many other research efforts.

Unsupervised Chinese Word Segmentation: [8] uses a soft counting version of EM to learn how to segment Chinese. To augment the influence of

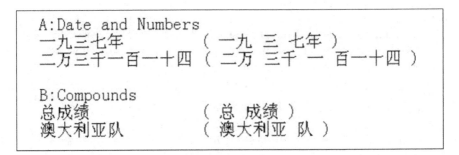

Fig. 3. Typical segmentation errors

important words, [8] shifts probability mass to likely words by soft counting. In our model, we shift half of the probability space to the core words by dividing the lexicon to two parts. Also, [8] does not employ any sort of lexicon pruning, which we have found is essential to improving performance. [12] uses a suffix tree word-based model and a bigram model to segment Chinese strings. This work takes the surrounding word information into consideration when constructing the lexicon. [14] uses a more complicated Hidden Markov Model (HMM) model that includes special recognizers for Chinese names and a component for morphologically derived words. As pointed out by [8], standard EM segmentation can be thought of as a zero order HMM.

Mutual Information Lexicon Optimization: Other researchers have considered using mutual information to build a lexicon. For example, [14] uses mutual information to build a lexicon, but only deals with words of up to 2 characters. [3,12] uses mutual information and context information to build a lexicon based on the statistics directly obtained from the training corpus. By contrast, we are using mutual information to prune a given lexicon. That is, instead of building a lexicon from scratch, we first add all possible words and then use mutual information to prune away illegal words after training with EM. Hence the statistics we use for calculating mutual information are more reliable than those directly obtained from corpus by frequency counting. Another difference is that we are using a probabilistic splitting scheme that sometimes just shifts probability between words, instead of completely discarding words.

7   Conclusion and Future Work

This paper describes a new unsupervised method for discovering Chinese words from an unsegmented corpus. Combined with an efficient mutual information based lexicon pruning scheme, we achieve competitive results.

However, there is much work left to be done. One problem is that we cannot detect the hierarchical structure of Chinese compounds, which is very useful in many NLP tasks. We are currently exploring a hierarchical unsupervised method to detect Chinese compounds. Also, instead of assuming the independence of each word, we should also consider context information, which has proven to be helpful [12]. Another problem with our self-supervised training procedure is that it puts equal weight on the core and candidate lexicons. One interesting idea would be to automatically estimate the weights of the two lexicons by using a mixture model.

8 Acknowledgments

We would like to thank Ge, Xianping [8] for kindly supplying us with the training corpus. Research supported by Bell University Labs, MITACS and NSERC.

References

1. Ando, R. and Lee, L.; Mostly-Unsupervised Statistical Segmentation of Japanese: Application to Kanji. ANLP-NAACL, 2000.
2. Brand, M.; Structure learning in conditional probability models via an entropic prior and parameter extinction. In Neural Computation, vol.11, page 1155-1182, 1999.
3. Chang, J.-S. and Su, K.-Y.; An Unsupervised Iterative Method for Chinese New Lexicon Extraction. International Journal of Computational Linguistics & Chinese Language Processing, 1997.
4. Dahan, D. and Brent, M.; On the discovery of novel word-like units from utterances: An artificial-language study with implications for native-language acquisition. Journal of Experimental Psychology: General, 128, 165-185, 1999.
5. Deligne, S. and Bimbot, F.; Language Modeling by Variable Length Sequences: Theoretical Formulation and Evaluation of Multigrams. ICASSP, 1995.
6. Dempster, A., Laird, N, and Rubin, D.; Maximum-likelihood from incomplete data via the EM algorithm. J. Royal Statist. Soc. Ser. B., 39, 1977.
7. Fung, P.; Extracting key terms from Chinese and Japnese text. The International Journal on Computer Processing of Oriental Language, Special Issue on Information Retrieval on Oriental Languages, 1998, 99-121.
8. Ge, X., Pratt, W. and Smyth, P.; Discovering Chinese Words from Unsegmented Text. SIGIR-99, pages 271-272.
9. Jin, W.; Chinese Segmentation and its Disambiguation. MCCS-92-227, Computing Research Laboratory, New Mexico State University, Las Cruces, New Mexico.
10. Manning, C. and Schütze, H.; Foundations of Statistical Natural Language Processing. MIT Press, Cambridge, Massachusetts, 1999, pages 66-68.
11. Palmer, D. and Burger, J.; Chinese Word Segmentation and Information Retrieval. AAAI Spring Symposium on Cross-Language Text and Speech Retrieval, Electronic Working Notes, 1997.
12. Ponte, J. and Croft, W.; Useg: A retargetable word segmentation procedure for information retrieval. Symposium on Document Analysis and Information Retrival 96 (SDAIR).
13. Rabiner, L.; A Tutorial on Hidden Markov Models and Selected Applications in Speech Recognition. Proceedings of IEEE, Vol.77, No.2, 1989.
14. Sproat, R., Shih, C., Gale, W. and Chang, N.; A stochastic finite-state word-segmentation algorithm for Chinese Computational Linguistics, 22 (3), 377-404, 1996.

Analyzing Data Clusters: A Rough Sets Approach to Extract Cluster-Defining Symbolic Rules

Syed Sibte Raza Abidi, Kok Meng Hoe, and Alwyn Goh

School of Computer Science, Universiti Sains Malaysia, 11800 Penang, Malaysia
{sraza, kmhoe, alwyn}@cs.usm.my

Abstract. We present a strategy, together with its computational implementation, to intelligently analyze the internal structure of inductively-derived data clusters in terms of symbolic cluster-defining rules. We present a symbolic rule extraction workbench that leverages *rough sets* theory to inductively extract CNF form symbolic rules from un-annotated continuous-valued data-vectors. Our workbench purports a hybrid rule extraction methodology, incorporating a sequence of methods to achieve data clustering, data discretization and eventually symbolic rule discovery via rough sets approximation. The featured symbolic rule extraction workbench will be tested and analyzed using biomedical datasets.

1 Introduction

Data clustering is a popular data analysis task that involves the distribution of 'un-annotated' data (i.e. with no a priori class information), in an inductive manner, into a finite sets of categories or clusters such that data items within a cluster are more similar in some respect and unlike those from other clusters. Notwithstanding the efficacy of traditional data clustering techniques, it can be argued that the outcome of a data clustering exercise does not necessarily explicates the intrinsic relationship between the various attributes of the dataset. From an intelligent data analysis perspective, the availability of (inductively-derived) cluster-defining knowledge, most preferably in a symbolic formalism such as deductive rules, can help identify both the structure of the emerged clusters and the cluster membership principles.

The featured work is motivated by the desirability of deriving cluster-defining knowledge for a priori defined data clusters [1]. We present a multi-strategy approach for the automated extraction of cluster-defining *Conjunctive Normal Form* (CNF) symbolic rules from un-annotated data-sets. The motivation for our work stems from the individual effectiveness of various data analysis mechanisms: (1) cluster formation via unsupervised clustering algorithms, (2) dataset simplification and attribute selection via attribute discretization, and (3) symbolic rule extraction via rough sets approximation [2]. The interestingness of the rough sets approach is primarily their capability to generate deterministic

F. Hoffmann et al. (Eds.): IDA 2001, LNCS 2189, pp. 248–257, 2001.

and *non-deterministic* rules of varying granularity and strength. We have implemented a generic *Symbolic Rule Extraction Workbench* (see Fig. 1) that can generate cluster-defining symbolic rules from continuous-valued data, such that the emergent rules are directly applicable to rule-based systems [3].

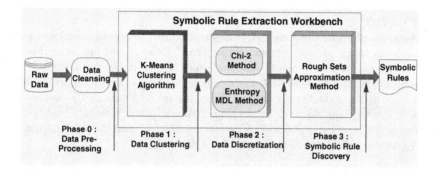

Fig. 1. The Functional Architecture of the Symbolic Rule Extraction Workbench

2   Rough Sets: A Brief Overview

The main objective of rough sets mediated data analysis is to form approximate concepts about the data based on available classification or decision information [2]. These data-defining approximate concepts generated via rough sets analysis are typically represented as succinct symbolic rules that provide an explanation about inter-attribute dependencies, attribute importance and topology-defining information vis-a-vis an annotated dataset.

In a rough sets framework, annotated data is represented as a decision system, Γ defined as a pair $\Gamma = (U, A)$, where U is a finite set of objects called the *universe* and A is a finite set of *attributes*. For every $a \in A$, $a : U \rightarrow V_a$, where $V_a \neq \emptyset$ is called the *values set* of a. Attributes in A are divided into two disjoint sets, $A = C \cup \{d\}$, where $\{d\} \notin C$ is the singular decision or class attribute and C is the set of condition attributes. Hence Γ can be denoted as $\Gamma = (U, C \cup \{d\})$. Rough sets based data analysis, leading to symbolic rule extraction involves the following processing steps:

- Definition of an approximate space by finding *indiscernibility relations* between objects in the universe. Two objects, $x, y \in U$ are indiscernible when they are equivalent with regards to their attributes and values. With any subset $B \subseteq A - \{d\}$ or $B \subseteq C$, an *indiscernibility relation* $I(B)$ partitions the universe into separate sets called *equivalence classes*, $[x]_{I(B)}$, which denotes the set of all objects equivalent to x in terms of all attributes in B.

The indiscernibility relations are used to reduce the size of the universe such that only a single element of the equivalence class is required to represent the characteristics of the entire class.

- Concept approximation is achieved through data reduction–i.e by retaining the minimum subset of attributes that can differentiate all equivalence classes in the universe. Such minimum subsets are called *reducts*, and are computed by eliminating attribute a when there exist $B' \subseteq B$ where $a \in B, a \notin B'$ and $I(B') = I(B)$.
- A decision rule provides a definitive description of the concepts within the universe in terms of a statement of the form "if *Conditions* are True then *Outcomes* are True". Symbolic decision rules can next be synthesized by superimposing the reducts with the decision system–i.e. by taking for each attribute in the reduct its corresponding value from objects in the dataset together with their decision values. For instance, given a reduct $\{a_1, a_3\}$, where $a_i \in C, i = 1, \ldots, |C|$ and the values set of a_i, V_a^i and the values set of D, V_d, where $d \in D$, the decision rule for Γ is

$$\text{IF } a_1 = V_a^1(j) \text{ and } a_3 = V_a^3(j) \text{ THEN } D = V_d(j), \quad \text{for } j = 1, 2, \ldots, |U| \ . \quad (1)$$

Finally, the decision rules are assessed in terms of their measure of *accuracy*–i.e. how well they perform in predicting the class or outcome of new data patterns.

2.1   The Efficacy of Rough Sets for Rule Discovery

We argue that rough sets is an efficient, non-invasive approach to rule discovery, as they use information provided by the data as opposed to assuming any static model about the data. Advantages of rough sets for rule extraction are:

- Decision tree based algorithms implicitly assume the availability of class-defining information in the data. The drawback being that *imprecise* objects–i.e. objects with the same description but of different classes–are treated as *noise* and hence discarded. On the contrary, rough set methods not only allow for the generation of *non-deterministic* rules from an entire 'un-annotated' dataset, but in addition they also associate each rule with a probability determined by the data itself.
- The rule synthesis process in rough sets performs *automatic* attribute reduction–i.e. a minimal number of attributes that are necessary to preserve the indiscernibility relation among objects are retained.
- Rough set methods can generate rules of varying granularity (length) and quality, in order to fulfill the requirements of different domains.

3   Extracting Cluster-Defining Rules: A Multi-strategy Approach

In our work, we intend to extract symbolic rules from un-annotated datasets comprising an undifferentiated collection of continuous-valued multi-component data-vectors $S = \{X_i : i \in [1, n]\}$, for which the classification attribute $c(X_i) = \alpha$ for $\alpha \in [1, k]$ is unknown. We have postulated a multi-strategy approach that systematically transforms un-annotated data-sets to deductive symbolic rule-sets via a sequence of phases, as described below:

3.1   Phase 1 - Data Clustering

Given an un-annotated dataset satisfying the above assumption, we first partition it into k clusters using the popular *K-Means* data clustering algorithm [5]. In practice, the clustering algorithm inductively derives the class information and next partitions the dataset into k number of clusters. The discovered clusters form the basis for subsequent discovery of symbolic rules that define the structure of the discovered clusters.

3.2   Phase 2 - Data Discretisation

The motivation for this phase is driven by the fact that ordinal or continuous valued attributes are proven to be rather unsuitable for the extraction of concise symbolic rules. Henceforth, the necessity to discretise continuous-valued attributes to discrete intervals–i.e. reduce the domain of values of an attribute to a small number of attribute-value ranges–where each interval can be represented by a label/token. More attractively, the discretization phase not only reduces the complexity and volume of the dataset, but also serves as a attribute filtering mechanism, whereby attributes that have minimum impact vis-a-vis classification are eliminated. We employ two data discretisation methods: (1) statistical discretization via Chi-2 [6] and (2) class information entropy reduction via MDL partitioning [7]; their respective results provide for an interesting contrast.

3.3   Phase 3 - Symbolic Rule Discovery

We use rough sets approximation–an interesting alternative to a variety of symbolic rule extraction methods [8,9]–to derive cluster-defining symbolic rules. We have devised a three step methodology for the generation of symbolic rules from un-annotated data.

Step 1: Construction of Dynamic Reducts. First, we randomly partition the discretized data into two disjoint sets: a *training set* (70% of the dataset) and a *testing set* (30% of the dataset). Next, we create 6 sub-samples of the training dataset by selecting 2 different random samples each comprising 90%, 80% and 70% of the training data. The rationale is to provide multiple perspectives of the training data in order to find frequently occurring reducts, i.e. dynamic reducts [10].

Finding minimal reducts is an NP-hard problem [10,11], hence we apply *genetic algorithms* to meet that end. Our work closely follows that of Wróblewski [11], in that individuals or *chromosomes* are represented as bit strings of length equal or less than the number of attributes. A *distinction table* is being constructed to transform the discernibility relations of pairs of different objects into bit strings for the initial population. We use the following fitness function:

$$F(r) = \frac{m - L_r}{m} + \frac{C_r}{K} \, , \tag{2}$$

where r = chromosome or possible reduct; m = total number of attributes; L_r = number of 1's in r; C_r = number of object pairs that r can discern between; K = total number of different pairs of objects i.e. $\frac{(n^2 - n)}{2}$; n = total number of objects in Γ. The first part of $F(.)$ rewards the possible reduct, r for being short (few 1's) and the second part determines the extent in which r can discern between all other pairs of objects.

For each sample the fittest chromosomes are selected as the minimal reduct set. Next, we accumulate the minimal reduct-sets (from the multiple data sub-samples) and proceed to select dynamic reducts–i.e. those reducts that have a high frequency of occurrence across the cumulative reduct-sets.

Step 2: Generate Symbolic Rules. We next proceed to generate symbolic rules from the set of dynamic reducts. We prefer to generate symbolic rules from the shortest possible length dynamic reducts; the reason being that shorter length dynamic reducts have been shown to yield concise rule-sets with high accuracy and better generalization capabilities [4]. Our rule generation strategy therefore involves: (1) the selection of dynamic reducts that have a short length and (2) the generation of rules that satisfy a user-defined accuracy level. Our strategy for generating symbolic rules is as follows:

1. Specify an acceptable minimum accuracy level for the rule-set.
2. Generate 6 random sub-samples as described in section 3.3.1.
3. Find dynamic reducts from the sub-samples and place in set DR. Note that DR will comprise reducts with varying lengths.
4. From the reducts in DR determine the shortest reduct length, SRL.
5. From DR, collect all reducts with length equal to SRL & put them in set $SHRED$.
6. Generate symbolic rules from the reducts placed in $SHRED$.
7. Determine the overall accuracy of the generated rules on the test data.
8. IF Overall accuracy is lower than the minimum accuracy level AND exist reducts in the DR set with length $> SRL$
   THEN Empty $SHRED$ & Update SRL to the next shortest reduct length in DR & Repeat from 5.
   ELSE Symbolic rules with the desired accuracy level cannot be generated.

At the conclusion of the above rule generation strategy we will ideally have a non-empty $SHRED$ that will contain reducts yielding a set of rules, $SHRED\_RULS$ that satisfy the acceptable accuracy level.

Step 3: Rule Filtering. The rule-set obtained from Step 2 (see Sect. 3.3.2) can be significantly large and can contain 'weak' rules. Hence, we next filter the rule-set, *SHRED_RULS*, to eliminate weak rules in order to achieve compact cluster-defining knowledge.

We define the rule filtering problem as follows: *Given a set of rules generated from a training set, find the least number of high quality rules of the rule-set that fulfils a pre-specified testing accuracy requirement.*

In our work, a rule is represented by an implication, $\alpha \rightarrow \beta$, which is true only when, both the conditions, α and the outcome β, are true. *Support, Consistency* and *Coverage* are key quality defining metrics for the rules. Support of a rule is the number of objects in the training set that matches the condition(s), outcome(s) or both. Consistency indicates how reliable or accurate a rule is. Coverage denotes how prevalent or powerful is the rule. Michalski [12] suggests that high consistency and coverage are requirements of good decision rules. Consistency and coverage are defined as,

$$\text{Consistency}(\alpha \rightarrow \beta) = \frac{\text{Support}(\alpha \cap \beta)}{\text{Support}(\alpha)} \ , \tag{3}$$

$$\text{Coverage}(\alpha \rightarrow \beta) = \frac{\text{Support}(\alpha \cap \beta)}{\text{Support}(\beta)} \ . \tag{4}$$

Numeric qualities of consistency and coverage can be combined into a single function or quality index (QI) for a rule. We employ the function of Brazdil & Torgo [13] which was derived based on empirical evidence and defined as,

$$\text{Quality}_{\text{Brazdil}}(\text{rule}) = \text{Consistency}(\text{rule}) \cdot e^{\text{Coverage}(\text{rule})-1} \ . \tag{5}$$

Our rule filtering heuristic employs a search algorithm guided by the QI median-the 50th percentile of the cumulated frequency distribution of QI values of the rules. The proposed heuristic tends to eliminate rules that have low QI values and hence do not contribute to sustaining a user-specified minimum accuracy level. In our work, the cut-off value for rule elimination is determined by the median of the QI.

4   Experimental Results

Experiments on several data-sets obtained from [14] were conducted. These data-sets were chosen for two reasons: (1) all their data vector components being continuous-valued and (2) all the class-subsets are well-separated–i.e. with inter-mean distances fairly large compared to the radii. Due to space constraints we will only present the results for the *Wisconsin Breast Cancer* (WBC) dataset here. The WBC dataset used comprised 699 patterns, divided into 2 classes–0=benign and 1=malignant. Each data-vector included 9 attributes: (1) Clump thickness, (2) Uniformity of cell size, (3) Uniformity of cell shape, (4) Marginal adhesion, (5) Single epithelial cell size, (6) Bare nuclei, (7) Bland chromatin, (8) Normal nucleoli, (9) Mitoses.

4.1   Phase 1: Data Clustering Using K-Means Clustering Algorithm

Prior to clustering the actual classification information is removed–i.e. we work with an un-annotated dataset. The K-Means algorithm is used to inductively cluster the data patterns. Upon completion of the clustering process the members of each cluster are associated with their respective class label (see Table 1).

Table 1. Results of K-Means clustering for the WBC dataset

CLASS	ACTUAL CLASS DISTRIBUTION (%)	K-MEANS CLUSTER DISTRIBUTION (%)	CLUSTERING ACCURACY (%)
Benign	65.01	66.33	96.1
Malignant	34.99	33.67	

4.2   Phase 2: Data Discretization

After the successful clustering of the dataset, we employ Chi-2 discretization [6] to (1) discretise the continuous data values into meaningful intervals; and (2) perform attribute elimination. Table 2 shows the discretization results. Note that two attributes have been eliminated and about 83% data reduction was achieved, thus resulting in a much smaller dataset for symbolic rules generation.

Table 2. Chi-2 discretization of the WBC dataset. $[x, y)$ means $\geq x$ but $< y$.

Attributes	Clump Thickness	Uniformity of Cell Size	Uniformity of Cell Shape	Marginal Adhesion	Single Epithelial Cell Size
# Intervals	2	2	2	0	2
Interval Value	1 = [_ , 7.0) 2 = [7.0, _)	1 = [_ , 3.0) 2 = [3.0, _)	1 = [_ , 3.0) 2 = [3.0, _)	Attribute Eliminated	1 = [_ , 3.0) 2 = [3.0, _)
Attributes	Bare Nuclei	Bland Chromatin	Normal Nucleoli	Mitoses	
# Intervals	3	2	2	0	
Interval Value	1 = [_ , 2.0) 2 = [2.0, 8.0) 3 = [8.0, _)	1 = [_ , 4.0) 2 = [4.0, _)	1 = [_ , 3.0) 2 = [3.0, _)	Attribute Eliminated	

4.3   Phase 3: Symbolic Rule Discovery

In this phase symbolic rule-sets of different lengths are generated from the dynamic reducts (see Sects. 3.3.1 & 3.3.2) of the training data for each sample. Based on the rule filtering heuristic (see Sect. 3.3.3), from each rule-set we extract all rules with QI equal or greater than the first 50th percentile of the QI frequency distribution. The QI of each rule is calculated using Brazdil's quality function (see (5)). The eventual rule-set to be used by the user can be selected as follows: (a) selecting the native rule-set derived from a sample x, most likely the rule-set with the highest accuracy; (b) performing an union of all the rules generated across the n samples; or (c) performing an intersection of all the rules generated across the n samples. We present the results from 3 samples of the WBC dataset.

In Table 3, we show the performance of three rule-sets with different antecedent lengths of 5 and 6 conditions. Table 4 shows an intersection of the symbolic rules (length=5) originating from three different rule-sets.

Table 3. Testing accuracy of 3 rule-sets generated from 3 samples of the WBC dataset including the union and the intersection of the rule-sets.

Rule-set	Rule Length	# Rules	Test Data Accuracy (%)
$s1$	5	119	95.12
$s1$	6	122	94.63
$s2$	5	67	94.15
$s2$	6	138	94.15
$s3$	5	109	95.12
$s3$	6	166	95.61
$s1 \cap s2 \cap s3$	5	186	96.10
$s1 \cap s2 \cap s3$	6	226	96.59
$s1 \cup s2 \cup s3$	5	15	87.32
$s1 \cup s2 \cup s3$	6	55	91.71

4.4   Experimental Observations

- Our method yields more rules of longer antecedent (or LHS). It may be noted that rule-sets of length 6 have more rules than rule-sets of length 5, although both share almost equal predictive accuracy on the test data. It is our belief that for practical purposes, rules with a longer LHS length are a better model of the dataset as they encode more complex relationships between data attributes.
- The union of the rule-sets $s1 \cup s2 \cup s3$ do not show a significant increase in accuracy compared to the individual rule-sets of similar lengths. We note that despite the increased size of the union rule-set–186 rules of length 5 and 226 rules of length 6–the accuracy improves by not more than 2%.
- The intersection of the rule-sets $s1 \cap s2 \cap s3$ yields a core, yet small, set of rules with reasonably high accuracy. The intersected rule-set comprises less than 10% (length 5) and 25% (length 6) of the corresponding union rule-set.
- There is a direct correlation between the number of rules and the accuracy level–as the number of rules increases in a rule-set the corresponding accuracy level also increases. Depending on the decision-support task, one can determine the acceptable accuracy level and choose the optimum rule-set.

Note that despite the higher accuracy of the larger-sized union rule-set, the smaller-sized individual rule-sets exhibit a reasonably high accuracy level.

Table 4. Rule-set $s1 \cap s2 \cap s3$ generated for the WBC dataset. The legend is: bn = bare_nuclei, bc = bland_chromatin, nn = normal_nucleoli, sez = sing_epi_cell_sz, ucp = uni_cell_shape; Cls = Class, Sup = Support, Con = Consistency, Cov = Coverage.

No	Attributes					Cls	Sup	Con	Cov
1	bn(1)	bc(1)	nn(1)	sez(1)	ucp(1)	0	225.75	1	0.7196
2	bn(3)	bc(2)	nn(2)	sez(2)	ucp(2)	1	65	1	0.3958
3	bn(2)	bc(2)	nn(2)	sez(2)	ucp(2)	1	36	1	0.2194
4	bn(3)	bc(2)	nn(1)	sez(2)	ucp(2)	1	19.25	1	0.1174
5	bn(2)	bc(1)	nn(1)	sez(1)	ucp(1)	0	18.25	1	0.0581
6	bn(3)	bc(1)	nn(2)	sez(2)	ucp(2)	1	10.5	1	0.0638
7	bn(1)	bc(1)	nn(1)	sez(1)	ucp(2)	0	18.75	1	0.0598
8	bn(1)	bc(1)	nn(1)	sez(2)	ucp(1)	0	12.25	1	0.0389
9	bn(3)	bc(2)	nn(2)	sez(1)	ucp(2)	1	5.5	1	0.0334
10	bn(3)	bc(2)	nn(1)	sez(1)	ucp(2)	1	2.75	1	0.0167
11	bn(3)	bc(1)	nn(2)	sez(1)	ucp(2)	1	1.25	1	0.0077
12	bn(2)	bc(1)	nn(2)	sez(1)	ucp(2)	0	2.5	1	0.0079
13	bn(1)	bc(1)	nn(2)	sez(1)	ucp(1)	0	4	1	0.0128
14	bn(2)	bc(1)	nn(1)	sez(2)	ucp(1)	0	5.5	1	0.0176
15	bn(3)	bc(1)	nn(1)	sez(2)	ucp(2)	1	2.5	1	0.0151

5   Concluding Remarks

We conclude that the systematic application of rough sets proves to be an effective tool for symbolic rule discovery as an alternate to traditional methods. More attractively, our experimental results may indicate that the rule-sets derived via the said rough sets approach have a classification accuracy greater than 90% (see Table 3), which indeed is a good measure of the efficacy of rough sets for symbolic rule extraction. In addition, we noted some other interesting observations:

1. The intersection and union operations on rules extracted through the rough sets framework provides an interesting exploration method of the rule-sets derived from different samples of a dataset (see Sect. 4.3). The intersection operator can be used to find the most common, and hence, more powerful and representative rules. In experiments with the WBC dataset, the intersection of rule-sets yielded only 15 rules with an accuracy of 87.3% in contrast to the average of relative accuracy of 95.12%, 94.15%, 95.12% for the individual samples $s1$, $s2$ and $s3$, respectively.

2. The rough sets approach favors the generation of rules with shorter lengths. In fact, our experiments show that, when comparing two rule-sets of length n

and $n-1$, the smaller rule-set with shorter rules achieved the same accuracy level as the larger and longer rule-set–i.e. concise rules are as accurate as long rules, which is a desirable effect for knowledge representation purposes.

3. As the number of rules eliminated from a rule-set increases, accuracy of the rule-set naturally reduces. This is in accordance with the theoretical assumptions i.e. as the number of rules reduces, less predictive power is available, hence the lower accuracy of the rule-set.

In conclusion we will like to point out that the proposed sequential application of multiple techniques–i.e. data clustering, data discretization and feature selection and finally, rough sets approximation–for knowledge extraction via symbolic rule generation, appears to be a pragmatic approach for the intelligent analysis of un-annotated data-vectors with continuous-valued attributes.

References

1. Agrawal R., Gehrke J., Gunopulos D., Raghavan P.: Automatic subspace clustering of high dimensional data for data mining applications. Proc. ACM-SIGMOD Int. Conf. Management of Data, Seattle, Washington (1998)
2. Pawlak, Z.: Rough Sets. In: Lin T.Y., Cercone N.(eds.): Rough Sets and Data Mining: Analysis of Imprecise Data. Kluwer Academic Publishers, Dordrecht (1997)
3. Abidi, S.S.R., Goh, A., Hoe, K.M.: Specification of Healthcare Expert Systems Using a Multi-Mechanism Rule Extraction Pipeline. Proc. Int. ICSC Congress on Intelligent Systems and Applications, Sydney (2000)
4. Abidi, S.S.R., Hoe, K.M., Goh, A.: Healthcare Simulation Model Specification Featuring Multi-Stage Neural Network Rule Extraction. Proc. 4th Int. Eurosim Congress, Netherlands (2001)
5. Bottou L., Bengio, Y.: Convergence Properties of the K-Means Algorithms. Proc. 7th Int. Conf. on Neural Information Processing Systems, Denver (1994)
6. Liu, H., Setiono, R.: Chi2: Feature Selection and Discretization of Numeric Attributes. Proc. 7th Int. Conf. on Tools with AI, Washington D.C (1995)
7. Kohavi, R., Sahami, M.: Error-based and Entropy-based Discretization of Continuous Features. Proc. Int. Conf. on Knowledge Discovery and Data Mining (1996)
8. Quinlan, R.: C4.5: Programs for Machine Learning. Morgan Kaufmann, CA (1993)
9. Tickle A., Andrews R., Golea M., Diederich J.: The Truth Will Come To Light: Directions and Challenges in Extracting the Knowledge Embedded Within Trained Artificial Neural Networks. In: IEEE Trans. on Neural Networks 9(6) (1998)
10. Bazan, J.G, Skowron, A.J., Synak, P.: Discovery of Decision Rules from Experimental Data. Proc. Int. W'shop on Rough Sets and Soft Computing, CA (1994)
11. Wróblewski, J.: Finding Minimal Reducts using Genetic Algorithms. Proc. 2nd Annual Joint Conf. on Information Sciences, Wrightsville Beach, NC. USA (1995)
12. Michalski, R.S.: A Theory and Methodology of Inductive Learning. In: Michalski, R.S., Carbonell, J.G. & Mitchell, T.M. (eds): Machine Learning, An Artificial Approach. Tioga Publishing, Palo Alto (1983)
13. Brazdil, P., Torgo, L.: Knowledge Acquisition Via Knowledge Integration. In: Current Trends in Knowledge Acquisition, IOS Press (1990)
14. Blake, C.L., Merz, C.J.: UCI Repository of machine learning databases (http://www.ics.uci.edu/~mlearn/MLRepository.html). Uni. of California Irvine.

Finding Polynomials to Fit Multivariate Data Having Numeric and Nominal Variables

Ryohei Nakano[1] and Kazumi Saito[2]

[1] Nagoya Institute of Technology
Goriso-cho, Showa-ku, Nagoya 466-8555 Japan
nakano@ics.nitech.ac.jp
[2] NTT Communication Science Laboratories
2-4 Hikaridai, Seika, Soraku, Kyoto 619-0237 Japan
saito@cslab.kecl.ntt.co.jp

Abstract. This paper proposes a new method for finding polynomials to fit multivariate data containing numeric and nominal variables. Each polynomial is accompanied with the corresponding nominal condition stating when to apply the polynomial. Such a nominally conditioned polynomial is called a rule. A set of such rules can be regarded as a single numeric function, and such a function can be closely approximated by a single three-layer neural network. After training single neural networks with different numbers of hidden units, the method selects the best trained network, and restores the final rules from it. Experiments using three data sets show that the proposed method works well in finding very succinct and interesting rules, even from data containing irrelevant variables and a small amount of noise.

1 Introduction

Finding a numeric relationship, e.g., Kepler's third law $T = kr^{3/2}$, from data is one important issue of computer-assisted data analysis. In AI field, the BACON systems [3] and many variants [2,7,13] have employed a recursive combination of two variables to find a polynomial-type relationship. This combinatorial approach, however, suffers from combinatorial explosion in search and lack of robustness. As an alternative, we have investigated a connectionist approach that employs a three-layer neural network to find a polynomial-type nonlinear relationship which fits multivariate data having only numeric variables[9,6].

In many real fields, however, data contains nominal variables as well as numeric variables. For example, Coulomb's law $F = 4\pi\epsilon q_1 q_2 / r^2$ relating the force of attraction F of two particles with charges q_1 and q_2, respectively, separated by a distance r depends on ϵ, the permittivity of surrounding medium; i.e., if substance is *"water"* then $F = 8897.352 q_1 q_2 / r^2$, if substance is *"air"* then $F = 111.280 q_1 q_2 / r^2$, and so on. Each polynomial is accompanied with the corresponding nominal condition stating when to apply it. In this paper we consider finding a set of such pairs to fit multivariate data having numeric and nominal variables. Hereafter such a pair is referred as a *rule*.

F. Hoffmann et al. (Eds.): IDA 2001, LNCS 2189, pp. 258–267, 2001.

To solve this type of problem, a "divide-and-conquer" approach may be a very natural one. For example, ABACUS [2] basically repeats the following: a numeric function describing some subset is sought by using numeric variables, a nominal condition for discriminating the subset is calculated by using nominal variables, and all the examples in the subset are removed. Clearly, in this approach the fitting reliability for a small subset will be necessarily degraded.

As an alternative approach, we propose a new connectionist method called *RF6.2* [1], where neural networks learning is employed and all training examples are used at once in the learning. Section 2 formalizes the background and shows the basic framework of our approach. Section 3 explains the first half of RF6.2, i.e., training neural networks with a smart regularizer. Section 4 explains the second half, i.e., a rule restoring procedure which finds coefficient representatives and generates nominal conditions to get the final rules. Section 5 reports experimental results using one artificial and two real data sets.

2   Basic Framework

Let $(q_1, \cdots, q_{K_1}, x_1, \cdots, x_{K_2}, y)$ or $(\boldsymbol{q}, \boldsymbol{x}, y)$ be a vector of variables describing an example, where q_k is a nominal explanatory variable[2], x_k is a numeric explanatory variable and y is a numeric target variable. Here, by adding extra categories, if necessary, without losing generality, we can assume that q_k exactly matches the only one category. Therefore, for each q_k we introduce a *dummy variable* expressed by q_{kl} as follows.

$$q_{kl} = \begin{cases} 1 & \text{if } q_k \text{ matches the } l\text{-th category} \\ 0 & \text{otherwise} \end{cases}$$

Here $l = 1, \cdots, L_k$, and L_k is the number of distinct categories appearing in q_k.
As a true model governing data, we consider the following set of rules.[3]

$$if \bigwedge_{q_{kl} \in Q^i} q_{kl} = 1 \quad then \quad y = y(\boldsymbol{x}; \boldsymbol{\Theta}^i) + \varepsilon, \quad i = 1, \cdots, I^* \tag{1}$$

where Q^i and $\boldsymbol{\Theta}^i$ denote a set of dummy variables and a parameter vector respectively used in the i-th rule. Moreover, ε is a noise term and I^* is the number of rules. As a class of numeric equations $y(\boldsymbol{x}; \boldsymbol{\Theta})$, we consider polynomials, whose power values are not restricted to integers, expressed by

$$y(\boldsymbol{x}; \boldsymbol{\Theta}) = w_0 + \sum_{j=1}^{J^*} w_j \prod_{k=1}^{K_2} x_k^{w_{jk}} = w_0 + \sum_{j=1}^{J^*} w_j \exp\left(\sum_{k=1}^{K_2} w_{jk} \ln x_k\right). \tag{2}$$

---

[1] RF6.2 denotes *Rule extraction from Facts, version 6.2.*
[2] An explanatory variable means an independent variable.
[3] Here a rule "if A then B" means "when A holds, apply B."

Here, each parameter w_j or w_{jk} is a real number, and J^* is an integer corresponding to the number of terms. Θ is a vector constructed by arranging parameters w_j and w_{jk}. Note that Eq. (2) can be regarded as a feedforward computation of a three-layer neural network [9].

This paper adopts the following basic framework: firstly, single neural networks are trained by changing the number of hidden units; secondly, among the trained neural networks selected is the one having the best criterion value; finally, a rule set is restored from the best trained neural network. The framework is explained in more detail in the remaining part of the paper.

To express a nominal condition numerically, we introduce a function $c(\boldsymbol{q}; \boldsymbol{V})$ defined by

$$c(\boldsymbol{q}; \boldsymbol{V}) = \exp\left(\sum_{k=1}^{K_1}\sum_{l=1}^{L_k} v_{kl} q_{kl}\right), \tag{3}$$

where $\boldsymbol{V}$ denotes a vector of parameters v_{kl}. For a nominal condition defined by Q, consider the values of $\boldsymbol{V}$ as follows:

$$v_{kl} = \begin{cases} 0 & if \ q_{kl} \in Q, \\ -\beta & if \ q_{kl} \notin Q, \ q_{kl'} \in Q \ for \ some \ l' \neq l, \\ 0 & if \ q_{kl'} \notin Q \ for \ any \ l', \end{cases} \tag{4}$$

where $\beta \gg 0$. Then $c(\boldsymbol{q}; \boldsymbol{V})$ is almost equivalent to the truth value of the nominal condition defined by Q, thus we can see that the following formula can closely approximate the final output value defined by Eq. (1). Thus, a set of rules can be merged into a single numeric function.

$$F(\boldsymbol{q}, \boldsymbol{x}; \boldsymbol{V}^1, \cdots, \boldsymbol{V}^{I^*}, \Theta^1, \cdots, \Theta^{I^*}) = \sum_{i=1}^{I^*} c(\boldsymbol{q}; \boldsymbol{V}^i) \, y(\boldsymbol{x}; \Theta^i). \tag{5}$$

On the other hand, with an adequate number J, the following can completely express Eq. (5).

$$y(\boldsymbol{q}, \boldsymbol{x}; \Theta) = w_0 + \sum_{j=1}^{J} w_j \exp\left(\sum_{k=1}^{K_1}\sum_{l=1}^{L_k} v_{jkl} \, q_{kl} + \sum_{k=1}^{K_2} w_{jk} \ln x_k\right), \tag{6}$$

where Θ is rewritten as an M-dimensional vector constructed by arranging all parameters w_j, v_{jkl} and w_{jk}. Therefore, a set of rules can be learned by using a single neural network defined by Eq. (6). Note that even when Eq. (6) closely approximates Eq. (1), the weights for the nominal variables are no more limited to such weights as those defined in Eq. (4). Incidentally, Eq. (6) can be regarded as the feedforward computation of a three-layer neural network.

3 Learning Neural Networks with Regularizer

Let $D = \{(\boldsymbol{q}^\mu, \boldsymbol{x}^\mu, y^\mu) : \mu = 1, \cdots, N\}$ be training data, where N is the number of examples. We assume that each training example $(\boldsymbol{q}^\mu, \boldsymbol{x}^\mu, y^\mu)$ is independent and identically distributed. Our goal in learning with neural networks is defined as a problem of minimizing the generalization error, that is, finding [4] the optimal estimator $\boldsymbol{\Theta}^*$ that minimizes

$$\mathcal{G}(\boldsymbol{\Theta}) = E_D E_T \left(y^\nu - y(\boldsymbol{q}^\nu, \boldsymbol{x}^\nu; \boldsymbol{\Theta}(D))\right)^2, \tag{7}$$

where $T = (\boldsymbol{q}^\nu, \boldsymbol{x}^\nu, y^\nu)$ denotes test data independent of the training data D. The *least-squares estimate* $\widehat{\boldsymbol{\Theta}}$ minimizes the error sum of squares

$$\mathcal{L}(\boldsymbol{\Theta}) = \frac{1}{2} \sum_{\mu=1}^{N} \left(y^\mu - y(\boldsymbol{q}^\mu, \boldsymbol{x}^\mu; \boldsymbol{\Theta})\right)^2. \tag{8}$$

However, for a small D this estimation is likely to overfit the noise of D when we employ nonlinear models such as neural networks. Thus, for a small training data by using Eq. (8) as our criterion we cannot obtain good results in terms of the generalization performance defined in Eq. (7) [1].

It is widely known that adding some penalty term to Eq. (8) can lead to significant improvements in network generalization [1]. To improve both the generalization performance and the readability of the learning results, we adopt a method, called the *MCV (Minimum Cross-Validation) regularizer* [11], to learn a distinct penalty factor for each weight as a minimization problem over the cross-validation error. Let $\boldsymbol{\Lambda}$ be an M-dimensional diagonal matrix whose diagonal elements are penalty factors λ_m, and $\boldsymbol{a}^T$ denotes a transposed vector of $\boldsymbol{a}$. Then, finding a numeric function subject to Eq. (6) can be defined as the following learning problem in neural networks, that is, finding the $\boldsymbol{\Theta}$ that minimizes the following objective function

$$\mathcal{E}(\boldsymbol{\Theta}) = \mathcal{L}(\boldsymbol{\Theta}) + \frac{1}{2}\boldsymbol{\Theta}^T \boldsymbol{\Lambda} \boldsymbol{\Theta}. \tag{9}$$

In order to efficiently and constantly obtain good learning results, we employ a second-order learning algorithm called *BPQ* [10]. The BPQ algorithm adopts a quasi-Newton method [5] as a basic framework, and calculates the descent direction on the basis of a partial BFGS (Broyden-Fletcher-Goldfarb-Shanno) update and then efficiently calculates a reasonably accurate step-length as the minimal point of a second-order approximation. In our experiments [12], the combination of the squared penalty and the BPQ algorithm drastically improves the convergence performance in comparison to other combinations and at the same time brings about excellent generalization performance.

4   Rule Restoring

Assume that we have already selected the best neural network from the trained ones which have different numbers of hidden units. In order to get the final rules as described in Eq. (1), the following procedure is proposed.

---

[4] Here the network structure is not fixed; i.e., among different numbers of hidden units the best one is selected on the basis of some adequate statistical measure.

rule restoring procedure :

  step 1. For $I = 1, 2, ...$, calculate the optimal I representative vectors which minimize vector quantization (VQ) error.

  step 2. Select the optimal number of representatives $\widehat{I}$ which minimizes cross-validation error.

  step 3. For the optimal $\widehat{I}$, generate nominal conditions to get the final rules.

4.1 Finding Coefficient Representatives

For each training example, we can extract a specific rule. The following c_j^μ is a coefficient of the j-th term in the polynomial for the μ-th training example:

$$c_j^\mu = \widehat{w}_j \exp\left(\sum_{k=1}^{K_1}\sum_{l=1}^{L_k} \widehat{v}_{jkl} q_{kl}^\mu\right), \tag{10}$$

where $\widehat{w}_j$ and $\widehat{v}_{jkl}$ denote the weights of the trained neural network. Coefficient vectors $\{c^\mu = (c_1^\mu, \cdots, c_J^\mu)^T : \mu = 1, \cdots, N\}$ are quantized into representative vectors $\{r^i = (r_1^i, \cdots, r_J^i)^T : i = 1, \cdots, I\}$, where I is the number of representatives. For VQ we employ the k-means algorithm [4] due to its simplicity.

  In the k-means algorithm, all of the coefficient vectors are assigned simultaneously to their nearest representative vectors, and then each representative vector is moved to its region's mean; this cycle is repeated until no further improvement. At last all of the coefficient vectors are partitioned into I disjoint subsets $\{G_i : i = 1, \cdots, I\}$ so that the following sum-of-squares error $\mathcal{VQ}$ is minimized. Here N_i is the number of coefficient vectors belonging to G_i.

$$\mathcal{VQ} = \sum_{i=1}^{I}\sum_{\mu \in G_i}\sum_{j=1}^{J}(c_j^\mu - r_j^i)^2, \text{ where } r_j^i = \frac{1}{N_i}\sum_{\mu \in G_i} c_j^\mu, \tag{11}$$

4.2 Selecting the Optimal Number of Representatives

For given data, since we don't know the optimal number I^* of polynomials in advance, we must evaluate it by changing I. For this purpose we employ the following S-fold *cross-validation*[14]. It divides data D at random into S distinct segments $\{D^s : s = 1, \cdots, S\}$. Then $S - 1$ segments are used for the training, and the remaining one is used for the test. This process is repeated S times by changing the remaining segment. The extreme case of $S = N$ is known as the *leave-one-out* method, which is often used for a small size of data [1].

  Here we introduce a function $i(q)$ that returns the index of the representative vector minimizing the distance, i.e.,

$$i(q^\mu) = \arg\min_i \sum_{j=1}^{J}(c_j^\mu - r_j^i)^2. \tag{12}$$

Then we get the following set of rules using the representative vectors.

$$if \ i(\boldsymbol{q}) = i \quad then \ \ \widehat{y} = \widehat{w}_0 + \sum_{j=1}^{J} r_j^i \prod_{k=1}^{K2} x_k^{\widehat{w}_{jk}}, \quad i = 1, \cdots, I, \qquad (13)$$

Then we get the following cross-validation error. The I minimizing $\mathcal{CV}$ is selected as the optimal number of representatives, $\widehat{I}$.

$$\mathcal{CV} = \frac{1}{N} \sum_{s=1}^{S} \sum_{\nu \in D_s} (y^\nu - \widehat{y}^\nu)^2 . \qquad (14)$$

4.3   Generating Conditional Parts

Finally, the indexing functions $\{i(\boldsymbol{q})\}$ described in Eq. (13) must be transformed into a set of nominal conditions as described in Eq. (1). One reasonable approach is to perform this transformation by solving a simple classification problem whose training examples are $\{(\boldsymbol{q}^\mu, i(\boldsymbol{q}^\mu)) : \mu = 1, \cdots, N\}$, where $i(\boldsymbol{q}^\mu)$ indicates the class label of the μ-th example. Here we employ the c4.5 decision tree generation program [8] due to its wide availability. From the generated decision tree, we can easily obtain the final rule set as described in Eq. (1).

5   Evaluation by Experiments

5.1   Experimental Settings

By using three data sets, we evaluated the performance of the proposed method RF6.2. Firstly, we summarize the common experimental settings for training neural networks. The initial values for the weights v_{jkl} and w_{jk} are independently generated according to a normal distribution with a mean of 0 and a standard deviation of 1; the initial values for the weights w_j are set to 0, but the bias value w_0 is initially set to the average output value of all training examples. The initial values for the penalty factors λ are set to 1, i.e., Λ is set to the identical matrix. The iteration is terminated when the gradient vector is sufficiently small, i.e., each elements of the gradient vector is less than 10^{-6}.

Next, we summarize the common experimental settings for rule restoring. In the k-means algorithm, initial representative vectors $\{r^i\}$ are randomly selected as a subset of coefficient vectors $\{c^\mu\}$. For each I, trials are repeated 100 times with different initial values, and the best result minimizing Eq. (11) is used. The cross-validation error of Eq. (14) is calculated by using the leave-one-out method, i.e., $S = N$. The candidate number I of representative vectors is incremented in turn from 1 until the cross-validation error increases. The c4.5 program is used with the initial settings.

Table 1. Performance for artificial data set

RMSE	$I = 1$	$I = 2$	$I = 3$	$I = 4$
training	2.090	0.828	0.142	0.142
cross-validation	2.097	0.841	0.156	0.160
generalization	2.814	1.437	0.320	0.322

5.2 Experiment Using Artificial Data Set

We consider the following artificial rules.

$$\begin{cases} if \ q_{21} = 1 \wedge (q_{31} = 1 \vee q_{33} = 1) \ then \ y = 2 + 3x_1^{-1}x_2^3 + 4x_3x_4^{1/2}x_5^{-1/3} \\ if \ q_{21} = 0 \wedge (q_{32} = 1 \vee q_{34} = 1) \ then \ y = 2 + 5x_1^{-1}x_2^3 + 2x_3x_4^{1/2}x_5^{-1/3} (15) \\ else \qquad\qquad\qquad\qquad\qquad\qquad y = 2 + 4x_1^{-1}x_2^3 + 3x_3x_4^{1/2}x_5^{-1/3} \end{cases}$$

Here we have three nominal and nine numeric explanatory variables, and the numbers of categories of q_1, q_2 and q_3 are set as $L_1 = 2$, $L_2 = 3$ and $L_3 = 4$, respectively. Clearly, variables $q_1, x_6, \cdots, x_9$ are irrelevant to Eq. (15). Each example is randomly generated with numeric variables $x_1, \cdots, x_9$ in the range of $(0, 1)$; The corresponding value of y is calculated to follow Eq. (15) with Gaussian noise with a mean of 0 and a standard deviation of 0.1. The number of examples is set to 400 ($N = 400$).

In this experiment, a neural network was trained by setting the number of hidden units J to 2. Table 1 compares the rule restoring results of RF6.2 with different numbers of representative vectors I, where the RMSE (root mean squared error) was used for the evaluation. The training error was evaluated as a rule set of Eq. (13); the cross-validation error was calculated by using Eq. (14); and the generalization error was evaluated by using new noise-free 10,000 test examples. This table shows that although the training error almost monotonically decreased, the cross-validation error was minimized when $I = 3$, which indicates that an adequate number of representative vectors is 3. The generalization errors supported the selection that $I = 3$. Incidentally, the generalization error of the trained neural network was 0.315.

Note that the case of $I = 1$ corresponds to the case where only numeric variables are used and all the nominal variables are ignored. Table 1 shows that nominal variables played a very important role since RMSEs decreased by one order of magnitude from $I = 1$ to $I = 3$.

By applying the c4.5 program, we obtained the following decision tree whose leaf nodes correspond to the following.

$q_{21} = 0$:
| $q_{34} = 1$: 2 (83) $\Leftrightarrow$ $r^2 = (+5.04, +2.13)$
| $q_{34} = 0$:
| | $q_{32} = 0$: 3 (129) $\Leftrightarrow$ $r^3 = (+3.96, +2.97)$
| | $q_{32} = 1$: 2 (53) $\Leftrightarrow$ $r^2 = (+5.04, +2.13)$

Table 2. Performance for financial data set

RMSE	$I = 1$	$I = 2$	$I = 3$	$I = 4$
training	562571.4	420067.5	388081.8	385114.0
cross-validation	572776.7	440071.6	404998.2	406696.5

$q_{21} = 1$:
$\quad q_{34} = 1$: 3 (36) $\qquad \Leftrightarrow \quad r^3 = (+3.96, +2.97)$
$\quad q_{34} = 0$:
$\quad \quad q_{32} = 0$: 1 (73) $\qquad \Leftrightarrow \quad r^1 = (+3.10, +4.07)$
$\quad \quad q_{32} = 1$: 3 (26) $\qquad \Leftrightarrow \quad r^3 = (+3.96, +2.97)$

Here the number of training examples arriving at a leaf node is shown in paren-thesis. Then, the following rule set was straightforwardly obtained. Although some of the weight values were slightly different, we can see that the rules al-most equivalent to the original were found.

$$\begin{cases} if \;\; q_{21} = 1 \wedge (q_{31} = 1 \vee q_{33} = 1) \\ \quad then \; y = 2.01 + 3.10x_1^{-1.00}x_2^{+3.01} + 4.07x_3^{+1.02}x_4^{+0.51}x_5^{-0.33} \\ if \;\; q_{21} = 0 \wedge (q_{32} = 1 \vee q_{34} = 1) \\ \quad then \; y = 2.01 + 5.04x_1^{-1.00}x_2^{+3.01} + 2.13x_3^{+1.02}x_4^{+0.51}x_5^{-0.33} \\ else \;\; y = 2.01 + 3.96x_1^{-1.00}x_2^{+3.01} + 2.97x_3^{+1.02}x_4^{+0.51}x_5^{-0.33}. \end{cases} \quad (16)$$

5.3 Experiment Using Financial Data Set

Our experiments used stock market data of 953 companies listed on the first section of the TSE (Tokyo Stock Exchange), where banks, and insurance, secu-rities and recently listed companies were excluded. The market capitalization of each company was calculated by multiplying the shares of outstanding stocks by the stock price at the end of October, 1999. As the first step in this study, we selected six fundamental items from all of the BS (Balance Sheet) items. Since market capitalization rules may differ according to the type of industry, 33 classifications of the TSE were used as a nominal variable.[5]

In order to intuitively understand the effect of the nominal variable, the number of hidden units was fixed at 1. Table 2 shows the rule restoring results of RF6.2, indicating $\hat{I} = 3$. Since RMSEs clearly decreased from $I = 1$ to $I = 3$, the nominal variable was effectively used. The final rule set obtained was

$$if \; \bigvee_{q_l \in Q^i} q_l = 1 \;\; then \;\; y = 12891.6 + r^i x_2^{+0.668} x_3^{+1.043} x_6^{-0.747}, \;\; i = 1, 2, 3 \quad (17)$$

$$r^1 = +1.907, \; r^2 = +1.122, \; r^3 = +0.657.$$

---

[5] In the experiment 30 classifications were used because three categories were excluded.

Table 3. Performance for automobile data set

RMSE	$I = 1$	$I = 2$	$I = 3$	$I = 4$
training	3725.07	1989.67	1522.44	1370.38
cross-validation	3757.56	2132.08	1665.48	1774.09

Each group of the nominal conditions was as follows:

Q^1 = {Pharmaceuticals, Rubber Products, Metal Products, Machinery, Electrical Machinery, Transport Equipment, Precision Instruments, Other Products, Communications, Services}.

Q^2 = {Foods, Textiles, Pulp & Paper, Chemicals, Glass & Ceramics, Nonferrous Metals, Maritime Transport, Retail Trade}.

Q^3 = {Fisheries, Mining, Construction, Oil & Coal Products, Iron & Steal, Electricity & Gas, Land Transport, Air Transport, Wearhousing, Wholesale, Other Financing Business, Real Estate}.

Since the second term of the polynomial appearing in Eq. (17) is always positive, each of the coefficient values r^i indicates the stock price setting tendency of industry groups; i.e., industries appearing in Q^1 are likely to have a high setting, while those in Q^3 are likely to have a low setting. More specifically, the Pharmaceuticals industry had the highest setting, while the Electricity & Gas industry had the lowest.

5.4   Experiment Using Automobile Data Set

The Automobile data set[6] contains data on the car and truck specifications in 1985, and was used to predict prices based on these specifications. The data set has 159 examples with no missing values ($N = 159$), and consists of 10 nominal and 14 numeric explanatory variables[7] and one target variable (price).

In this experiment, since the number of examples was small, the number of hidden units was set to 1. Table 3 shows the experimental results of RF6.2, indicating that the adequate number of representatives was 3. Since RMSEs decreased by 2.3 times from $I = 1$ to $I = 3$, the nominal variable again played an important role. The polynomial part of the final rules was as follows:

$$y = 1163.16 + r^i x_2^{+1.638} x_4^{+0.046} x_5^{-1.436} x_6^{+0.997} x_9^{-0.245} x_{13}^{-0.071} \tag{18}$$
$$r^1 = +1.453, \ r^2 = +1.038, \ r^3 = +0.763.$$

The relatively simple nominal conditions were obtained.[8] Similarly as for the financial data set, since the second term of Eq. (18) is always positive, the coefficient value r^i indicates the car price setting tendency for similar specifications. In the final rules the following price setting groups are found:

---

[6]   from the UCI repository of machine learning databases

[7]   We ignored one nominal variable (engine location) having the same value.

[8]   We used the c4.5 rules program to obtain the rule set.

High price setting: {5-cylinder ones, BMW's, convertibles, VOLVO turbos, SAAB turbos, 6-cylinder turbos}.

Middle price setting: {PEUGOT's, VOLVO non-turbos, SAAB non-turbos, HONDA 1bbl-fuel-system ones, MAZDA fair-risk-level ones, non-BMW non-turbos & 6-cylinder ones, non-5-cylinder turbos & fair-risk-level ones}.

Low price setting: other cars.

6   Conclusion

To find polynomials to fit multivariate data having numeric and nominal variables, we have proposed a new method employing neural networks learning. Our experiment using artificial data showed that the method can successfully restore rules equivalent to the original despite the existence of irrelevant variables and a small noise. Other experiments using real data showed that it can find interesting rules in the domains, making good use of nominal variables. In the future we plan to carry out further experiments to evaluate the proposed method.

References

1. C. M. Bishop. *Neural networks for pattern recognition.* Clarendon Press, Oxford, 1995.
2. B. C. Falkenhainer and R. S. Michalski. Integrating quantitative and qualitative discovery in the abacus system. In *Machine Learning: An Artificial Intelligence Approach (Vol. 3)*, pages 153–190. Morgan Kaufmann, 1990.
3. P. Langley, H. A. Simon, G. Bradshaw, and J. Zytkow. *Scientific discovery: computational explorations of the creative process.* MIT Press, 1987.
4. S. P. Lloyd. Least squares quantization in pcm. *IEEE Trans. on Information Theory*, IT-28(2):129–137, 1982.
5. D. G. Luenberger. *Linear and nonlinear programming.* Addison-Wesley, 1984.
6. R. Nakano and K. Saito. Computational characteristics of law discovery using neural networks. In *Proc. 1st Int. Conference on Discovery Science, LNAI 1532*, pages 342–351, 1998.
7. B. Nordhausen and P. Langley. An integrated framework for empirical discovery. *Machine Learning*, 12:17–47, 1993.
8. J. R. Quinlan. *C4.5: programs for machine learning.* Morgan Kaufmann, 1993.
9. K. Saito and R. Nakano. Law discovery using neural networks. In *Proc. 15th International Joint Conference on Artificial Intelligence*, pages 1078–1083, 1997.
10. K. Saito and R. Nakano. Partial BFGS update and efficient step-length calculation for three-layer neural networks. *Neural Computation*, 9(1):239–257, 1997.
11. K. Saito and R. Nakano. Discovery of relevant weights by minimizing cross-validation error. In *Proc. PAKDD 2000, LNAI 1805*, pages 372–375, 2000.
12. K. Saito and R. Nakano. Second-order learning algorithm with squared penalty term. *Neural Computation*, 12(3):709–729, 2000.
13. C. Schaffer. Bivariate scientific function finding in a sampled, real-data testbed. *Machine Learning*, 12(1/2/3):167–183, 1993.
14. M. Stone. Cross-validatory choice and assessment of statistical predictions (with discussion). *Journal of the Royal Statistical Society B*, 64:111–147, 1974.

Fluent Learning: Elucidating the Structure of Episodes

Paul R. Cohen

Department of Computer Science, University of Massachusetts
cohen@cs.umass.edu

Abstract. Fluents are logical descriptions of situations that persist, and composite fluents are statistically significant temporal relationships between fluents. This paper presents an algorithm for learning composite fluents incrementally from categorical time series data. The algorithm is tested with a large dataset of mobile robot episodes. It is given no knowledge of the episodic structure of the dataset (i.e., it learns without supervision) yet it discovers fluents that correspond well with episodes.

1 Introduction

The problem addressed here is unsupervised learning of structures in time series. When we make observations over time, we effortlessly chunk the observations into episodes: I am driving to the store, stopping at the light, walking from the parking lot into the store, browsing, purchasing, and so on. Episodes contain other episodes; purchasing involves receiving the bill, writing a check, saying thank you, and so on. What actually happens, of course, is a continuous, extraordinarily dense, multivariate stream of sound, motion, and other sensor data, which we somehow perceive as events and processes that start and end. This paper describes an incremental algorithm with which a robot learns to chunk processes into episodes.

2 The Problem

Let x_t be a vector of sensor values at time t. Suppose we have a long sequence of such vectors $S = x_0, x_1, \ldots$. Episodes are subsequences of S, and they can be nested hierarchically; for example, a robot's approach-and-push-block episode might contain an approach episode, a stop-in-contact episode, a push episode, and so on. Suppose one does not know the boundaries of episodes and has only the sequence S: How can S be chunked into episodes? A model-based approach assumes we have models of episodes to help us interpret S. For example, most people who see a robot approach a block, make contact, pause, and start to push would interpret the observed sequence by matching it to models of approaching, touching, and so on. Where do these models come from? The problem here is to learn them. We wish to find subsequences of S that correspond to episodes, but

F. Hoffmann et al. (Eds.): IDA 2001, LNCS 2189, pp. 268–277, 2001.

we do not wish to say (or do not know) what an episode is, or to provide any other knowledge of the generator of S.

This problem arises in various domains. It is related to the problem of finding changepoints in time series and motifs in genomics (motifs are repeating, meaningful patterns). In our version of the problem, a robot must learn the episodic structure of its activities.

Note that episode is a semantic concept, one that refers not to the observed sensor data S but to what is happening—to the *interpretation* of S. Thus, episodes are not merely subsequences of S, they are subsequences that correspond to qualitatively different things that happen in a domain. To qualify as an episode in S, a subsequence of S should span or cover one or more things that happen, but it should not span part of one and part of another. Suppose we know the processes P_s that generate S, labelled a,b,c,f, and we have two algorithms, X and Y, that somehow induce models, labelled 1, 2, and 3, as shown here:

```
Pₛaaaaabbbbbbaaaacccfffffffffaaaaaffffffffffffaaaa
X  11111122222221122223333333311111333333333111
Y  2221111111111111133333111111122222222233333333
```

The first five ticks of S are generated by process a, the next six by b, and so on. Algorithm X does a pretty good job: its models correspond to types of episodes. When it labels a subsequence of S with 1, the subsequence was generated entirely or mostly by process a. When it labels a subsequence with 2, the subsequence was generated by process b or c. It's unfortunate that algorithm X doesn't induce the distinction between processes of type b and c, but even so it does much better than algorithm Y, whose model instances show no correspondence to the processes that generate S.

3   Fluents and Temporal Relationships

In general, the vector x_t contains real-valued sensor readings, such as distances, RGB values, amplitudes, and so on. The algorithm described here works with binary vectors only. In practice, this is not a great limitation if one has a perceptual system of some kind that takes real-valued sensor readings and produces propositions that are true or false. We did this in our experiments with the Pioneer 1 robot. Sensor readings such as translational and rotational velocity, the output of a "blob vision" system, sonar values, and the states of gripper and bump sensors, were inputs to a simple perceptual system that produced the following nine propositions: STOP, ROTATE-RIGHT, ROTATE-LEFT, MOVE-FORWARD, NEAR-OBJECT, PUSH, TOUCH, MOVE-BACKWARD, STALL.

Nine propositions permit $2^9 = 512$ world states, but many of these are impossible (e.g., moving forward and backward at the same time) and only 35 unique states were observed in the experiment, below. States are not static: the robot can be in the state of moving forward. Moving forward near an object is a state in the sense that it remains true that the robot is moving forward and near an object.

States with persistence are called fluents [4]. They have beginnings and ends. Allen [1] gave a logic for relationships between the beginnings and ends of fluents. We use a nearly identical set of relationships:

SBEB X starts before Y, ends before Y; Allen's "overlap"
SWEB Y starts with X, ends before X; Allen's "starts"
SAEW Y starts after X, ends with X; Allen's "finishes"
SAEB Y starts after X, ends before X; Allen's "during"
SWEW Y starts with X, ends with X; Allen's "equal"
SE   Y starts after X ends; amalgamating Allen's "meets" and "before"

In Allen's calculus, "meets" means the end of X coincides exactly with the beginning of Y, while "before" means the former event precedes the latter by some interval. In our work, the truth of a predicate such as SE or SBEB depends on whether start and end events happen within a window of brief duration; for example, SE(XY) is true if Y starts within a few ticks of the end of X; these events can coincide, but they needn't. Similarly, SBEB(XY) is true if Y does not start within a few ticks of the start of X; if it did, then the appropriate relationship would be SWEB. Said differently, "starts with" means "starts within a few ticks of" and "starts before" means "starts more than a few ticks before" The reason for this window is that on a real robot, it takes time for events to show up in sensor data and be processed perceptually into propositions, so coinciding events will not necessarily produce propositional representations at exactly the same time.

4 Learning Composite Fluents

As noted, a fluent is a state with persistence. A composite fluent is a statistically significant temporal relationship between fluents. Suppose that every time the robot pushed an object, it eventually stalled. This relationship might look like this:

```
touch   ------------------------
push    ----------
stall             ----------
```

Three temporal relationships are here: SWEB(touch,push), SAEW(touch,stall) and SE(stall,push). But there are other ways to represent these relationships, too; for example, the relationship SAEW(stall,SWEB(touch,push)) says, "the relationship between touch and push begins before and ends with their relationship with stall." In what follows, we describe how to learn representations like these that correspond well to episodes in the life of a robot.

Let $\rho \in$ [SBEB,SWEB,SAEW,SAEB,SWEW,SE], and let f be a proposition (e.g., MOVING-FORWARD). Composite fluents have the form:

$$F \leftarrow f \text{ or } \rho(f, f)$$
$$CF \leftarrow \rho(F, F)$$

That is, a fluent F may be a proposition or a temporal relationship between propositions, and a composite fluent is a temporal relationship between fluents. As noted earlier, a situation has many alternative fluent representations, we want a method for choosing some over others. The method will be statistical: We will only accept $\rho(F, F)$ as a representation if the constituent fluents are statistically associated, if they "go together."

An example will illustrate the idea. Suppose we are considering the composite fluent SE(jitters,coffee), that is, the start of the jitters begins after the end of having coffee. Four frequencies are relevant:

	jitters	*no jitters*
coffee	a	b
no coffee	c	d

Certainly, a should be bigger than b, that is, I should get the jitters more often than not after drinking coffee. Suppose this is true, so $a = kb$. If the relative frequency of jitters is no different after I drink, say, orange juice, or talk on the phone (e.g., if $c = kd$) then clearly there's no special relationship between coffee and jitters. Thus, to accept SE(jitters,coffee), I'd want $a = kb$ and $c = md$, and $k \gg m$. The chi-square test (among others) suffices to test the hypothesis that the start of the jitters fluent is independent of the end of the drinking coffee fluent.

It also serves to test hypotheses about the other five temporal relationships between fluents. Consider a composite fluent like SBEB(brake,clutch): When I approach a stop light in my standard transmission car, I start to brake, then depress the clutch to stop the car stalling; later I release the brake to start accelerating, and then I release the clutch. To see whether this fluent— SBEB(brake,clutch)—is statistically significant, we need two contingency tables, one for the relationship "start braking then start to depress the clutch" and one for "end braking and then end depressing the clutch":

	s(x=clutch)	*s(x!=clutch)*			*e(x=clutch)*	*e(x!=clutch)*
s(x=brake)	a1	b1		*e(x=brake)*	a2	b2
s(x!=brake)	c1	d1		*e(x!=brake)*	c2	d2

Imagine some representative numbers in these tables: Only rarely do I start something other than braking and then depress the clutch, so c1 is small. Only rarely do I start braking and then start something other than depressing the clutch (otherwise the car would stall), so b1 is also small. Clearly, a1 is relatively large, and d1 bigger, still, so the first table has most of its frequencies on a diagonal, and will produce a significant statistic. Similar arguments hold for the second table. When both tables are significant, we say SBEB(brake,clutch) is a significant composite fluent.

5 Fluent Learning Algorithm

The fluent learning algorithm incrementally processes a time series of binary vectors. At each tick, a bit in the vector x_t is in one of four states:

$$\text{Still off: } x_{t-1} = 0 \wedge x_t = 0$$
$$\text{Still on: } x_{t-1} = 1 \wedge x_t = 1$$
$$\text{Just off: } x_{t-1} = 1 \wedge x_t = 0$$
$$\text{Just on: } x_{t-1} = 0 \wedge x_t = 1$$

The fourth case is called *opening*; the third case *closing*. Recall that the simplest fluents f are just propositions, i.e., bits in the vector x_t, so we say a simple fluent f closes or opens when the third or fourth case, above, happens; and denote it open(f) or close(f). Things are slightly more complicated for composite fluents such as SBEB(f_1,f_2), because of the ambiguity about which fluent opened. Suppose we see open(f_1) and then open(f_2). It's unclear whether we have just observed open(SBEB(f_1,f_2)), open(SAEB(f_1,f_2)), or open(SAEW(f_1,f_2)). Only when we see whether f_2 closes after, before, or with f_1 will we know which of the three composite fluents opened with the opening of f_2.

The fluent learning algorithm maintains contingency tables that count co-occurrences of open and close events. For example, the tables for SBEB(f_1,f_2) are just:

	open(f = f2,t+m)	open(f != f2,t+m)
open(f=f1,t)	a1	b1
open(f! =f1,t)	c1	d1

	close(f = f2,t+m)	close(f != f2,t+m)
close(f=f1,t)	a2	b2
close(f! =f1,t)	c2	d2

That is, f_2 must open after f_1 and close after it, too. We restrict the number of ticks, m, by which one opening must happen after another: m must be bigger than a few ticks, otherwise we treat the openings as simultaneous; and it must be smaller than the length of a *short-term memory*. The short term memory has two kinds of justification. First, animals do not learn associations between events that occur far apart in time. Second, if every open event could be paired with every other (and every close event) over a long duration, then the fluent learning system would have to maintain an enormous number of contingency tables.

At each tick, the fluent learning algorithm first decides which simple and composite fluents have closed. With this information, it can disambiguate which composite fluents opened at an earlier time (within the bounds of short term memory). Then, it finds out which simple and composite fluents have just opened, or might have opened (recall, some openings are ambiguous). To do this, it consults a list of accepted fluents, which initially includes just the simple fluents—the bits in the time series of bit vectors— and later includes statistically-significant composite fluents. This done, it can update the open and close contingency tables for all fluents that have just closed. Next, it updates the χ_2 statistic for each table and it adds the newly significant composite fluents to the list of accepted fluents.

The algorithm is incremental because new composite fluents become available for inclusion in other fluents as they become significant.

6 An Experiment

The dataset is a time series of 22535 binary vectors of length 9, generated by a Pioneer 1 mobile robot as it executed 48 replications of a simple approach-and-push plan. In each trial, the robot visually located an object, oriented to it, approached it rapidly for a while, slowed down to make contact, attempted to push the object, and, after a variable period of time, stalled and backed up. In one trial, the robot got wedged in a corner of its playpen.

Data from the robot's sensors were sampled at 10Hz and passed through a simple perceptual system that returned values for nine propositions: STOP, ROTATE-RIGHT, ROTATE-LEFT, MOVE-FORWARD, NEAR-OBJECT, PUSH, TOUCH, MOVE-BACKWARD, STALL. The robot's sensors are noisy and its perceptual system makes mistakes, so some of the 35 observed states contained semantic anomalies (e.g., 55 instances of states in which the robot is simultaneously stalled and moving backward).

Because the robot collected data vectors at 10Hz and its actions and environment did not change quickly, long runs of identical states are common. In this application, it is an advantage that fluent learning keys on temporal relationships between open and close events and does not attend to the durations of the fluents: A push fluent ends as a stall event begins, and this relationship is significant irrespective of the durations of the push and stall.

Each tick in the time series of 22353 vectors was marked as belonging to exactly one of seven episodes:

A : start a new episode, orientation and finding target
B1 : forward movement
B2 : forward movement with turning or intruding periods of turning
C1 : B1 + an object is detected by sonars
C2 : B2 + an object is detected by sonars
D : robot is in contact with object (touching, pushing)
E : robot stalls, moves backwards or otherwise ends D

This markup was based on our knowledge of the robot's controllers (which we wrote). The question is how well do the induced fluents correspond to these episodes.

7 Results

The composite fluents involving three or more propositions discovered by the
fluent learning system are shown below. (This is not the complete set of such
fluents, but the others in the set are variants of those shown, e.g., versions of
fluent 4, involving two and four repetitions of SWEW(PUSH,MOVE-FORWARD)
respectively.) In addition, the system learned 23 composite fluents involving two
propositions. Eleven of these involved temporal relationships between MOVE-
FORWARD, ROTATE-RIGHT and ROTATE-LEFT. Let's begin with the fluents shown
below. The first captures a strong regularity in how the robot approaches an
obstacle. Once the robot detects an object visually, it moves toward it quite
quickly, until the sonars detect the object. At that point, the robot immediately
stops, and then moves forward more slowly. Thus, we expect to see SAEB(NEAR-
OBJECT,STOP), and we expect this fluent to start before MOVE-FORWARD, as
shown in the first composite fluent. This fluent represents the bridge between
episodes of types B and C.

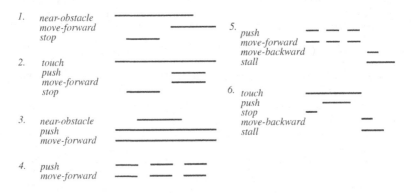

The second fluent shows that the robot stops when it touches an object but
remains touching the object after the stop fluent closes (SWEB(TOUCH,STOP))
and this composite fluent starts before and ends before another composite fluent
in which the robot is simultaneously moving forward and pushing the object.
This is an exact description of episodes of type D, above.

The third fluent is due to the noisiness of the Pioneer 1 sonars. When the
sonars lose contact with an object, the NEAR-OBJECT fluent closes, and when
contact is regained, the fluent reopens. This happens frequently during the pushing phase of each trial because, when the robot is so close to a (relatively small)
box, the sonar signal off the box is not so good.

The fourth fluent represents episodes of type D, pushing the object. The robot
often stops and starts during a push activity, hence the SE fluents. The fifth fluent represents the sequence of episodes of type D and E: The robot pushes, then
after the pushing composite fluent ends, the MOVE-BACKWARD and STALL fluents begin. It is unclear why this composite fluent includes SWEW(STALL,MOVE-
BACKWARD), implying that the robot is moving while stalled, but the data do

indeed show this anomalous combination, suggesting a bug in the robot's perceptual system.

The last fluent is another representation of the sequence of episodes D and E. It shows the robot stopping when it touches the object, then pushing the object, and finally moving backward and stalled.

At first glance, it is disappointing that the fluent learning algorithm did not find higher-order composite fluents— involving two or more temporal relationships between fluents—for episodes of type A and B. During episode A the robot is trying to locate the object visually, which involves rotation; and during episodes B1 and B2 it moves quickly toward the object. Unfortunately, a bug in the robot controller resulted in a little "rotational kick" at the beginning of each forward movement, and this often knocked the robot off its chosen course, and sometimes required it to visually reacquire the object. Consequently, during episodes of type A and B2, we see many runs of combinations of moving forward, rotating left, rotating right, and sometimes backing up. This is why 15 of 23 fluents of two propositions involve these propositions. For example, we have SAEB, SAEW, SWEB, and SBEB fluents relating MOVE-FORWARD and ROTATE-LEFT.

None of these fifteen fluents was eventually combined into higher order fluents. Why not? The reason is simply that during episodes of type A and B, it is common to see two things happening simultaneously or sequentially, but it is uncommon to see systematic associations between three or more things.

In sum, the fluents above represent the episodic structure of episodes C, D and E; while episodes of types A and B are represented by composite fluents of two propositions, typically moving forward, rotating left, and rotating right. Qualitatively, then, these fluents are not bad representations of episodes and sequences of episodes in the robot data set. Results of a more quantitative nature follow.

Recall that each of 22535 ticks of data belongs to one of seven episode types, so we can obtain a time series of 22535 episode labels in the set A,B1, B2, C1, C2, D, E. Similarly, we can "tile" the original dataset with a set of fluents. Each element in this fluent tiling series will contain the labels of zero or more open fluents. Then, we can put the episode-label series next to the fluent tiling series and see which fluents opened and closed near episode boundaries.

A particularly interesting result is that two fluents occurred nowhere near episode boundaries. They are SAEB(SWEW(MOVE-FORWARD, PUSHING), NEAR-OBSTACLE) and SAEB(PUSHING, NEAR-OBSTACLE). Is this an error? Shouldn't fluent boundaries correspond to episode boundaries? In general, they should, but recall from the previous section that these fluents are due to sonar errors during a pushing episode (i.e., episodes of type D). A related result is that these were the only discovered fluents that did not correlate with either the beginning or the end of episodes.

When the fluent tiling series and the episode-label series are lined up, tick-by-tick, one can count how many unique episode labels occur, and with what frequency, during each occurrence of a fluent. Space precludes a detailed description of the results, but they tend to give quantitative support for the earlier qualitative conclusions: The composite fluents above generally span episodes of type D and E. For example, the fifth fluent in the figure above spans 1417 ticks labeled D and 456 labeled E (these are not contiguous, of course, but distributed over the 48 trials in the dataset). And the sixth fluent covers 583 ticks labeled D and 33 ticks labeled E. The third fluent, in which the robot loses and regains sonar contact with the object, spans 402 ticks of episode D.

Not all the higher-order composite fluents are so tightly associated with particular types of episodes. The first fluent in the figure above, in which the robot stops and then begins to move forward, all while near an object, spans 405 ticks of episodes labeled C1, 157 ticks of episodes labeled D, 128 ticks of episodes labeled C2, and two ticks of episodes labeled B1. Although this fluent is statistically significant, it is not a good predictor of any episode type.

This story is repeated for fluents involving just two propositions from the set MOVING-FORWARD, MOVING- BACKWARD, ROTATE-LEFT, ROTATE-RIGHT. Each of these fluents covers a range of episode types, mostly B2, B1, C2 and C1. These fluents evidently do not correspond well with episodes of particular types.

In sum, higher-order composite fluents involving more than one temporal relationship tend to be very strongly predictive of episodes of types D and E (or the sequence D,E). Some low-order composite fluents, involving just two propositions, are also very strongly predictive of episodes (e.g., SWEW(ROTATE-LEFT,MOVE-FORWARD) occurs almost exclusively in episodes of type B2; but other low-order composite fluents are not strongly associated with a particular episode type. Finally, it appears that the corpus of fluents learned by the algorithm contained none that strongly predict episodes of type A.

8   Discussion

Fluent learning works for multivariate time series in which all the variables are binary. It does not attend to the durations of fluents, only the temporal relationships between open and close events. This is an advantage in domains where the same episode can take different amounts of time, and a disadvantage in domains where duration matters. Because it is a statistical technique, fluent learning finds common patterns, not all patterns; it is easily biased to find more or fewer patterns by adjusting the threshold value of the statistic and varying the size of the fluent short term memory. Fluent learning elucidates the hierarchical structure of episodes (i.e., episodes contain episodes) because fluents are themselves nested. We are not aware of any other algorithm that is unsupervised, incremental, multivariate, and elucidates the hierarchical structure of episodes.

Fluent learning is based on the simple idea that random coincidences of events are rare, so the episodic structure of a time series can be discovered by counting these coincidences. Thus, it accords with psychological literature on neonatal abilities to detect coincidences [9], and it has a strong statistical connection to causal induction algorithms [6]; though we do not claim that the algorithm discovers causal patterns. Our principal claim is that the algorithm discovers patterns (a syntactic notion) that correspond with episodes (a semantic notion) without knowledge of the latter. In discovering patterns—the "shape" of episodes—it differs from techniques that elucidate only probabilistic structure, such as autoregressive models [3], HMMs [2], and markov-chain methods such as MBCD [7]. Clustering by dynamics and time-warping also discover patterns [5,8], but require the user to first identify episode boundaries in time series.

Acknowledgments

Prof. Cohen's work was supported by Visiting Fellowship Research Grant number GR/N24193 from the Engineering and Physical Sciences Research Council (UK). Additional support was provided by DARPA under contract(s) No.s DARPA/USASMD-CDASG60-99-C-0074 and DARPA/AFRLF30602-00-1-0529. The U.S. Government is authorized to reproduce and distribute reprints for governmental purposes notwithstanding any copyright notation hereon. The views and conclusions contained herein are those of the authors and should not be interpreted as necessarily representing the official policies or endorsements either expressed or implied, of DARPA or the U.S. Government.

References

1. James F. Allen. 1981. An interval based representation of temporal knowledge. In IJCAI-81, pages 221–226. IJCAI, Morgan Kaufmann, 1981.
2. Charniak, E. 1993. Statistical Language Learning. MIT Press.
3. Hamilton, J..D. 1994. Time Series Analysis. Princeton University Press.
4. McCarthy, J. 1963. Situations, actions and causal laws. Stanford Artificial Intelligence Project: Memo 2; also, http://wwwformal.stanford.edu/jmc/mcchay69/mcchay69.html
5. Oates, Tim, Matthew D. Schmill and Paul R. Cohen. 2000. A Method for Clustering the Experiences of a Mobile Robot that Accords with Human Judgements. Proceedings of the Seventeenth National Conference on Artificial Intelligence.pp. 846-851. AAAI Press/The MIT Press: Menlo Park/Cambridge.
6. Pearl, J. 2000.Causality: Models, Reasoning and Inference. Cambridge University Press.
7. Ramoni, Marco, Paola Sebastiani and Paul R. Cohen. 2000. Multivariate Clustering by Dynamics. Proceedings of the Seventeenth National Conference on Artificial Intelligence,pp. 633-638. AAAI Press/The MIT Press: Menlo Park/Cambridge.
8. David Sankoff and Joseph B. Kruskal (Eds.) Time Warps, String Edits, and Macromolecules: Theory and Practice of Sequence Comparisons. Addison-Wesley. Reading, MA. 1983
9. Spelke. E. 1987. The Development of Intermodal Perception. The Handbook of Infant Perception. Academic Press

An Intelligent Decision Support Model for Aviation Weather Forcasting

Sérgio Viademonte* and Frada Burstein**

School of Information Management and Systems
Monash University, Australia
sergio.viademonte@sims.monash.edu.au; frada.burstein@sims.monash.edu.au

Abstract. Ability to collect and utilize historical data is very important for efficient decision support. On the other hand this data often requires significant pre-processing and analysis in order to bring any value for the user-decision-maker. Knowledge Discovery in Databases (KDD) can be used for these purposes in exploiting massive data sets. This paper describes a computational architecture for decision support system, which comprises an artificial neural network component for the KDD purposes. It integrates mining data set stored in databases, knowledge base produced by data mining and artificial neural network components, which serve the role of an intelligent interface for producing recommendations for decision-maker. The architecture is being implemented in the context of aviation weather forecasting. The proposed architecture can serve as a model for a KDD-based intelligent decision support for any complex decision situations where large volume of historical data is available.

1   Introduction

Information about past decisions and their outcomes can help decision-makers identify potential successes and failures in dealing with current decision situations. The Knowledge Discovery in Databases (KDD) is concerned with the theory and applications of the technologies for supporting analysis and decision-making involving, normally massive, data sets. Soft computing technologies such as artificial neural networks (ANN) are being used for KDD purposes in developing intelligent systems for decision-making.

This paper describes a research project, which is concerned with the integration of artificial neural network components for the KDD purposes in decision support context. The paper discusses the basic components of the computational architecture we propose and describes the computational technologies that are involved in this integration.

The architecture is being implemented in the context of aviation weather forecasting. The preliminary results show that the proposed computational architecture is considered a powerful technology for intelligent data analysis. In

---

* Doctoral Candidate
** Associate Professor, Director of the Monash Knowledge Management Laboratory

F. Hoffmann et al. (Eds.): IDA 2001, LNCS 2189, pp. 278–288, 2001.

this section we presented the introduction to the paper. The next section 2, describes the application domain; section 3 presents the proposed architecture for intelligent decision support, its components and functionality; section 4 presents the conclusion and final comments.

2 The Problem of Aviation Forecasting – Fog Phenomenon

The research addressed in this paper is applied in the context of rare event weather forecasting at airport terminals. Some of the most hazardous weather events are severe thunderstorms, low cloud and fog. This research project is particularly concerned with fog forecasting at the Tullemarine Airport, Melbourne, Australia.

Rare event forecasts are problematic because, by definition, forecasters do not have extensive experience in forecasting such events. A fog phenomenon is a rare event, which is particularly difficult to forecast. By international convention, fog is defined as restricting visibility to equal or less than 1000 meters [1]. Dense fog represents one of the greatest hazards to aviation and to nearly all forms of surface transport. Aircraft are generally not allowed to take off or land if the visibility is less than 400 meters. The most relevant information for short-term aviation forecasting can be found from synoptic observations and aircraft reports within about 150 km of the airport, satellite pictures and radar [2]. In addition, several other weather observations must be evaluated when forecasting fog. For example: localization of the area (Melbourne Airport in the case of this research project), precipitation, seasonality, time, wind velocity and direction, temperature, dew point depression, previous afternoon dew point, low cloud amount, etc [1, 3].

The complexity of the weather patterns and weather phenomenon occurrences implies serious problems for forecasters trying to came up with correlation models, besides, characteristics of weather observations can change often, as new instruments are used to read, measure and report weather observations. Consequently the area is a potential candidate for KDD purposes.[1]

3 A Hybrid Intelligent Decision Support System Architecture

This research project proposes a hybrid architecture for Intelligent Decision Support System (IDSS) to support decision-making. The aim of this architecture

[1] We would like to thank the Central Operations and Service Branch and the Regional Forecasting Centre from Australian Bureau of Meteorology, Victorian Regional Office for providing meteorological data and supporting in understanding and pre-processing this large amount of information. We also thank expert-forecasters from Victorian Regional Forecasting Centre for their help in providing data and explanations about aviation weather forecasting.

is to support decision-making processes by recalling relevant past experience, extracting "chunks" of knowledge from those past experiences and performing reasoning upon this knowledge in order to reach conclusions in a given new situation. The elements of this architecture can be divided in tow layers, the data and process layer. The data layer includes all the data repositories used during the various stages of decision support; it includes a database (ideally a dataware-house), a case base and a knowledge (rule) base. The process layer includes a data-mining component for automatic knowledge acquisition by extracting knowledge from transactional historical databases and an ANN (artificial neural network) based system applied as an advisory intelligent system interface.

3.1   The Proposed Architecture and Its Components

The proposed model extends the results of the KDD to fit the requirements of intelligent advisory system [4, 5]. Figure 1 presents the proposed decision-making scenario, its components and interaction processes. The basic computational elements of the proposed architecture are:

- A decision-oriented data repository, such as a data warehouse or datamart;
- A case base;
- Data mining technology;
- A knowledge base;
- An ANN based decision support interface.

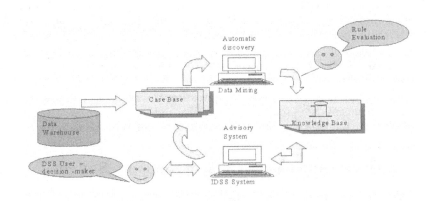

Fig. 1. Scenario for the proposed architecture for DSS

This research project is concerned with the extension of the KDD process towards a decision support level. The proposed architecture combines several computational components to derive possible solutions for the decision-maker's consideration. This IDSS does not intend to replace the human decision-maker, but provides advice based on the expert knowledge extracted from the historical data through the KDD processing. The architecture for IDSS, its elements and their roles are described in next sections, through the experiment done in the domain of aviation weather (fog) forecast.

Database The database in this architecture is the primary source of information; ideally it should be a datawarehouse. The database for the particular study of aviation weather forecasting was generated from ADAM (Australian Data Archive for Meteorology) data repository. It is a historical database with daily weather observations stored. The database used in this project has weather observations from July 1970 until June 2000. It contains 49,901 rows of data and 17 attributes. The most relevant observations are the time and month when a particular event was observed, and specific weather observations like dry-bulb temperature, dew point temperature, cloud amount, sea level pressure, amount of rainfall, visibility at the airport area, wind speed and direction. Besides that, also information about current weather conditions is provided in the form of codes, for instance, code 99 indicates a "Heavy Thunderstorm with Hail". There are nearly 70 codes expressing present and past weather conditions [1].

Case Base The case base was generated from the database mentioned above. A case base contains selected instances of significant and representative cases about the problem being addressed (fog occurrence). Each case is normally described by a number of attributes and their values. In this experiment, rows of historical weather observations were transformed in meaningful cases for analysis. Table 1 illustrates a reported case when fog was observed:

Table 1. A reported observation of fog.

Case: Weather observations when Fog was observed on morning, February.
Wind speed 4.1 m/s
Dryer moderate to fresh westerly winds during the day before.
Dew Point dropping from 15 C in the morning to 5 C degrees - one day before
Easterly wind in the evening
Low cloud in the morning = 1 h.
Dew point temperature = 10C
Sea level pressure = 1022.0

Meteorological data is usually stored without necessary care for KDD purposes. The original database was submitted to an intensive process of data analysis and pre-processing. After the data pre-processing stage, the resulting case base had 48,993 rows, divided into two classes. One class is related with the fog phenomenon observations and the other class - when fog phenomenon was not observed. The final case base has 938 fog cases (Fog) and 47,995 not fog cases

(Not Fog). Because Fog and Not Fog classes have a heterogeneous distribution was decided to split the dataset into two, one data set with fog cases and the other with not fog cases. Each class was split into training (mining), testing and evaluation data sets. The case bases have been built through the mining data sets.

To generate the data sets samples from the whole population were extracted. The sampling approach used in this study is the stratified sampling. This approach is suitable when the classes in the population are heterogeneously distributed [6]. Fog class has 938 instances in the whole population. All population was used, being 85% of the whole population randomly selected for mining data set, with 807 rows of data. In addition, 7% of the population was selected for testing and for evaluation, with 63 and 68 rows of data respectively. All samples were randomly selected without replacement.

Not Fog class has 47,995 instances. Two samples were generated from Not Fog population, one sample being 10% of the whole population, named Model1, with 4,763 instances. The second sample being 20% of the whole population, named Model2, with 9,572 instances. From Model1, three data sets were generated. A mining data set being 60% of the sample, called Model1-60, an evaluation data set and a test data set, both of them being 10% out of the Model1 sample. Again, all generated samples were randomly selected. A second mining data set was generated, being 80% of the initial sample (Model1) and new evaluation data set and test data set were generated, both of them being 10% out of the Model1 sample (named Model1-80).

The same approach was used with Model2. A mining data set being 60% of the sample, an evaluation data set and a test data set, both of them being 10% were generated from sample Model2. Those data sets are named Model2-60. The final case bases were obtained by joining the Fog class mining set with each Not Fog class mining set, therefore three versions of case bases were generated. Table 2 shows the scope of the generated case bases:

Table 2. Case base models

	Case Base Model 1-60	Case Base Model 1-80	Case Base Model 2-60
Case Base	Fog + Not Fog Model 1-60	Fog + Not Fog Model 1-80	Fog + Not Fog Model 2-60
Rows in Fog class	807	807	807
Rows in Not Fog class	2.836	3.869	5.699
Total number of rows	3.643	4.676	6.506

The samples generation was done using random sampling without replacement, using SPSS package. Analyses of variability were performed between the samples and the population to certify the samples had captured the population variability within a 90 % confidence degree. The case bases were used as the training sets for a data-mining component, described in the next section.

Data Mining The data-mining component is used to extract relevant relations out of the case base. We apply data mining to the case base to find "chunks" of knowledge about the problem domain. Specifically, this technologies, data mining and case base, are applied to perform automatic knowledge acquisition for the proposed decision support model. Consequently, handling one of the bottlenecks in intelligent system and expert systems development, which is the knowledge acquisition from human experts. We used an associative rules generator algorithm for data mining, based on AIS algorithm [7]. An association rule is an expression X→Y, where X and Y are sets of predicates; being X the precondition of the rule in disjunctive normal form and Y the target post condition.

The used algorithm is a data mining oriented version of the Combinatorial Neural Model (CNM) algorithm [8, 9]. The CNM artificial neural network is a feed-forward network with a topology of 3 layers. The input layer of a CNM network represents the set of evidences about the examples. That is, each input is in the [0,1] interval, indicating the pertinence of the training example to a certain concept. The combinatorial layer represents combination of evidences. The output layer corresponds to the possible classes an example could belong to [8], Fog and Not Fog classes in the scope of this research project.

Some conceptual differences from the original model were introduced in the algorithm, as deterministic and heuristic mechanisms for evidence selection. Next section illustrates the algorithm's output, i.e., a set of relations - rules - extracted from the case bases described in the Case Base item above. Those rules form the content of the knowledge base.

Knowledge Base Knowledge base contains structured knowledge that corresponds to relations extracted from the case bases, and describes the domain being addressed. Experiments, using the data mining algorithm, were carried out with the case base models described in the Case Base item above. The output attribute, named Fog Type, was discretised in two classes: "F"(for Fog) and "NF"(for Not Fog); this corresponds to fog phenomenon observed or not observed, respectively.

In a first approach, rules with 70%, 80% and 90% of confidence degree levels were generated and have been appraised to select the most significant ones. All the experiments were done using a minimum rule support of 10%. These parameters were arbitrarily selected. The support represents the ratio of the number of the records in the database for which the rule is true to the total number of the records. The confidence degree represents the belief in the consequent being true for a record once the antecedent is known to be true, it means, the conditional probability of the consequent given the antecedent [10, 11]. Due to the fact that fog occurrence is a rare event the level of confidence has a dramatic effect to the number of rules that can be derived from the same case base. Table 3 shows the number of rules generated for each data models for a particular confidence degree level.

Table 3. Amount of generated rules for each case base

Case Base	Rule Confidence Degree								
	70%			80%			90%		
	Fog	NotFog	Total	Fog	NotFog	Total	Fog	NotFog	Total
Model 1-60	108	134	242	74	130	204	11	86	97
Model 1-80	101	126	227	61	126	187	13	90	103
Model 2-60	83	189	272	45	189	234	11	180	191

For the confidence level of 90% too few rules were obtained; with a maximum amount of 13 rules for fog class when using the case base Model1-80. This amount of rules is unlikely to be enough for a satisfactory description of the fog phenomenon. Experiments with the confidence factor of 70% and 80% resulted in several rules some of which are very trivial and not really add much to decision support knowledge. For example, one of the obtained rules is showed below:

```
Rule 1 for Fog Class:
  If DryBulb temperature is <= 7.7 Celsius degree
    And Total Low Cloud Amount is > 7 Oktas
    And Rainfall = 0
  Then Fog Type = F
  (Confidence: 96.10%, Number of cases: 74, Support: 9.17%)
```

The same structure applies to all discovered rules. A set of obtained rules will populate the knowledge bases. At the moment this paper has been written, the set of rules generated through the data mining experiments are exported to a MS Access table, consequently, the knowledge base is stored as relational tables. Different formats are being considered to store the knowledge base, besides a Relational Model as MS Access, XLM document type definition and PMML, an XML-based language defined by the Data Mining Group (http://www.dmg.org/), formats are under consideration for further handling by the intelligent interface.

Intelligent Advisory System One of the purposes of this research project is to provide the KDD process with the following capabilities:

– Automatic reasoning upon discovered knowledge; and
– Justification about recommended solutions.

These capabilities are provided by the intelligent advisory system, identified as IDSS component in this project (see Figure 1). It is called advisory, as it does not take any decision or action by itself, but offers suggested choices to the user decision-maker together with its respective justifications. The architecture of this advisory system combines ANN [12] and a symbolic object-based mechanism for knowledge representation [12, 13]. This IDSS component is capable of learning from a training data set through its neural network learning mechanism and is able to justify its reasoning through its symbolic knowledge representation mechanism. Figure 2 shows an overview of the architecture of such an IDSS, its main components and processes [12].

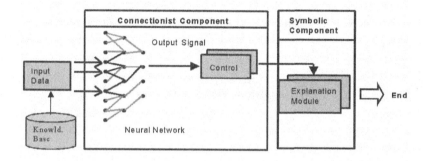

Fig. 2. IDSS component architecture.

The IDSS component uses knowledge stored in the knowledge base to learn about a particular problem being addressed. Preliminary experiments have been completed using the IDSS component. A training data set with 25 cases extracted from the case base Model1-60 and a test data set with 12 cases were generated to simulate the system behavior, to verify the neural networks training and learning times and accuracy. The observations Visibility, Rainfall, Wind Speed, Wind Compass Points, Sea Level Pressure and Dew Point temperature were selected in this first analysis. The selected observations and their possible attribute values perform 32 input neurons in the neural network input layer. The experiment showed a good performance and accuracy, all the cases in the test dataset were correctly evaluated; the learning time was 1.292 seconds with combination order 3. Although it as a small experiment, the preliminary results shows the suitability and applicability of the IDSS component in the proposed architecture for intelligent decision support. 3.2 - Functionality of the IDSS Architecture The proposed model for IDSS works in two levels, first at the rule generation level where data mining technology is applied. Second, at the ANN learning and consulting level, where an ANN framework is applied.

The proposed model extracts relations from the cases stored in the case base using an associative rules generator algorithm; and uses those relations as input into the IDSS component, which implements an inductive learning algorithm. The inductive learning is performed through a punishment and reward algorithm and an incremental learning algorithm implemented in the ANN model [9, 13]. The IDSS component pursues the following strategy to come up with a decision for a specific case (e.g. fog forecasting). The ANN evaluates the given case and calculates a confidence degree value for each hypothesis. The inference mechanism finds the winning hypothesis and returns the corresponding set of evidences. It must be noted that a case is modeled and presented to the IDSS as a set of evidences, where each pair evidence/attribute value is mapped into an input neuron in the ANN topology. Figure 3 illustrates this mapping mechanism. The rule's evidences are mapped into the ANN input layer as input neurons E1, E2 and E3, with their respective confidence degrees. The bold arrows indicate the combination of these evidences, it means a rule, with its calculated confidence degree, W1. The combination appoints to Fog hypothesis, represented by the output neuron named fog in Figure 3.

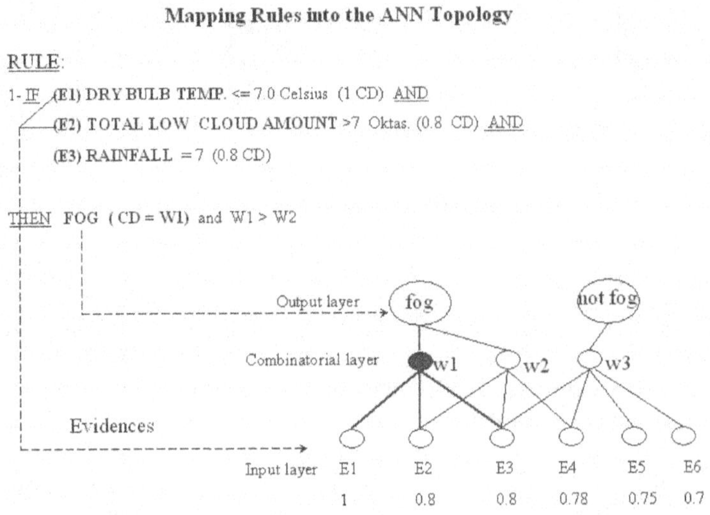
Mapping Rules into the ANN Topology
RULE:
1- IF (E1) DRY BULB TEMP. <= 7.0 Celsius (1 CD) AND
(E2) TOTAL LOW CLOUD AMOUNT >7 Oktas. (0.8 CD) AND
(E3) RAINFALL = 7 (0.8 CD)
THEN FOG (CD = W1) and W1 > W2
Output layer
fog
not fog
Combinatorial layer
w1
w2
w3
Evidences
Input layer
E1
E2
E3
E4
E5
E6
1
0.8
0.8
0.78
0.75
0.7

Fig. 3. Mapping rules into the ANN topology

The system follows three stages of decision support: rules generation, case consult and case base consult. The rule generation step produces a set of relationship generated in a data mining session. Case consult and case base consult are provided by the IDSS component. Cases consult function presents to the decision maker a selection of evidences, and their respective evaluation of relevance to the situation at hand. The IDSS component sets up a set of hypotheses based on the presented input data about the current decision situation. It evaluates the selected evidences and calculates a confidence degree for each hypothesis.

The inference mechanism appoints the hypothesis with the higher confidence degree as the most suitable solution to the problem. A case base consult stage is similar to the case consult, however, instead of presenting one single case each time several cases are simultaneously presented.

4   Conclusion and Comments

This research project aims to develop a methodology for decision support through the integration of knowledge discovery in databases and soft computing technologies. As a final result we intend to propose an efficient way of integration of the required computational elements and technologies in a hybrid model. To verify the efficiency of the proposed architecture it has been applied in a specific decision problem, aviation weather forecasting. As part of the valuation process we intend to assess the level of generalizability of the proposed model to the other domains.

In this paper we propose and discuss an approach for such a hybrid computational architecture. The proposed architecture and results obtained so far were presented to the forecasters from Australian Bureau of Meteorology, and assessed as potentially useful in supporting their tasks. The generated set of rules shows a good level of accuracy when appraised by the forecasters. Also, if compared with the statistical methods currently used for severe weather forecaster, the generated set of rules showed a higher level of capacity to generalize relations stored in case base due to the application of ANN methods. One of the main conclusion achieved in this project concerns with the data preparation for KDD. A significant amount of work was required in preparing row data, in building the case bases and training data sets. Among the most difficult questions that arose were those that related to the amount of cases that were enough for training, to define suitable sampling strategies and suitable sample models. Apparently these questions are empirically defined, depending on the nature of the problem being addressed, the knowledge about the problem and the amount of available information. To automatically perform those tasks, seems to be a difficult and encouraging challenge.

As a future work as a result of thorough analysis of data modeling, the authors came up with the suggestion to use fuzzy logic to represent semantic variables such as wind speed and wind direction. In addition, others data mining algorithms are under evaluation to be used in a comparison study. Besides that, there is the aim to extend the CANN framework [12, 13] by plug into the framework a new class hierarchy for data mining algorithms. Potentially, this is the best approach to achieve a fully integrated software system to the architecture discussed in this research paper. Some others future tasks are the implementation of a class to store and export the CANN output, in order to facilitate the system integration with other systems and to store the trained neural network topology, it means, to make the neural network knowledge explicit.

The system will be available to the weather forecasters at Australian Bureau of Meteorology for evaluation.

References

[1] A. H. Auer, J., Guidelines for Forecasting Fog. Part 1: Theoretical Aspects, 1992, Meteorological Service of New Zealand. p. 32.

[2] Love, G., Tropical weather forecasting: A pratical viewpoint., 1985, Australian Bureau of Meteorology.: Darwin, Northern Territory.

[3] Keith, R., Results And Recommendations Arising From An Investigation Into Forecasting Problems At Melbourne Airport., 1991, Bureau of Meteorology, Meteorological Office.: Townsville. p. 19.

[4] Fayyad, U.M., &, G. Piatetsky-Shapiro, and P. Smyth, From Data Mining to Knowledge Discovery: An Overview., in Advances in Data Mining and Knowledge Discovery, A. Press, (Ed.), 1996, AAAI/MIT Press: Cambridge. p. 1-34.

[5] Han, J. Data Mining: An Overview from Databases Perspective, in PAKDD 1998. 1998. Melbourne, Australia.

[6] Catlett, J., Megainduction: Machine learning on very large databases., in School of Computer Science. 1991, University of Technology: Sydney, Australia.

[7] Agrawal, R., Imielinski T., and A. Swami. Mining association rules between sets of items in large databases, in Conference Management of Data, 1993.

[8] Machado, R.J., V.C. Barbosa, and P.A. Neves, Learning in the Combinatorial Neural Model., in IEEE Transactions on Neural Networks, 1998.

[9] Machado, R.J. and A.F. Rocha, Handling Knowledge in High Order Neural Networks: The Combinatorial Neural Model, 1989, IBM Rio Scientific Center: Rio de Janeiro.

[10] Buchner, A.G., et al., A meteorological knowledge-discovery environment., in Knowledge Discovery and Data Mining, 1998. pp. 204–226.

[11] Piatetsky-Shapiro, G. and W. Frawley, Knowledge Discovery in Databases., ed. (Eds). 1991: MIT Press.

[12] Beckenkamp, F., S.I.V. Rosa, and W. Pree. OO Design & Implementation of a Flexible Software Architecture for Decision Support Systems. in The 9th. International Conference on Software Engineering & Knowledge Engineering - SEKE'97, 1997, Madrid, Spain.

[13] Beckenkamp, F. and W. Pree, Neural Network Framework Components., in Object-Oriented Application Framework: Applications and Experiences., S.D.C. a.J.R. Fayad M., (Ed.). 1999, John Wiley.

MAMBO: Discovering Association Rules Based on Conditional Independencies

Robert Castelo, Ad Feelders, and Arno Siebes

Utrecht University, Institute of Information and Computing Sciences,
P.O. Box 80089, 3508TB Utrecht, The Netherlands,
{roberto, ad, arno}@cs.uu.nl

Abstract. We present the Mambo algorithm for the discovery of association rules. Mambo is driven by conditional independence relations between the variables instead of the minimum support restrictions of algorithms like Apriori. We argue that making use of conditional independencies is an intuitively appealing way to restrict the set of association rules considered. Since we only have a finite sample from the probability distribution of interest, we have to deal with uncertainty concerning the conditional independencies present. Bayesian methods are used to quantify this uncertainty, and the posterior probabilities of conditional independence relations are estimated with the Markov Chain Monte Carlo technique. We analyse an insurance data set with Mambo and illustrate the differences in results compared to Apriori.

1 Introduction

The prototypical application of association rules [1] is that of market basket analysis. An association rule $X \to Y$ indicates cross-sell effects with as *support* the fraction of customers that bought all articles in X and Y and as *confidence* the fraction of all customers that bought X that also bought Y. With the Apriori algorithm [2], all association rules exceeding user defined thresholds for support and confidence can be discovered.

The support pruning by Apriori poses a serious problem when applying association rules in practice. Rules with high support are in general already known, while using a (very) low support threshold causes two problems. Firstly, it results in very many rules, most of which are uninteresting. The second is that the lower the support, the longer the running time.

For the first problem, many interestingness measures, such as the lift (the number of times buyers of X are more likely to buy Y compared to all customers), have been defined, that can be used to filter or order the resulting association rules. The computational problem, however, has proven to be more difficult.

One approach to speed-up the discovery of low support but interesting association rules is to devise algorithms that do not rely on support pruning to find the interesting association rules. Two recent papers in this direction are [5,11], this paper presents another algorithm in this line of research.

F. Hoffmann et al. (Eds.): IDA 2001, LNCS 2189, pp. 289–298, 2001.

If one doesn't use support to prune the set of possible association rules, one needs a different measure to achieve this. Like in [11] our pruning is based on dependencies between attributes. However, instead of searching for minimally dependent sets of attributes [11], we use *conditional independence* for pruning. More precisely, *Mambo* discovers all association rules $X \to Y$ such that Y is a singleton and X is a subset of a *Markov Blanket* of Y.

A Markov blanket of an attribute A is a minimal set of attributes MB such that A is conditionally independent of any other attribute B given MB. A more detailed explanation is given in section 2.

The rationale for our approach is that A only depends on MB. More precisely, the rationale is that for any attribute $B \notin MB \cup \{A\}$, we have that $P(A|MB, B) = P(A|MB)$. This is, again, discussed in more detail in section 2.

In section 3 we present the Mambo algorithm, which yields the association rules described above, sorted on lift. To compute the rules, Mambo needs to know the Markov blankets of each attribute. For these blankets, it consults an *oracle*. This oracle is computed using the Markov Chain Monte Carlo method on a class of graphical models called decomposable models (DEC). The details of this are given in section 3.2. Not all results discovered by Mambo are equally interesting. In particular, we may find long(er) rules that describe nearly the same subset of the database as short(er) rules. Mambo filters out these long rules, the details are given in section 3.3.

Clearly, Mambo will find a different rule set than Apriori. In section 4 we present results of an experiment in which both algorithms were run on the same data set. We discuss some of the differences in result sets.

In section 5, we formulate our conclusions.

2   Association Rules and Conditional Independencies

Let db be a database with schema $\mathbf{A} = \{A_1, \ldots, A_n\}$ in which each A_i has a finite domain D_i. For a subset $\mathbf{X} \subseteq \mathbf{A}$, the domain $D_{\mathbf{X}}$ is simply the cartesian product of the domains of the elements of $\mathbf{X}$. For disjoint $\mathbf{X}, \mathbf{Y} \subseteq \mathbf{A}$ and $\mathbf{x} \in D_{\mathbf{X}}, \mathbf{y} \in D_{\mathbf{Y}}$, the association rule

$$\mathbf{Y} = \mathbf{y} \to \mathbf{X} = \mathbf{x}$$

has:

support $sup(\mathbf{Y} = \mathbf{y} \to \mathbf{X} = \mathbf{x}) = \frac{|\{t \in db | \pi_{\mathbf{X}}(t) = \mathbf{x} \wedge \pi_{\mathbf{Y}}(t) = \mathbf{y}\}|}{|db|}$

confidence $conf(\mathbf{Y} = \mathbf{y} \to \mathbf{X} = \mathbf{x}) = \frac{|\{t \in db | \pi_{\mathbf{X}}(t) = \mathbf{x} \wedge \pi_{\mathbf{Y}}(t) = \mathbf{y}\}|}{|\{t \in db | \pi_{\mathbf{Y}}(t) = \mathbf{y}\}|}$

lift $lift(\mathbf{Y} = \mathbf{y} \to \mathbf{X} = \mathbf{x}) = \frac{conf(\mathbf{Y} = \mathbf{y} \to \mathbf{X} = \mathbf{x})}{conf(\emptyset \to \mathbf{X} = \mathbf{x})}$

As stated in the introduction, finding association rules with low support but high lift using the standard support pruning approach, such as Apriori, is difficult. Therefore, the Mambo algorithm doesn't rely on support pruning but on conditional independencies.

To motivate our approach, first recall the definition of conditional independence:

Definition 1. *Conditional Independence (CI)*
Let **A** *be a finite set of discrete random variables in which each variable has a finite domain. Let* $p(\mathbf{A})$ *be a joint probability distribution over the random variables in* **A**, *that belongs to some family of probability distributions* P. *Let* $\mathbf{X}, \mathbf{Y}, \mathbf{Z}$ *be subsets of* **A** *such that* $\mathbf{X}, \mathbf{Y}$ *are disjoint and non-empty. We say that* **X** *is* conditionally independent *of* **Y** *given* **Z**, *denoted by* $\mathbf{X} \perp\!\!\!\perp \mathbf{Y}|\mathbf{Z}\ [P]$, *if* $\forall \mathbf{x} \in D_{\mathbf{X}}, \mathbf{y} \in D_{\mathbf{Y}}, \mathbf{z} \in D_{\mathbf{Z}}$:

$$p(\mathbf{Y} = \mathbf{y}, \mathbf{Z} = \mathbf{z}) > 0 \Rightarrow p(\mathbf{X} = \mathbf{x}|\mathbf{Y} = \mathbf{y}, \mathbf{Z} = \mathbf{z}) = p(\mathbf{X} = \mathbf{x}|\mathbf{Z} = \mathbf{z})$$

In other words, conditional independence is a form of irrelevance. If we know **Z**, any further information about **Y** cannot enhance the current state of information about **X**, i.e. given **Z**, **Y** becomes irrelevant to **X**.

If we reformulate association rules in the context of random variables and probability distributions, the support of the rule $\mathbf{Y} = \mathbf{y} \rightarrow \mathbf{X} = \mathbf{x}$ becomes $p(\mathbf{Y} = \mathbf{y}, \mathbf{X} = \mathbf{x})$, the confidence becomes $p(\mathbf{X} = \mathbf{x}|\mathbf{Y} = \mathbf{y})$, and the lift becomes $p(\mathbf{X} = \mathbf{x}|\mathbf{Y} = \mathbf{y})/p(\mathbf{X} = \mathbf{x})$.

If we know that $\mathbf{X} \perp\!\!\!\perp \mathbf{Y}|\mathbf{Z}\ [P]$, we have that:

$$p(\mathbf{X} = \mathbf{x}|\mathbf{Y} = \mathbf{y}, \mathbf{Z} = \mathbf{z})/p(\mathbf{X} = \mathbf{x}) = p(\mathbf{X} = \mathbf{x}|\mathbf{Y} = \mathbf{y})/p(\mathbf{X} = \mathbf{x})$$

In other words, the lift doesn't rise by adding knowledge about **Y**, **Y** is irrelevant to **X** given **Z**. Or, if we have an association rule for **X** with **Y** and **Z** on the lefthand side, we might as well filter **Y** out.

This is interesting if **Z** *shields* **X** from all other variables, i.e., if $\mathbf{X} \perp\!\!\!\perp \mathbf{A} \setminus (\mathbf{Z} \cup \mathbf{X})|\mathbf{Z}\ [P]$. Because then we only have to consider association rules whose lefthand side is within **Z**. All of this is even more interesting if **Z** is *minimal*, i.e, if we remove an element from **Z** it no longer shields **X** from the rest. Such a minimal set is called a *Markov Blanket* of **X**.

There are two obstacles to use this idea to discover association rules. Firstly, we have a database rather than a probability distribution. Secondly, **X** may have many Markov Blankets.

Clearly, the database is a sample from a probability distribution and hence we will use statistical techniques to decide on conditional independence. This gives an uncertainty as to whether a set **Z** is a Markov Blanket of **X**; there is a *probability* that **Z** is a Markov blanket of **X**. Hence, we attack both problems at the same time: Mambo uses the n most probable Markov blankets of **X**. Where Apriori prunes on support, Mambo prunes on the posterior probability of the Markov blanket.

To keep our discussions simple, we will assume in the remainder of this paper that **X** is a singleton set denoted by X.

3   Mambo

For each attribute X, Mambo computes association rules using its (most probable) Markov blankets. Hence, it needs to know these blankets. The algorithm uses an oracle for this; hence the name: Maximum A posteriori Markov Blankets using an Oracle. First we specify Mambo, then we explain how the Oracle is computed. Finally, we discuss how the result set is filtered on lift.

3.1   The Algorithm

Given a Markov blanket MB for an attribute X, Mambo simply computes all association rules of the form $\mathbf{Y} = \mathbf{y} \rightarrow X = x$ in which $\mathbf{Y} \subseteq MB$. The results are filtered before they are given to the user. hence, we have:

Mambo
> $nmb :=$ number of blankets
> $Res := \emptyset$
> **For all $A_i \in \mathbf{A}$ do**
> > $Pres := \emptyset$
> > $O := Oracle(A_i, nmb)$
> > > **For all $I \in O$ Do**
> > > > **For all $\mathbf{Y} \subseteq I$ Do**
> > > > > $Pres := Pres \cup \{(\mathbf{Y} = \mathbf{y} \rightarrow A_i = v_i, lift) | \mathbf{y} \in D_{\mathbf{Y}}, v_i \in D_i\}$
> > $Pres := Filter(Pres)$
> > $Res := Res \cup Pres$
> Return Res

3.2   The Oracle

There are various ways in which one can build an oracle for Mambo. Here we chose a Bayesian approach. The oracle provides a posterior distribution over the Markov blankets given the data. For any given Markov blanket, the higher the posterior, the more certain we are about the validity of that conditional independence. The Bayesian approach for the computation of such a posterior is by integrating over a nuisance parameter θ

$$p(I|D) = \int_{\theta} p(I|\theta, D)p(\theta|D)$$

This nuisance parameter should, of course, handle conditional independencies in general and Markov blankets in particular. Graphical Markov models (see for instance [14]) have this property. They use a graph, in which the nodes correspond to the random variables, to summarize and represent a set of conditional independencies. In particular we will use DEC (decomposable Markov) models, whose graph is an undirected chordal graph. That is, an undirected graph in which a chordless cycle involves at most three vertices.

The graph of a DEC model encodes conditional independencies as follows. For any triple of disjoint subsets $\mathbf{X}, \mathbf{Y}, \mathbf{Z} \subset \mathbf{A}$, if all paths between $\mathbf{X}$ and $\mathbf{Y}$ intersect $\mathbf{Z}$, $\mathbf{X}$ is conditionally independent of $\mathbf{Y}$ given $\mathbf{Z}$. In particular, *the*

Markov blanket of an attribute A in such a graph is simply the set of all its neighbours; for the full details see [14].

Using this finite set of models we get:

$$p(I|D) = \sum_{M \in DEC} p(I|M, D)p(M|D) \tag{1}$$

To compute the posterior in this way, we have to compute $p(M|D)$. Using Bayes rule, we have $p(M|D) = p(M, D)/p(D)$, where $P(D) = \sum_M p(M, D)$. The term $p(M, D)$ is known as the *likelihood* of the model and the data, and it has a closed formula under certain assumptions [7]. However, the summations in (1) and of $p(D)$ have a number of terms which is exponential in the number of variables, which renders their exact computation infeasible.

So, rather than exact computations, we approximate $p(M|D)$ and $p(I|D)$ using the Markov Chain Monte Carlo (MCMC) method; see, e.g. [13]. The MCMC method was already adapted for DEC models by Madigan and York [8] and we use the implementation described in [6], where posteriors on CI restrictions were also computed to analyze high-order interactions among random variables.

The state space of the Markov chain is the set of all DEC models. For the transitions we define a *neighbourhood*. For $M, M' \in DEC$, we say that M' is a neighbour of M if M' has one arc more or one arc less than M, and is chordal. The transition probability of M to M' is defined by the transition matrix q, in which $q(M \to M') = 0$ if $M' \notin nbd(M)$ and $q(M \to M') > 0$ if $M' \in nbd(M)$. Using the transition matrix q we can build a Markov chain $\mathbf{M}(t, q)$, $t = 1, 2, \ldots, N$, with state space DEC.

In particular, if the chain is in state M, we randomly generate an $M' \in nbd(M)$ and M' is accepted as the next state with probability:

$$min\left\{1, \frac{p(M'|D)}{p(M|D)}\right\}$$

Because $p(M|D) \propto p(D|M)p(M)$ this ratio corresponds to the Bayes factor $p(D|M)/p(D|M')$, which only involves efficient local computations [7].

With this definition, we have that the chain has $p(M|D)$ as its equilibrium distribution, that q is irreducible and, that the chain is aperiodic. Hence, we know that for a function $g(M)$, after the simulation of $\mathbf{M}(t, q)$, the average

$$\hat{G} = \frac{1}{N}\sum_{t=1}^{N} g(\mathbf{M}(t, q))$$

will converge to E(g(M)), i.e. the expectation of $g(M)$, as $N \to \infty$ [4]. As argued in [8], by setting $g(M) = p(I|M, D)$, we will obtain the posterior $p(I|D)$.

To define $P(I|M, D)$, let $I = X \perp\!\!\!\perp \mathbf{A}\backslash(X \cup \mathbf{Y})|\mathbf{Y}$, $p(I|M, D) = 1$ if $\mathbf{Y}$ is the Markov blanket of X in M, and 0 otherwise. In other words, all we have to do is to count the number of times each different model and each different Markov blanket are visited, and divide those counts by the number of iterations of the Markov chain.

3.3   Filter

There is a natural inclusion relation on association rules. We say that R_1 is a subrule of R_2 and R_2 is a superrule of R_1 iff

$$R_1 : \mathbf{Y} = \mathbf{y} \rightarrow X = x \wedge R_2 : \mathbf{Y} = \mathbf{y}, \mathbf{Z} = \mathbf{z} \rightarrow X = x$$

Given a rule, not all of its superrules are necessarily of interest. In particular, if R_1 and R_2 have (more or less) the same lift, R_2 is not very interesting. In other words, for the valuations $\mathbf{Y} = \mathbf{y}$ and $X = x$, the additional information that $\mathbf{Z} = \mathbf{z}$ is irrelevant. Hence, we filter the results of Mambo for such uninteresting rules.

4   Experimental Results

The experiments have been performed on a P-III 650Mhz with 512Mbytes of RAM, and the software has been implemented as a module on top of the main memory database system Monet [9].

In order to build the oracle $\mathcal{O}$ and obtain the posterior $p(I|D)$ we have run the Markov chain for $N = 10^6$ iterations, without burn-in, and starting from the empty graph.

For the experiments we used the insurance company benchmark of the COIL 2000 challenge, available from the UCI KDD archive [3]. This data set contains information about the customers of an insurance company, and consists of 86 variables concerning product ownership and socio-demographic characteristics derived from postal area codes. For the association rule discovery we ignored the socio-demographic variables (1-43) in the data set. The remaining variables consist of the target variable of the original challenge (whether or not someone owns a caravan policy) and variables concerning the ownership (number of policies) and contribution (money amount) for 21 other insurance products. In order to make the data suited for analysis with binary association rules, they were discretized in the obvious way, i.e. new value = I(old value > 0). After this discretization the variables concerning amounts contain exactly the same information as those concerning numbers of policies so we discarded the variables concerning the contribution amount. Finally then, this leaves us with a data set consisting of 22 binary variables about a total of 5822 customers. Table 1 provides a description of the variables and the corresponding relative frequencies of product ownership.

Below we describe the analysis of this data set with Mambo, and compare the results with a minimum support driven approach like Apriori. The comparison with Apriori is not intended to show that one is "better" than the other, but rather to illustrate where the differences lie with our conditional independence driven approach.

Table 1. Description of insurance data set.

Variable	Description	Mean
1	private third party insurance	0.402
2	third party insurance (firms)	0.014
3	third party insurane (agriculture)	0.021
4	car policy	0.511
5	delivery van policy	0.008
6	motorcycle/scooter policy	0.038
7	lorry policy	0.002
8	trailer policy	0.011
9	tractor policy	0.025
10	agricultural machines policy	0.004
11	moped policy	0.068
12	life insurance policy	0.050
13	private accident insurance policy	0.005
14	family accidents insurance policy	0.007
15	disability insurance policy	0.004
16	fire policy	0.542
17	surfboard policy	0.001
18	boat policy	0.006
19	bicycle policy	0.025
20	property insurance policy	0.008
21	social security insurance policy	0.014
22	mobile home insurance policy	0.060

4.1   Analysis of Results

As a point of reference we analysed the insurance data with the Apriori algorithm using a minimum support of 0.01 (59 records). All rules with 1 item on the right-hand side were generated from the large itemsets found. This yielded a total of 102 rules.

We have performed the analysis with Mambo with a number of variations in the parameter values in order to illustrate the effect of the different parameters.

In the most basic analysis we selected for each variable $X \in \mathbf{A}$ only the Markov blanket with highest posterior probability $p(I|D)$, i.e $nmb = 1$. Furthermore we only consider positive rules for the moment, i.e. rules of type $\{\mathbf{Y} = 1\} \rightarrow \{X = 1\}$. In general however, values in the antecedent and consequent of the rule may be either 0 or 1. For an initial comparison with the Apriori results we furthermore selected only those rules with support at least 0.01. It is important to note however that this was done afterwards, since the minimal support is not used in Mambo to constrain the search.

In order to consider the effect of filtering the superrules that do not improve the lift enough, we performed the initial analysis without filtering. This yielded a total of 72 rules, 30 less than Apriori (since we selected rules with support at least 0.01, all 72 rules were also found by Apriori). To illustrate why some rules generated by Apriori were not considered by Mambo we look at an example. For variable 9 we found that the Markov blanket with highest posterior probability (0.4) is $\{3,10\}$. One of the rules found with Apriori is $\{3,16\} \to \{9\}$ (we use this notation as an abbreviation of $\{3 = 1, 16 = 1\} \to \{9 = 1\}$) with confidence 0.64, lift 26 and support 0.013. This rule was not found with Mambo since the antecedent contains variable 16 which does not belong to the best Markov blanket. The rule $\{3\} \to \{9\}$ with confidence 0.62, lift 25 and support 0.013 *is* found by Mambo however, since its antecedent is a subset of the best Markov blanket.

If we filter superrules that do not improve the lift by at least 0.1, the number of rules is reduced from 72 to 60. For example, one of the 72 rules found is $\{4,3\} \to \{16\}$ with a lift of 1.83. There exists a subrule $\{3\} \to \{16\}$ however, with a lift of 1.80. Therefore the superrule is filtered from the ruleset.

Now, as mentioned, with Mambo we can find interesting rules regardless of their support. For example, the best rule for product 4 found with Apriori is $\{1, 16, 22\} \to \{4\}$ with a confidence of 0.88 and lift of 1.7. With Mambo (without any support restrictions) we find the rule $\{21, 1\} \to \{4\}$, with confidence 0.96, lift 1.9, and support 0.008 (46 records). This rule has higher confidence and lift than the best rule found with Apriori, but does not meet the minimum support requirement of 0.01.

It is interesting to note that setting nmb to 5 instead of 1 does not lead to a dramatic increase of the number of rules (see Table 2). For example, if we consider only positive rules and remove superrules that do not meet the minimum improvement requirements, we obtain a total of 78 rules with nmb set to 1 (selecting only the rules that have support at least 0.005). If we use the same settings, but select the 5 most probable Markov blankets we obtain a total of 85 rules. The 7 additional rules found are of high quality in the sense that they all have a lift larger than 1. One of the additional rules found with $nmb = 5$ is $\{16, 21\} \to \{1\}$, with confidence 0.72, lift 1.8 and support 0.007. This rule was not found with $nmb = 1$ since the Markov blanket for 1 with highest posterior probability (0.78) is $\{2,3,4,11,12,16\}$, i.e. does not contain 21. However, the Markov blanket with third highest posterior probability (0.036) is the same set with 21 added (see figure 1). Furthermore, $\{16, 21\} \to \{1\}$ is not pruned by a subrule since it improves the lift more than required on the subrules $\{16\} \to \{1\}$ (lift 1.6) and $\{21\} \to \{1\}$ (lift 1.5).

Table 2. The number of rules found by Mambo for different values of nmb, with removal of superrules. Only rules with support at least 0.005 have been counted. The last row also counts rules with zero values in the antecedent or consequent.

	nmb		
ruletype	1	2	5
positive	78	78	85
all	518	549	750

Looking at figure 1 we can understand why not very many rules are added when nmb is increased from 1 to 5. For example, the set at the top is a subset of the best blanket, and can therefore not generate any additional rules. The sets at the bottom have one element additional to the best set and could generate additional rules. Many of these rules are removed however because they are only slightly better, or even worse, than one of their subrules. For example, $\{4, 16\} \rightarrow \{1\}$ (lift: 1.9) is generated from the best Markov blanket. Later the rule $\{4, 16, 21\} \rightarrow \{1\}$ (lift: 1.9) is considered through the third best Markov blanket but immediately filtered because it has the same lift as the subrule just mentioned.

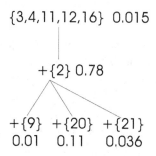

Fig. 1. The 5 best Markov blankets for variable 1. Only additional elements to set above are shown. Posterior probability is shown to the right of or below a set.

5   Summary and Concluding Remarks

We have presented an algorithm for the discovery of association rules that is driven by conditional independence relations between the variables rather than minimum support restrictions. We consider this to be an interesting alternative because for some applications minimum support is a rather artificial constraint primarily required for the efficient computation of large itemsets.

Making use of conditional independencies is an intuitively appealing way to restrict the set of association rules considered, because if $\mathbf{X}$ is conditionally independent of $\mathbf{Y}$ given $\mathbf{Z}$, then the association rule $\{\mathbf{YZ} = \mathbf{yz}\} \rightarrow \{\mathbf{X} = \mathbf{x}\}$ has the same lift and confidence as $\{\mathbf{Z} = \mathbf{z}\} \rightarrow \{\mathbf{X} = \mathbf{x}\}$ for any values $\mathbf{x}, \mathbf{y}$ and $\mathbf{z}$. Furthermore, the information concerning conditional independencies between variables, as provided by the Mambo algorithm, may be of interest in its own right.

The algorithm for the generation of all possible rules from the best Markov blankets can be improved. Currently it does not exploit the subset relations that exist between them. For example, if $p(I_i|D) > p(I_j|D)$ (where $I_k = X \perp\!\!\!\perp$

$\mathbf{A} \backslash (X \cup \mathbf{Y}_k)|\mathbf{Y}_k)$ and $\mathbf{Y}_j \subset \mathbf{Y}_i$ then clearly I_j will not add any new rules. Other efficiency improvements are possible, and will be dealt with in future work.

We have analysed an insurance data set with Mambo and illustrated the differences with a minimum support driven approach. It is ultimately an empirical question whether rules based on conditional independencies or rules based on minimum support restrictions are to be preferred for a particular application. If the rules are used exclusively for descriptive purposes, it makes sense to consult domain experts for the assessment of rule quality. If the rules are used for predictive purposes as well one can compare the predictive accuracy of the rules obtained with the different approaches.

References

1. Agrawal, R., Imilinski, T., Swami, A.: Mining Association Rules between Sets of Items in Large Databases. In *Proc. ACM SIGMOD Conf. Management of Data.* (1993) 207–216
2. Agrawal, R., Srikant, R.: Fast Algorithms for Mining Association Rules. In *Proc. 20th Int'l Conf Very Large Databases.* (1994)
3. Bay, S. D.: *The UCI KDD Archive [http://kdd.ics.uci.edu].* Irvine, CA: University of California, Department of Information and Computer Science. (1999)
4. Chung, K.L.: *Markov Chains with Stationary Transition Probabilities (2nd ed).* Springer-Verlag (1967)
5. Cohen, E., Datar, M., Funjiwara, S., Gionis, A., Indyk, P., Motwani, R., Ullman, J.D., Yang, C.: *Finding Interesting Associations without Support Pruning.* In *IEEE Transactions on Knowledge and Data Engineering.* **13**(1) (2001) 64–78
6. Giudici, P. and Castelo, R.: Improving Markov Chain Monte Carlo model search for data mining. Technical Report 117, Department of Economics and Quantitative Methods, University of Pavia, Italy. (2000)
7. Dawid, A.P. and Lauritzen, S.L.: Hyper Markov laws in the statistical analysis of decomposable graphical models. *Annals of Statistics.* **21** (1993) 1272–1317
8. Madigan, D. and York, J.: Bayesian graphical models for discrete data. *International Statistical Review.* **63** (1995) 215–232
9. http://www.cwi.nl/~monet
10. Pearl, J.: *Probabilistic Reasoning in intelligent systems.* Morgan Kaufmann (1988)
11. Silverstein, C., Brin, S., Motwani, R.: Beyond Market Baskets: Generalizing Association Rules to Dependence Rules. *Data Mining and Knowledge Discovery.* **2** (1998) 69–96
12. Lauritzen, S.L. and Spiegelhalter, D.J.: Local computations with probabilities on graphical structures and their application to expert systems (with discussion). *Journal of the Royal Statistical Society, Series B.* **50** (1988) 157–224
13. Tanner, M.A.: *Tools for Statistical Inference 3ed.* Springer (1996)
14. Whittaker, J.: *Graphical Models in Applied Mathematical Multivariate Statistics.* Wiley (1990)

Model Building for Random Fields

R.H. Glendinning

Defence Evaluation and Research Agency, Gt. Malvern, WR14 3PS, UK
rhg@signal.dera.gov.uk

Abstract. Random fields are used to model spatial data in many application areas. Typical examples are image analysis and agricultural field trials. We focus on the relatively neglected area of model building [1]and draw together its widely dispersed literature, which reflects the aspirations of a wide range of application areas. We include a spatial analogue of predictive least squares which may be of independent interest.

1 Introduction

We are concerned with the analysis of data collected on a $t_1 \times t_2$ rectangular subset D_T of the two dimensional lattice Z^2. Data of this type is collected in many application areas [10]. Typical examples are image analysis [13] and agricultural field trials [32]. We assume that the data can be described by a random field $X^{D_T} = (X_i, i \in D_T)$, where $T = (t_1, t_2)$. The probabilistic properties of X^{D_T} can be described by a wide variety of models [13]. Their key characteristics are described by the dependence of X_i on neighbouring random variables $(X_{i+l}, l \in N)$. We focus on stationary random fields with zero mean, where two classes of models are widely used. The first are *simultaneous* processes and includes the *kth* order autoregression (SAR)

$$X_i = \sum_{j \in N} \phi_j X_{i+j} + Z_i, \ i \in Z^2, \tag{1}$$

where the innovations $(Z_i, i \in Z^2)$ are independent and identically distributed random variables [47] with zero mean and variance σ_z^2. An important group of models (known as causal or unilateral) have neighbourhoods satisfying $N \subset H = (j_1 < 0, j_2 < \infty) \cup (j_1 = 0, j_2 < 0)$ with $j = (j_1, j_2) \in Z^2$. Models of this type are of particular interest as they are tractable approximations for more complex processes [47]. The second class of models describe spatial Markov processes. These are defined by

$$P(X_i|X_j, j \neq i) = P(X_i|X_j, j \in N), \ i \in Z^2 \tag{2}$$

where $P(Y)$ is the distribution of the random variable Y and N a symmetric *finite* subset of Z^2. While this class includes conditional autoregressions $(CAR's)$ under Gaussian assumptions, the latter differs from (1) as $(Z_i, i \in Z^2)$ are now *dependent* random variables when $N \neq \emptyset$ [26].

F. Hoffmann et al. (Eds.): IDA 2001, LNCS 2189, pp. 299–308, 2001.

We focus on the relatively neglected area of model building for random fields and bring together key contributions from its widely dispersed literature. We view model building as an iterative process of knowledge discovery using the data and prior knowledge. Its key steps are: 1) exploratory analysis to identify a family of candidate models

$$M^T = (M_k(\phi), \phi^{(k)} \in \Phi^{(k)}, k = 1, \ldots, K_T), \tag{3}$$

where the parameters of the kth model are $\phi^{(k)} \in \Phi^{(k)}$ with $\epsilon(k)$ freely adjustable parameters; 2) model selection, the process of identifying a member $\hat{k}_T$ of M^T using the data (although an ensemble of models may be desirable in some applications); 3) an assessment of the selected model or models using diagnostic procedures; 4) if the latter are unsatisfactory, update M^T and repeat steps 2 to 4. Model building may be carried out using the Box-Jenkins paradigm [32] (based on the examination of summary statistics) or the use of model selection criteria and diagnostics in the frequentist or Bayesian [6] paradigms.

There are a number of theoretical and practical difficulties associated with model building for random fields. These include the intractable nature of the likelihood associated with important models, the effect of boundaries, the prevalence of non-nested families of models, the computational demands of large data sets, and the difficulty in choosing an appropriate loss function when spatial prediction is inappropriate. This survey is organized by approach: hypothesis testing, pattern recognition techniques, Bayesian methods, criterion based techniques, step-wise procedures and comparative performance.

2   Hypothesis Testing

Hypothesis tests are used to identify suitable families of models, to directly select a model and generate model diagnostics. A typical example of the first activity is described in Guo and Billard [20], who derive an expression for the asymptotic distribution of a generalized score statistic for data generated by a causal SAR with $N = ((0, -1), (-1, -1), (0, -1))$. This is used to test whether the model generating the data is separable. The latter have tractable properties as their covariance generating functions (cgf's) are proportional to the product of two one-dimensional cgf's [32]. An important class of separable models are generated by $ARMA$ components and denoted by $ARMA * ARMA$. Testing the assumption of separability against non-separability using data generated by an $AR(p) * AR(1)$ is considered by Shitan and Brockwell [42] using a vector autoregressive representation of the data generating mechanism. A number of authors advocate the use of hypothesis testing to select spatial models, see [29] or [22]. A popular paradigm is to test whether model alterations give significant improvements in fit using a likelihood ratio test, although approximations to the likelihood are typically required. The properties of this approach are well known for nested time series models, although the control of type I errors and the selection of null hypotheses for non-nested families of candidate models remain problematic. While this approach is still advocated for spatial model selection

[10] (and more general graphical models), difficulties of this type are circumvented by criterion based methods. Finally we consider diagnostic tests. These include hypothesis tests against specific alternative models or *whiteness* tests for the residuals generated by $M_{\hat{k}_T}$. Rigorous whiteness tests are developed for $ARMA * ARMA$ models in [11], with modifications for separately assessing row and column models described in [14].

3   Pattern Recognition Techniques

The first group of pattern recognition techniques are based on a spatial analogue of the auto, partial and inverse autocorrelation functions. These statistics play a key role in the exploratory phase of model building, where the aim is to suggest a plausible family of models [32], and obtain gross estimates of their neighbours (N), see [25]. Statistics of this type are of particular value for separable $ARMA * ARMA$ models. Here the sample autocorrelation function can be used to suggest the order of the moving average component of $MA * ARMA$ processes, while the partial autocorrelation function is used to suggest the order of the autoregressive component of $AR * ARMA$ processes [15]. The latter also use the sample partial autocorrelation function to test whether the data is generated by an $AR(p) * ARMA$ against $AR(p + 1) * ARMA$ alternatives. The latter is based on the asymptotic properties of the sample auto and partial autocorrelation functions for $AR * ARMA$ or $MA * ARMA$ processes. Next we consider the inverse auto-correlation function [50]. This is the collection of autocorrelations associated with $1/f(x)$, where $f(x)$ is the spectral density of $(X_i, i \in Z^2)$. When the data are generated by a spatial Markov process, it is easy to see that the collection of inverse auto-correlations are given by the parameters of this model. Yuan and Subba Rao [50] estimate the inverse auto-correlation function by replacing $f(x)$ with an appropriate estimate and suggest the informal examination of these estimates as a means of identifying the neighbourhood associated with a spatial Markov process. Formal procedures based on multiple hypothesis testing are unavailable as the asymptotic properties of the sample inverse auto-correlation function are not sufficiently well developed. The second group of procedures are based on the singular value decomposition of matrices constructed from the sample autocorrelation function, see Martin [32] for separable models, Guyon [21], who suggests the use of the *corner* method for spatial $ARMA$ model selection, and Aksasse, Badidi and Radouane [1], who describe a minimum eigenvalue technique for unilateral $SAR's$ with finite quarter plane support $N \in (j : (j_1 < 0, j_2 \leq 0) \cup (j_1 = 0, j_2 < 0))$.

4   The Bayesian Approach to Model Selection

We introduce a *prior* probability over (Φ, χ), where $\chi = (1, \ldots, K_T)$ and $\Phi = \bigcup_{k \in \chi} \Phi^{(k)}$. Put $P(\phi, k) = P(k)P(\phi^{(k)}|k)$ for $k \in \chi$ and $\phi^{(k)} \in \Phi^{(k)}$, where $P(k)$ and $P(\phi^{(k)}|k)$ have the usual meaning. A popular approach is to use the decision rule which selects the model giving the largest posterior probability $P(\hat{k}_T | X^{D_T})$. As analytic expressions for this decision rule are generally unavailable, techniques for numerically estimating $P(k | X^{D_T})$ have been proposed, see Raftery and Banfield in the discussion of [6] or Barker and Rayner [4]. The key step is to draw samples from the posterior distribution $P(\phi, k | X^{D_T})$ using an appropriate scheme. One approach is reversible jump Markov Chain Monte Carlo (MCMC) [3], where earlier work by Grenander is described. A number of authors criticize the selection of a single model and advocate an ensemble of models as a means of propagating model uncertainty (model averaging) [31].

The Bayesian paradigm is well established in many image processing applications [38], where spatial Markov models are commonly used as image priors. However, the large data sets generated by many remote sensing applications currently favour the use of approximations for posterior quantities of interest [30], rather than computationally intensive sampling techniques. This contrasts with other imaging [3] and artificial intelligence applications, see [19] or [31].

5   Criterion Based Techniques

These are decision rules of the form $\hat{k}_T = argmin_k W(X^{D_T}, k)$, where $k = 1, \ldots, K_T$ and $W(.)$ describes the suitability of M_k given the data.

5.1   Akaike's Approach (AIC)

Larimore [29] suggests the following spatial analogue of Akaike's criterion

$$AIC(k) = -2 \log P(X^{D_T}; \hat{\phi}^{(k)}(X^{D_T}), M_k) + 2\epsilon(k), \tag{4}$$

where $P(X^{D_T}; \hat{\phi}^{(k)}(X^{D_T}), M_k)$ is the joint distribution of X^{D_T} evaluated at the maximum likelihood estimate $\hat{\phi}^{(k)}(X^{D_T})$. This gives [5]

$$AIC_G(k) = \log \hat{\sigma}_z^2(k) + 2 \frac{\epsilon(k)}{\chi(D_T)} \tag{5}$$

for complex-valued Gaussian $SAR's$ with finite non-symmetric half plane support $N_k = (j : (p_k \leq j_1 < 0, -q_k \leq j_2 \leq r_k) \cup (j_1 = 0, 0 > j_2 \geq -q_k))$ and $\chi(D_T)$ the cardinality of D_T. The value of the complexity penalty associated with AIC is derived for a wide range of Gaussian Markov processes on Z^d in Künsch [28], where $P(X^{D_T}; \hat{\phi}^{(k)}(X^{D_T}), M_k)$ is replaced by Whittle's spectral approximation. Künsch associates AIC with an estimate of the information gain or Kullback-Leibler divergence between $P(X^{D_T}; \phi^{(k)}, M_k)$ and $P(X^{D_T}; \phi^*, M_*)$, where $M_*(\phi^*)$ is the model generating the data. The value of the penalty term in (5) is known under weaker conditions for real-valued time series using an interpretation of AIC as an estimate of the Kullback-Leibler divergence between predictive distributions [7]. The asymptotic distribution of $\hat{k}_T$ selected by AIC is not known in the spatial case, although the results of [47], [39], [21] or [23] point to the inconsistency of $\hat{k}_T$ under various conditions. However, the use of AIC may be desirable for model based spectral estimation, see [48] for a time series perspective. Note that AIC can be used with more general models [34].

5.2   Final Prediction Error (FPE)

Criteria of this type are used in applications where prediction is a meaningful measure of performance. By analogy with time series procedures, Bello [5] estimates the mean square error associated with the one step ahead prediction error $Y_i - \hat{Y}_i(Y^{B_i \cap D_T}; \hat{\phi}^{(k)}(X^{D_T}), M_k)$ for large T, where $B_i = i + H$ and $\hat{\phi}^{(k)}(X^{D_T})$ is an estimate of $\phi^{(k)}$ using X^{D_T}. Prediction is assessed in Y^{D_T}, an independent sample with the same probabilistic structure as X^{D_T}. After correcting for the bias associated with $\hat{\sigma}_z^2(k)$, the conditional maximum likelihood estimate of σ_z^2 using $\Omega_k(D_T) = (j : ((N_k + j) \cap \bar{D}_T) = \emptyset)$, gives

$$FPE(k) = \hat{\sigma}_z^2(k) \frac{\Omega_k(D_T) + \epsilon(k)}{\Omega_k(D_T) - \epsilon(k)} \qquad (6)$$

for unilateral $SAR's$ with non-symmetric half plane support and complex-valued Gaussian innovations. Tjøstheim [46] suggests a similar criterion with $\Omega_k(D_T)$ replaced by $\chi(D_T)$ for unilateral real-valued $SAR's$ on Z^d with quarter plane support. Parameter estimation is carried out using the spatial Yule-Walker equations. It is easy to see that (6) is asymptotically equivalent to AIC_G.

5.3   Cross-Validation (CV)

While there are numerous versions of CV for non-spatial data, all techniques partition X^{D_T} into two disjoint sets, with parameter estimation carried out in the first and assessment in the second. Here

$$CV_1(k) = \sum_{i \in D_T} \rho \left(X_i - \hat{X}_i(X^{D_T} \setminus i; \hat{\phi}^{(k)}(X^{D_T} \setminus i), M_k) \right) \quad \text{with} \qquad (7)$$

$$\hat{\phi}^{(k)}(X^{D_T} \setminus i) = argmin_{\phi^{(k)}} \sum_{j \in (D_T \setminus i)} \rho \left(X_j - \hat{X}_j(X^{D_T} \setminus i; \phi^{(k)}, M_k) \right) \qquad (8)$$

where $(X^{D_T} \setminus i) = (X_j \in X^{D_T}, j \neq i)$ and $\rho(x)$ a loss function. When $\rho(x) = x^2$, (7) (with (8)) is asymptotically equivalent to AIC for time series data, although its value for spatial model building is the subject of debate [10].

5.4   The Asymptotic Bayesian Approach (BIC)

The form of the decision rule considered in section 4 simplifies as $T \to \infty$. The resulting rule is known as BIC and is derived for Gaussian $SAR's$ and Markov processes with toroidal edge corrections in [26]. Other Gaussian Markov processes are considered in [41], with Gibbs random fields taking values in a finite set (typically grey levels) in [40]. In all cases

$$BIC(k) = -2 \log P(X^{D_T}; \hat{\phi}^{(k)}(X^{D_T}), M_k) + \epsilon(k) \log \chi(D_T) \qquad (9)$$

and $\hat{\phi}^{(k)}(X^{D_T})$, the maximum likelihood estimate of $\phi^{(k)}$ using the kth model. Note that BIC does not depend on the priors for k and $\phi^{(k)}|M_k$. For further work on Bayesian decision rules for spatial data, see Denny and Wright [12], with the time series case described in [48]. As maximum likelihood estimation is intractable for many spatial processes, Seymour and Ji [40] suggest MCMC based estimation and show that the resulting criterion is an asymptotic approximation for the Bayes decision rule described in section 4. An alternative approach is to replace the likelihood term in (9) by the pseudo-likelihood, see [24] and [43]. The asymptotic consistency of the resulting estimate $\hat{k}_T$ is a consequence of two additional propositions [24]. Results describing the asymptotic properties of criteria with general complexity penalties are described in section 5.6.

5.5   Complexity and Coding (MDL and PLS)

Here parameter estimation and model selection are associated with efficient data transmission, see [36] and [49]. This is achieved by associating a binary code with the data X^{D_T} and transmitting an appropriate code. Data values which occur with high probability can be given short codes. An asymptotically optimal coding scheme associates X^{D_T} with a code of approximately $-\log_2 P(X^{D_T})$ digits, where $P(.)$ is the joint distribution of X^{D_T}. Then one model is superior to another, if it leads to a shorter code for X^{D_T} on average. We focus on two part codes which transmit the data and parameter values $\hat{\phi}^{(k)}(X^{D_T})$, where all quantities of interest are expressed with finite precision. We determine the *strict minimum message length* estimates $\hat{\phi}^{(k)}(X^{D_T})$ and $\hat{k}_T$, which minimizes the expected value of

$$L(X^{D_T}, \phi^{(k)}(X^{D_T}), M_k) = L(X^{D_T}|\phi^{(k)}(X^{D_T}), M_k) + L(\phi^{(k)}(X^{D_T}), M_k) \quad (10)$$

where $L(X^{D_T}|\phi^{(k)}(X^{D_T}), M_k)$ is the length of code used to describe X^{D_T} given $\phi^{(k)}(X^{D_T})$, and $L(\phi^{(k)}(X^{D_T}), M_k)$ the analogous quantity for $(\phi^{(k)}(X^{D_T}), M_k)$ [49]. This approach is analytically and computationally intractable due to the discreteness of the sample space when X^{D_T} is recorded with finite precision [48]. The following approximation is used to provide computationally feasible estimates for image processing applications [16]. These are the minimum description estimates (MDL)

$$(\hat{k}_T, \hat{\phi}^{(k)}(X^{D_T})) = argmin_{(\phi^{(k)}, k)} L(X^{D_T}, \phi^{(k)}, M_k) + L(\phi^{(k)}, M_k) \quad (11)$$

where $L(X^{D_T}, \phi^{(k)}, M_k)$ is based on $P(X^{D_T}; \phi^{(k)}, M_k)$, the joint distribution of X^{D_T} given $M_k(\phi^{(k)})$, rather than $P(X^{D_T}|\phi^{(k)}(X^{D_T}), M_k)$, the conditional distribution of X^{D_T}. Further approximations follow from the fact that $\hat{\phi}^{(k)}(X^{D_T})$ is approximately the maximum likelihood estimate for fixed k. Asymptotic considerations lead to a criterion of the same form as BIC. Rissanen [36] demonstrates the equivalence of BIC and MDL using an upper bound for stochastic complexity. This is the marginal distribution of the data after integrating out priors for $\phi^{(k)}|M_k$ and k. Similar results are determined explicitly for $ARMA$ models in [48]. This approach is applied to Gibbs random fields taking values in a finite set [43] and unilateral $SAR's$ [2]. The use of MDL and BIC are now widespread in the image processing community [16]. Complexity based criteria for incomplete data X^{D_T} are described in [9] and [16], with model selection for noise corrupted fields using maximum entropy ideas in [8]. The use of complexity in modeling hierarchical (multi-scale) images is discussed in [33]. An alternative approach for sequentially generated data is described in Rissanen [35]. Here parameter values are not coded as they can be calculated from previously transmitted data. Glendinning [18] uses the lexical ordering implied by B_i to develop a range of sequential criteria. These include

$$PLS(X^{D_T}, k) = \sum_{i \in \Omega_{max}(D_T)} \left(X_i - \hat{X}_i(X^{B_i \cap D_T}; \hat{\phi}(k)(X^{B_i \cap D_T}), M_k) \right)^2, \quad (12)$$

$$\hat{X}_i(X^{B_i \cap D_T}, \hat{\phi}^{(k)}(X^{B_i \cap D_T}), M_k) = \sum_{j \in N_k} \hat{\phi}_j^{(k)}(X^{B_i \cap D_T}) X_{i+j} \quad (13)$$

where $\hat{\phi}_j^{(k)}(X^{B_i \cap D_T})$ are estimates of the parameters of M_k using $X^{B_i \cap D_T}$ and $(i \in \Omega_{max}(D_T))$ the set of lattice points which give unique parameter estimates for the *largest* model contained in M^T. The experiments in [18] show that PLS is a competitive alternative to BIC and HQ (section 5.6).

5.6 Penalized Goodness of Fit (HQ)

Tjøstheim [47] proposes an analogue of the Hannan-Quinn criterion for Gaussian $SAR's$ with non-symmetric half plane support on Z^d. In the spatial case

$$HQ(k) = \log \hat{\sigma}_z^2(k) + \epsilon(k)v(\chi(D_T)) \quad (14)$$

where $v(\chi(D_T)) = 2c\chi(D_T)^{-1} \log \log \chi(D_T)$, $c > 1$ and parameter estimation carried out using least squares. This is the fastest rate of decay of $v(.)$ which ensures that $\hat{k}_T$ is strongly consistent for time series data [48]. Tjøstheim [47] demonstrates weak consistency for $\hat{k}_T$ when $v(\chi(D_T))\chi(D_T)^{-1} \to 0$ and $v(\chi(D_T)) \to \infty$ as $T \to \infty$ for models with finite non-symmetric half plane support. Criteria of this form are applied to cepstral modeling [45], where a truncated Fourier series expansion of the log spectrum is used as a model for Gaussian random fields. More general criteria of the form

$$W(X^{D_T}, k) = U(X^{D_T}; \hat{\phi}^{(k)}, M_k) + \epsilon(k)v(\chi(X^{D_T})) \quad (15)$$

are considered by Senoussi [39], who demonstrates the strong consistency of $\hat{k}_T$ when the complexity penalty decays at an appropriate rate and the true model lies in M^T. Here $U(.)$ is a loss function satisfying certain conditions and $\hat{\phi}^{(k)} = argmin_\phi U(X^{D_T}; \phi^{(k)}, M_k)$. Typically $U(.)$ is some approximation to the likelihood. Similar results are demonstrated in the time series case in [48]. Senoussi's results are applicable to random fields defined in terms of an associated Gibbs random field, with $U(.)$ replaced by the pseudo-likelihood or similar quantities. These results are extended to certain non-ergodic processes in [21]. The conditions required by these results can be difficult to demonstrate.

6   Step-wise Procedures

Spatial sub-set models (typically SAR's) are defined in an analogous way to their time series counterparts. Khotanzad and Bennett [27] suggests a forward selection procedure with time domain whiteness test as a stopping rule and [25] forward selection followed by backward selection, with BIC used as a stopping rule. The latter restrict the size of M^T by excluding implausible models by an initial hypothesis test using a spatial analogue of the multiple partial correlation function. For an alternative approach, see Smyth [44], who restricts attention to models with lower complexity than the data itself. Sarkar, Sharma and Sonak [37] use a singular value decomposition to generate candidate models followed by the application of BIC to the remaining candidate models. Upper bounds for the probability of selecting under or over-determined models using an all sub-set search with (15) are developed in [23].

7   Comparative Performance

In spite of the importance of model selection [17], there are few studies describing the finite sample performance of spatial model selection criteria, see [29], [46], [41], [2]. A notable early study [46] shows that FPE tends to suggest over-determined models using $SAR's$ with quarter plane support, with FPE and BIC performing poorly for model based spectral estimation in data contaminated by period components. We must treat the conclusions of these small scale studies with caution as performance typically depends on the structure of M^T and the parameters of the true model [18]. The experiments described in [18] show that the relative performance of a number of criterion based techniques are similar to their time series analogues for moderately large T. Similarly, little can be drawn from the numerical studies of pattern recognition techniques [2] or hypothesis testing [29], although the former are generally inferior to criterion based methods for time series problems.

References

1. Aksasse, B., Badidi, L., Radouane, L.: A rank test based approach to order estimation-Part I: 2-D AR models application. IEEE Trans. Signal Proc. **47** (1999) 2069–2072
2. Akasse, B., Radouane, L.: Two-dimensional autoregressive (2-D AR) model order estimation. IEEE Trans. Signal Proc. **47** (1999) 2072–2074
3. Andrieu, C., Djurić, P. M., Doucet, A.: Model selection by MCMC computation. Signal Proc. **81** (2001) 19-37.
4. Barker, S. A., Rayner, P. J. W.: Unsupervised image segmentation using Markov random field models. Patt. Rec. **33** (2000) 587-602.
5. Bello, M. G.: A random field model based algorithm for anomalous complex image pixel detection. IEEE Trans. Image Proc. **1** (1992) 186–196
6. Besag, J., York, J., Mollié, A.: Bayesian image restoration with two applications in spatial statistics. Ann. Inst. Statist. Math. **43** (1991) 1–59

7. Bhansali, R. J.: A derivation of the information criteria for selecting auto-regressive models. Adv. Appl. Probab. **18** (1986) 360–387

8. Bouzouba, K., Radouane, L.: Image identification and estimation using the maximum entropy principle. Patt. Rec. Lett. **21** (2000) 691–700

9. Bueso, M. C., Qian, G., Angulo, J. M.: Stochastic complexity and model selection from incomplete data. J. Statist. Plann. Inf. **76** (1999) 273-284

10. Cressie, N.: Spatial Statistics. John Wiley, New York (1992)

11. Cullis, B. R., Gleeson, A. C.: Spatial analysis of field experiments-An extension to two dimensions. Biometrics **47** (1991) 1449–1460

12. Denny, J. L., Wright, A. L.: Inference about the shape of neighboring points in fields. In: Passolo, A. (ed.): Spatial Statistics and Imaging. IMS Lecture Notes Vol. 20 (1991) 46–54

13. Dubes, R. C., Jain, A. K.: Random field models in image analysis. J. Appl. Statist. **16** (1989) 131–164

14. Etchison, T., Brownie, C., Pantula, S. G.: A portmanteau test for spatial ARMA models. Biometrics **51** (1995) 1536–1542

15. Etchison, T., Pantula, S. G., Brownie, C.: Partial autocorrelation function for spatial processes. Statist. Probab. Lett. **21** (1994) 9–19

16. Figueiredo, M. A. T., Leitaõ, J. M. N.: Unsupervised image restoration and edge location using compound Gauss-Markov random fields and the MDL principle. IEEE Trans. Image Proc. **6** (1997) 1089-1102

17. Glendinning, R. H.: An evaluation of the ICM algorithm for image reconstruction. J. Statist. Comput. Simul. **31** (1989) 169–185

18. Glendinning, R. H.: Robust model selection for random fields. *Submitted*

19. Guidici, P., Green, P. J.: Decomposable graphical Gaussian model determination. Biometrika **86** (1999) 785-801

20. Guo, J-H., Billard, L.: Some inference results for causal autoregressive processes on a plane. J. Time Ser. Anal. **19** (1998) 681–691

21. Guyon, X.: Champs aléatoires sur un réseau, modélisations, statistique et applications. Masson, Paris (1993).

22. Guyon, X., Hardouin, C.: The chi-square coding test for nested Markov random field hypotheses. In: Lecture Notes in Statistics Vol. 74, Springer (1992) 165-176

23. Guyon, X., Yao, J-F.: On the under fitting and overfitting sets of models chosen by order selection criteria. J. Mult. Anal. **70** (1999) 221-249

24. Ji, C., Seymour, L.: A consistent model selection procedure for Markov random fields based on penalized pseudolikelihood. Ann. Appl. Probab. **6** (1996) 423–443

25. Kartikeyan, B., Sarkar, A.: An identification approach for 2-d autoregressive models in describing textures. CVGIP: Graphical Models and Image Processing **53** (1991) 121–131

26. Kashyap, R.L., Chellappa, R.: Estimation and choice of neighbors in spatial interaction models of images. IEEE Trans. Inform. Theory **29** (1983) 60–72

27. Khotanzad, A., Bennett, J.: A spatial correlation based method for neighbor set selection in random field image models. IEEE Trans. Image Proc. **8** (1999) 734–740

28. Künsch, H.: Thermodynamics and statistical analysis of Gaussian random fields. Z. Wahrsch. verw. Geb. **58** (1981) 407–421

29. Larimore, W. E.: Statistical inference on stationary random fields. Proc. IEEE **65** (1977) 961–970

30. LaValle, S. M., Moroney, K. J., Hutchinson, S. A.: Methods for numerical integration of high-dimensional posterior densities with application to statistical image models. IEEE Trans. Image Proc. **6** (1997) 1659–1672

31. Madigan, D., Raftery, A. E.: Model selection and accounting for model uncertainty in graphical models using Occam's window. J. Amer. Statist. Assoc. **89** (1994) 1535–1546

32. Martin, R. J.: The use of time series models and methods in the analysis of agricultural field trials. Commun. Statist.-Theory Meth. **19** (1990) 55–81

33. Mobasseri, B.: Expected complexity in hierarchical representations for random fields. IEEE 10th Int. Conf. Patt. Rec. (1990) 508–512.

34. Nishii, R., Kusanobu, S., Nakaoka, N.: Selection of variables and neighborhoods for spatial enhancement of thermal infrared images. Commun. Statist.-Theory Meth. **28** (1999) 965–976.

35. Rissanen, J.: A predictive least squares principle. IMA J. Math. Control Inform. **3** (1986) 211–222

36. Rissanen, J.: Stochastic complexity (with discussion). J. Roy. Statist. Soc. **49** (1987) 223–265.

37. Sarkar, A., Sharma, K. M. S., Sonak, R. V.: A new approach for subset 2-d AR model identification for describing textures. IEEE Trans. Image Proc. **6** (1997) 407–413

38. Schröder, M., Rehrauer, H., Seidel, K., Datcu, M.: Spatial information retrieval from remote sensing images-part II: Gibbs-Markov random fields. IEEE Trans. Geoscience and Remote Sensing **36** (1998) 1446–1455

39. Senoussi, R.: Statistique asymptotique presque súre de modélés statistiques convexes. Ann. Inst. Henri Poincaré **26** (1990) 19–44

40. Seymour, L., Ji, C.: Approximate Bayes model selection procedures for Gibbs-Markov random fields. J. Statist. Plann. Inf. **51** (1996) 75–97

41. Sharma, G., Chellappa, R.: A model based approach for estimation of two dimensional maximum entropy power spectra. IEEE Trans. Inform. Theory **31** (1985) 90–99

42. Shitan, M., Brockwell, P. J.: An asymptotic test for separability of a spatial autoregressive model. Commun. Statist.- Theory Meth. **24** (1995) 2027–2040

43. Smith, K. R., Miller, M. I.: A Bayesian approach incorporating Rissanen complexity for learning Markov random field texture models. In Proc. 15th Int. Conf. on Acoustics, Speech, Signal Proc. (1990) 2317–2320

44. Smyth, P.: Admissible stochastic complexity models for classification problems. Statist. Comput. **2** (1992) 97–104

45. Solo, V.: Modeling of two dimensional random fields by parametric cepstrum. IEEE Trans Inform. Theory **32** (1986) 743–750

46. Tjøstheim, D.: Autoregressive modeling and spectral analysis of array data in the plane. IEEE Trans. Geoscience and Remote Sensing **19** (1981) 15–24

47. Tjøstheim, D.: Statistical spatial series modeling II: some further results on unilateral lattice processes. Adv. Appl. Probab. **15** (1983) 562–584

48. Veres, S. M.: Structure Selection of Stochastic Dynamic Systems. The Information Criterion Approach. Gordon and Breach, New York (1991).

49. Wallace, C. S., Freeman, P. R.: Estimation and inference by compact coding (with discussion). J. Roy. Statist. Soc. **B 49** (1987) 240–265

50. Yuan, J., Subba Rao, T.: Spectral estimation for random fields with applications to Markov modelling and texture classification. In: Chellappa, R., Jain, A. (eds.): Markov Random Fields, Theory and Applications. Academic Press, Boston (1993) 179–209

---

[1] This work was carried out as part of Technology Group 10 of the MOD Corporate Research Programme. ©British Crown Copyright 2001/DERA

Active Hidden Markov Models for Information Extraction

Tobias Scheffer[1,2], Christian Decomain[1,2], and Stefan Wrobel[1]

[1] University of Magdeburg, FIN/IWS, PO Box 4120, 39016 Magdeburg, Germany
[2] SemanticEdge, Kaiserin-Augusta-Allee 10-11, 10553 Berlin, Germany
{scheffer, decomain, wrobel}@iws.cs.uni-magdeburg.de

Abstract. Information extraction from HTML documents requires a classifier capable of assigning semantic labels to the words or word sequences to be extracted. If completely labeled documents are available for training, well-known Markov model techniques can be used to learn such classifiers. In this paper, we consider the more challenging task of learning hidden Markov models (HMMs) when only *partially (sparsely) labeled* documents are available for training. We first give detailed account of the task and its appropriate loss function, and show how it can be minimized given an HMM. We describe an EM style algorithm for learning HMMs from partially labeled data. We then present an active learning algorithm that selects "difficult" unlabeled tokens and asks the user to label them. We study empirically by how much active learning reduces the required data labeling effort, or increases the quality of the learned model achievable with a given amount of user effort.

1 Introduction

Given the enormous amounts of information available only in unstructured or semi-structured textual documents, tools for *information extraction* (IE) have become enormously important (see [5,4] for an overview). IE tools identify the relevant information in such documents and convert it into a structured format such as a database or an XML document [1]. While first IE algorithms were hand-crafted sets of rules (*e.g.,* [7]), researchers soon turned to learning extraction rules from hand-labeled documents (*e.g.,* [9,11]). Unfortunately, rule-based approaches sometimes fail to provide the necessary robustness against the inherent variability of document structure, which has led to the recent interest in the use of hidden Markov models (HMMs) [15,12] for this purpose.

Markov model algorithms that are used for part-of-speech tagging [2], as well as known hidden Markov models for information extraction [12] require the training documents to be labeled *completely, i.e.,* each token is manually given an appropriate semantic label. Clearly, this is an expensive process. We therefore concentrate on the task of learning information extraction models from *partially* labeled texts, and develop appropriate EM-style HMM learning algorithms.

We develop an *active* hidden Markov model that selects *unlabeled* tokens from the available documents and asks the user to label them. The idea of

F. Hoffmann et al. (Eds.): IDA 2001, LNCS 2189, pp. 309–318, 2001.

active learning algorithms (*e.g.,* [3]) is to identify unlabeled observations that would be most useful when labeled by the user. Such algorithms are known for classification (*e.g.,* [3]), clustering [8], and regression [10]; here, we present the first algorithm for active learning of hidden Markov models.

The paper is organized as follows. We give formal account of task and loss function in Section 2, followed by a description of how it can be minimized given an HMM. We then present in detail our EM style algorithm for learning HMMs from partially labeled data (Section 3). We then extend this algorithm to perform active learning in Section 4. We report on our experiments with active hidden Markov models in Section 5.

2   Information Extraction Problem and HMMs

We begin by giving a definition of the task considered in this paper. A document is a sequence of observations, $O = (O_1, \ldots, O_T)$. The observations O_t correspond to the tokens of the document. Technically, each token is a vector of attributes generated by a collection of NLP tools. Attributes may include the word stem, the part of speech, the HTML context, and many other properties of the word, sentence, or paragraph.

The IE task is to attach a semantic *tag* X_i to some of the tokens O_t. Observations can also be left untagged (special tag *none*). An extraction algorithm f maps an observation sequence $O_1, \ldots, O_T$ to a single sequence of tags $(\tau_1, \ldots, \tau_T); \tau_t \in \{X_1, \ldots, X_n, none\}$ (multi-valued assignments would have to be handled by using several IE models, one per label). An IE problem is defined by a joint distribution $P(\tau_1, \ldots, \tau_T, O_1, \ldots, O_T)$ on documents (observation sequences) and their corresponding tag sequences. We can now define the error rate of an extraction algorithm for this IE problem.

Definition 1 (Per-token error). *Assuming that all documents are* finite *token sequences, we can define the* per-token error *of* f *as the probability of tokens with false tags (Equation 1). We write* $f(O)[i]$ *for the ith tag returned by* f *for* O.

$$E_{token}(f) = \int \frac{1}{T} \sum_{t=1}^{T} \delta(f(O)[t], \tau_t) dP(\tau_1, \ldots, \tau_T, O_1, \ldots, O_T) \qquad (1)$$

We can then state the task of *learning* IE models from partially labeled documents as follows.

Definition 2 (Task). *Let D be a distribution* $P(\tau_1, \ldots, \tau_T, O_1, \ldots, O_T)$, $\tau_i \in \{X_1, \ldots, X_n, none\}$ *over observations and tags, H a space of possible IE models (hypotheses).* **Given** *a set E of example vectors drawn according to D in which however some or all of the* τ_i *may be hidden, the task is to* **find** *the IE model* $f \in H$ *to minimize the per-token error.*

Hidden Markov models (see, [13] for an introduction) are a very robust statistical method for structural analysis of temporal data. An HMM $\lambda = (\pi, a, b)$

consists of finitely many states $\{S_1, \ldots, S_N\}$ with probabilities $\pi_i = P(q_1 = S_i)$, the probability of starting in state S_i, and $a_{ij} = P(q_{t+1} = S_j | q_t = S_i)$, the probability of a transition from state S_i to S_j. Each state is characterized by a probability distribution $b_i(O_t) = P(O_t | q_t = S_i)$ over observations. In the information extraction context, an observation is a token. The tags X_i correspond to the n target states $S_1, \ldots, S_n$ of the HMM. Background tokens without tag are emitted in all HMM states $S_{n+1}, \ldots, S_N$ which are not one of the target states. We might, for instance, want to convert an HTML phone directory into a database using an HMM with four nodes (labeled *name, firstname, phone, none*). The observation sequence "John Smith, extension 7343" would then correspond to the state sequence (*firstname, name, none, phone*).

How can the IE task be addressed with HMMs? Let us first consider what we would do if we already knew an HMM which can be assumed to have generated the observations (in Section 3, we will then discuss how learn HMMs from partially labeled texts). Given an observation sequence $O = O_1, \ldots, O_T$, which tag sequence should we return to minimize the per-token error? According to Bayes' principle, for each observation O_t, we have to return the tag X_i which maximizes the probability $P(\tau_t = X_i | O)$. This means that we have to identify a sequence of states $q_1, \ldots, q_T$ which maximize $P(q_t = S_i | O, \lambda)$ and return that tag X_i that corresponds to the state S_i for each token O_t.

We briefly describe some elements that we need for our HMM algorithms and refer to [13] for a more detailed discussion. $\alpha_t(i) = P(q_t = S_i, O_1, \ldots, O_t | \lambda)$ is the forward variable; it quantifies the probability of reaching state S_i at time t and observing the initial part $O_1, \ldots, O_t$ of the observation sequence. $\beta_t(i) = P(O_{t+1} \ldots O_T | q_t = S_i, \lambda)$ is the backward variable and quantifies the chance of observing the rest sequence $O_{t+1}, \ldots, O_T$ when in state S_i at time t.

$\alpha_t(i)$ and $\beta_t(i)$ can be computed recursively as in steps 2, 3, and 4 of the *forward-backward* procedure (Table 1). In step 5, we calculate $P(O|\lambda)$, the probability of observation sequence O given the model λ. We can now express the probability of being in state S_i at time t given observation sequence O (called $\gamma_t(i) = P(q_t = S_i | O, \lambda)$) in terms of α and β, as in step 6 of Table 1. Given an observation sequence $(O_1, \ldots, O_T)$ that has been generated by HMM λ, we can minimize the *per-token* error by returning the sequence of states $q_t^* = S_i$ that maximize $\gamma_t(i)$. Table 2 shows how we can assign semantic tags to the tokens of a document such that the per-token error is minimized.

3   Learning HMMs from Partially Labeled Documents

We have now seen how we can use a given HMM for information extraction, but we still have to study how we can learn the parameters λ of an HMM from a set $\mathcal{O} = \{O^{(1)}, \ldots, O^{(n)}\}$ of example sequences. We assume that the user labels some of the *tokens* (not whole sequences) with the appropriate tag. We express this by means of a labeling function $l : \{O_t^{(s)}\} \rightarrow \{X_1, \ldots, X_n, unknown, nolabel\}$. The user may label a token $(l(O_t^{(s)}) = X_i)$ which means that the HMM must have been in state S_i while emitting $O_t^{(s)}$. Tokens may be labeled as not possessing

Table 1. Forward-Backward algorithm

---

1. **Input** Observation sequence $(O_1, \ldots, O_T)$, HMM λ.
2. **Let** $\alpha_1(i) = \pi_i b_i(O_1)$. **Let** $\beta_T(i) = 1$.
3. **For all** $t = 1 \ldots T$ and $i = 1 \ldots N$ **Let** $\alpha_{t+1}(j) = \left(\sum_{i=1}^{N} \alpha_t(i) a_{ij} \right) b_j(O_{t+1})$.
4. **For all** $t = T \ldots 1$ and $i = 1 \ldots N$ **Let** $\beta_t(i) = \sum_{j=1}^{N} a_{ij} b_j(O_{t+1}) \beta_{t+1}(j)$.
5. **Let** $P(O|\lambda) = \sum_{i=1}^{n} \alpha_T(i)$
6. **For** $t = 1 \ldots T$, $i = 1 \ldots N$, **Let** $\gamma_t(i) = \frac{\alpha_t(i)\beta_t(i)}{P(O|\lambda)}$.

Table 2. Information extraction using hidden Markov models

---

1. **Input** Document $Doc = (w_1, \ldots w_T)$; HMM λ; set of tags $X_1, \ldots X_n$ corresponding to target HMM states $S_1, \ldots, S_n$.
2. **Call** the tokenizer, POS tagger, parser, etc, to generate sequence $O = (O_1, \ldots, O_T)$ of augmented tokens, where each O_t is a vector containing word w_t and attributes. The attributes refer to properties of the word w_t, the sentence, or the paragraph in which w_t occurs.
3. Call Forward-Backward and **Let** $q_t^* = \max_{S_i} \gamma_t(i)$ for all $t = 1 \ldots T$.
4. **For** $t = 1 \ldots T$
   (a) **If** $q_t^* = S_i \in \{S_1, \ldots, S_n\}$ (target state) **Then Output** "$\langle X_i \rangle w_t \langle /X_i \rangle$".
   (b) **Else** (background state) **Output** w_t.

---

a tag $(l(O_t^{(s)}) = nolabel)$ which implies that the HMM must have been in one of the background states, or may be left unlabeled $(l(O_t^{(s)}) = unknown)$ which says nothing about the state.

The Baum-Welch algorithm finds the *maximum likelihood (ML) hypothesis* λ which maximizes $P(\mathcal{O}|\lambda)$. Unfortunately, there is no obvious relation between the ML hypothesis and the Bayes hypothesis that minimizes the expected error given sample $\mathcal{O}$. The ML hypothesis does not take the prior probability of hypotheses $P(\lambda)$ into account. In practice, ad-hoc regularization mechanisms have to be applied in order to avoid finding ML hypotheses that explain the data well but are very unlikely a priori. We use Laplace smoothing with an additive value of 0.0001; maximum entropy is a suitable regularization technique, too [12].

Baum-Welch is an instantiation of the EM algorithm. The main difficulty that this algorithm addresses is that, in order to estimate the transition and emission probabilities, we have to know the state sequence that corresponds to each of the observation sequences. But, unless all observations are labeled with one of the tags, we need to know the transition and emission probabilities in order to calculate the probability of a state sequence, given one of the observation sequences. The algorithm starts with a random model, and then interleaves calculation of the state probabilities (the E-step) with estimation of the transition and emission probabilities based on the calculated state probabilities (the M-step). This bootstrapping procedure converges towards a stable (at least local) optimum with small error rate. We will only elaborate on those elements of the Baum-Welch algorithm that are required to explain our modifications; we refer the reader to [13] for detailed explanations.

Two more variables are needed to define the algorithm; we have to modify their definition and introduce a third to adapt the algorithm to partially labeled sequences. $\xi_t(i, j)$ is the probability of a transition from S_i to S_j at time t. When O_t and O_{t+1} are unlabeled, it is defined in Equation 2 and can be calculated as in Equation 4. In Equation 3, we split ξ into two parts; the right hand side of Equation 3 is equal to our definition of $\gamma_t(i)$. In Equation, 5, we factor this definition of $\gamma_t(i)$ and introduce a new term $\gamma_t'(i, j)$ (defined in Equation 6) for the residual of Equation 5.

$$\xi_t(i, j) = P(q_t = S_i, q_{t+1} = S_j | O, \lambda) \tag{2}$$

$$= P(q_t = S_i | O, \lambda) P(q_{t+1} = S_j | q_t = S_i, O, \lambda) \tag{3}$$

$$\xi_t(i, j) = \frac{\alpha_t(i) a_{ij} b_j(O_{t+1}) \beta_{t+1}(j)}{P(O|\lambda)} \tag{4}$$

$$= \left(\frac{\alpha_t(i) \beta_t(i)}{P(O|\lambda)} \right) \left(\frac{a_{ij} b_j(O_{t+1}) \beta_{t+1}(j)}{\beta_t(i)} \right) = \gamma_t(i) \gamma_t'(i, j) \tag{5}$$

$$\gamma_t'(i, j) = P(q_{t+1} = S_j | q_t = S_i, O, \lambda) = \left(\frac{a_{ij} b_j(O_{t+1}) \beta_{t+1}(j)}{\beta_t(i)} \right) \tag{6}$$

Let us now consider the case in which a token is labeled X_k. In this case, we have to force $\gamma_t(i)$ to one for $i = k$ and to zero for all other i (step 3(c)iii of Table 3). Similarly, $\gamma_t'(i, j) = 1$ if O_{t+1} is labeled X_j and zero for other indices (step 3(c)vi). When $l(O_t) = nolabel$, then we we know that the HMM cannot be in any of the target states. In this case, we have to set these probabilities to zero and renormalize (step 3(c)ii for γ and step 3(c)v for γ'). Table 3 shows the resulting Baum-Welch algorithm which maximizes the likelihood of the token sequences while obeying the constraints imposed by the manually added labels.

Theorem 1. *When the documents are* completely unlabeled *($l(O_t^{(s)} = unknown$ for all t, s), then $P(O|\lambda)$ increases at each iteration or stays constant in which case it has reached a (local) maximum. When all documents are completely labeled ($l(O_t^{(s)} = X_i$ for some i and all t, s), then the algorithm stabilizes after the first iteration and λ maximizes $P(l(O_1^1), \ldots, l(O_{T_m}^{(m)})|\lambda)$, the likelihood of the labels.*

4 Active Revision of Hidden Markov Models

Unlabeled documents can be obtained very easily for most information extraction problems; only labeling tokens in a document imposes effort. An active learning approach to utilizing the available amount of user effort most effectively is to select, from the available unlabeled tokens, the "difficult" ones of which to know the labels that would be most interesting.

Table 3. Baum-Welch Algorithm for partially labeled observation sequences

---

1. **Input** Set $\mathcal{O}$ of m token sequences $\mathcal{O} = \{(O_1^{(1)}, \ldots, O_{T_1}^{(1)}), \ldots, (O_1^{(m)}, \ldots, O_{T_m}^{(m)})\}$.
   Labeling function $l : \{O_t^{(s)}\} \to \{X_1, \ldots, X_n, unknown, nolabel\}$. Number of states
   N. Optionally an initial parameter set λ.
2. **If** no initial model λ is provided **Then** initialize (π, a, b) at random, but making
   sure that $\sum_{i=1}^{N} \pi_i = 1$, $\sum_o b_i(o) = 1$ (for all i), $\sum_{j=1}^{N} a_{ij} = 1$ (for all i), and that
   $0 < p < 1$ for all probabilities $p \in \{\pi, a, b\}$.
3. **Repeat**
   (a) **For** all $s = 1 \ldots m$ Call the forward-backward procedure to calculate the
       $\alpha_t^{(s)}(i)$, $\beta_t^{(s)}(i)$, and $\gamma_t^{(s)}(i)$.
   (b) **For** all $s = 1 \ldots m$ **Let** $P(O^{(s)}|\lambda) = \sum_{i=1}^{N} \alpha_T^{(s)}(i)$.
   (c) **For** $s = 1 \ldots m$, $t = 1 \ldots T_s$, and $i = 1 \ldots N$ **Do**
       i. **If** $l(O_t^{(s)}) = unknown$ **Then Let** $\gamma_t(i) = \frac{\alpha_t^{(s)}(i)\beta_t^{(s)}(i)}{P(O^{(s)}|\lambda)}$.
       ii. **Else If** $l(O_t^{(s)}) = nolabel$ **Then Let** $\gamma_t^{(s)}(i) = \frac{\alpha_t^{(s)}(i)\beta_t^{(s)}(i)}{\sum_{j=n+1}^{N} \alpha_t^{(s)}(j)\beta_t(j)}$ for
          $i > n$, and 0 for all $i \le n$.
       iii. **Else Let** $\gamma_t^{(s)}(i) = 1$ for $l(O_t^{(s)}) = X_i$ and 0 for all other i.
       iv. **If** $l(O_{t+1}^{(s)}) = unknown$ and $t < T_s$ **Then Let** $\gamma_t'^{(s)}(i,j) = \frac{a_{ij}b_j(O_{t+1})\beta_{t+1}^{(s)}(j)}{\beta_t^{(s)}(i)}$ for all $j = 1 \ldots N$.
       v. **Else If** $l(O_{t+1}^{(s)}) = nolabel$ and $t < T_s$ **Then Let** $\gamma_t'^{(s)}(i,j) = \frac{a_{ij}b_j(O_{t+1})\beta_{t+1}^{(s)}(j)}{\sum_{k=n+1}^{N} a_{ik}b_k(O_{t+1})\beta_{t+1}^{(s)}(k)}$ for all $j > n$, and 0 for all $j \le n$.
       vi. **Else If** $t < T_s$ **Then Let** $\gamma_t'^{(s)}(i,j) = 1$ for $l(O_{t+1}^{(s)}) = X_j$, and 0 for all
           other j.
       vii. **If** $t < T_s$ **Then Let** $\xi_t^{(s)}(i,j) = \gamma_t^{(s)}(i)\gamma_t'^{(s)}(j)$ (frequency of a transition
            from S_i to S_j at time t).
   (d) **Let** $\pi_i = \frac{1}{m} \sum_{s=1}^{m} \gamma_1^{(s)}$ (expected frequency of starting in state S_i).
   (e) **For** $i = 1 \ldots N$, $j = 1 \ldots N$ **Let** $a_{ij} = \frac{1}{m} \sum_{s=1}^{m} \frac{\frac{1}{m}\sum_{s=1}^{m}\sum_{t=1}^{T_s-1}\xi_t^{(s)}(i,j)}{\frac{1}{m}\sum_{s=1}^{m}\sum_{t=1}^{T_s-1}\gamma_t^{(s)}(i)}$ (ex-
       pected number of transitions from S_i to S_j / number of transitions from S_i).
   (f) **For** $i = 1 \ldots N$ and all observation values o **Let** $b_i(o) = \frac{1}{m} \sum_{s=1}^{m} \frac{\sum_{t=1}^{T_s}\gamma_t^{(s)}\cdot\delta(O_t^{(s)}=o)}{\sum_{t=1}^{T_s}\gamma_t^{(s)}}$ (expected frequency of observing o when in state
       S_i; δ returns 1 if $O_t^{(s)} = o$, 0 otherwise).
4. **Until** the probabilities a_{ij} and $b_j(O_t)$ stay nearly constant over an iteration.
5. **Output** parameters $\lambda = (\pi, a, b)$.

When our objective is to minimize the *per-token error*, then a Bayes-optimal
extraction algorithm has to label each token $O_t^{(s)}$ with the tag X_i that maxi-
mizes $P(S_i|O^{(s)}, \lambda) = \gamma_t^{(s)}(i)$ (or *none*, if $i > n$). We deviate from this optimal
strategy when our parameter estimates λ' differ from the true parameters λ such
that some state S_j seems to be most likely although state S_i really is most likely

($\max_i P(q_t = S_i | O, \lambda) \neq \max_i P(q_t = S_i | O, \lambda')$) We can see the difference between the probability of the most likely and that of the second most likely state as the confidence of the state given O. When such confidence values or margins can be defined for instances, then we often see empirically that instances with low margins are most relevant to adjust the hypothesis parameters (*e.g.*, [16,3]).

In analogy, we define the *margin* $M(q_t | O, \lambda)$ of the token that we read to time t as the difference between the highest and second highest probability for a state (Equation 8).

$$M(q_t | O, \lambda) = \max_i \{ P(q_t = S_i | O, \lambda) \} - \max_{j \neq i} \{ P(q_t = S_j | O, \lambda) \} \qquad (7)$$

$$= \max_i \{ \gamma_t(i) \} - \max_{j \neq i} \{ \gamma_t(j) \} \qquad (8)$$

Intuitively, the margin can be seen as quantifying how "difficult" (low margin) or "easy" (large margin) an example is. Our active HMM learning algorithm (Table 4) first learns an initial model λ_1 from a set of partially labeled documents. It then determines the margins of all tokens and starts asking the user to label those tokens which have a particularly low margin. The Baum-Welch algorithm is restarted, using the previous parameters λ_{k-1} as initial model and adapting λ_k to the new data.

Table 4. Active Revision of hidden Markov Models

---

1. **Input** Set $\mathcal{O}$ of m Sequences $\mathcal{O} = \{ (O_1^{(1)}, \ldots, O_{T_1}^{(1)}), \ldots, (O_1^{(m)}, \ldots, O_{T_m}^{(m)}) \}$ of tokens; Labeling function $l : O_t^{(s)} \mapsto \{ S_1, \ldots, S_n, unknown, nolabel \}$; Number of states N; query parameter n_q.
2. **Call** the Baum-Welch algorithm to determine initial parameters λ_1.
3. **For** $k = 2 \ldots \infty$ **Repeat**
   (a) **For** $s = 1 \ldots m$ **Call** the forward-backward algorithm.
   (b) **For** $s = 1 \ldots m$ and $t = 1 \ldots T_s$ **Let** $M(q_t | O^{(s)}) = \max_i \{ \gamma_t(i) \} - \max_{j \neq i} \{ \gamma_t(j) \}$.
   (c) **Ask** the user to label the n_q unlabeled tokens $O_t^{(s)}$ ($l(O_t^{(s)}) = nolabel$) with smallest margin $M(q_t | O^{(s)})$. Update the labeling function l.
   (d) **Estimate** the new model parameters λ_k using the Baum-Welch algorithm with initial model λ_{k-1}.
4. **Until** the user gets tired of labeling data.
5. **Return** HMM parameters $\lambda_i = (\pi, a, b)$.

---

5   Experiments

For our experiments, we generated HMMs with variable numbers of background and target states at random. We used these HMMs to generate unlabeled observation sequences; we label a number of initial tokens drawn at random. We

then study how the error developes with the number of additional labels added
to the observation sequences according to three strategies. The "random" strag-
egy is to label randomly drawn observations; the "margin" strategy is to label
"difficult" tokens with smallest margins; this corresponds to our active hidden
Markov model. As a control strategy ("large margins"), we also try selecting
"easy" tokens that have the largest margins. If "margins" is really better than
"random", we expect "large margins" to perform worse.

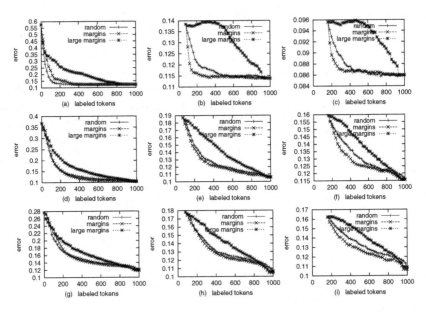

Fig. 1. Error rate of active and regular learning over number of labeled tokens.
(a)-(c), easy HMM; initial sample contains (a) no (b) 80 (c) 160 labeled tokens
drawn at random. (d)-(f) medium size HMM; initial sample with (d) no (e) 80
(f) 160 initial labels. (g)-(i) large HMM; (g) no (h) 80 (i) 160 initial labels.

We used three different HMM sizes. The "easy" HMM consists of one background and two target states. Each state emits three out of 20 observations
with randomly drawn probabilities. We generated 50 sequences of 20 initially
unlabeled observations. The "medium size" HMM possesses 10 nodes, and the
"large" HMM consists of 15 states. The curves in Figure 1 are averages over 50
leaning problems. The initial sample contains only unlabeled tokens (Figures 1a
for the easy, 1d for the medium size and 1g for the hard learning problem), labels
of 80 (Figures 1b, 1e, and 1h, respectively) and 160 tokens (Figures 1c, 1f, and
1i) drawn at random.

In Figures 1a, 1d, and 1g we see a slight but significant advantage of random
token selection over selecting tokens with small margins. Using only difficult

tokens from the beginning is not beneficial on average. In the later phase of learning, token selection by small margins gains a small advantage. The benefit of the margin strategy becomes more clearly visible, when the initial sample contains the labels of 80 (Figures 1b, 1e, and 1h) or 160 (Figures 1c, 1f, and 1i) tokens drawn at random, and only from then on tokens with smallest margins are selected. For the small HMM learning problem (when much unlabeled data relative to the problem complexity is available), the bottom line error is reached after about 300 labels under the margin strategy and after 600 labeled tokens under the random strategy.

Using active learning with small margin examples after 70 initial random tokens seems to be most beneficial. In this case (Figure 1b), the base level error is reached after less than 200 examples for active and after 1000 examples regular learning – *i.e.*, five times fewer labels are needed. Choosing only the most "easy" examples (large margins) is clearly a bad strategy in all cases. Our experiments show that, at least for the classes of HMM that we generated, using only difficult low-margin tokens for learning from the beginning results in higher error rates. However, when the training is started by labeling randomly drawn tokens and the active HMM chooses difficult low-margin tokens after that initial phase, then a significant improvement is achiever over regular HMMs that can result in the sufficiency of many times fewer labeled examples.

6   Discussion and Related Work

We defined a hidden Markov model that operates on partially labeled observation sequences. We defined the margin of tokens as a measure of their difficulty. Our experiments show that it is particularly important to know the labels of difficult examples with low margins. We observed that active HMMs sometimes require three times fewer examples than regular HMMs to achiev a high level of accuracy.

An alternative hidden Markov model for information extraction has been described by [12]. Instead of using the usual distributions a_{ij} and $b_i(O_t)$, the alternative model uses a conditional probability $P(q_{t+1}|q_t, O_{t+1})$. This appears appropriate at first blush but has considerable consequences.

Firstly, the number of probabilities to be estimated is $|Q|^2 \times |O|$ as opposed to $|Q|^2 + |Q| \times |O|$ in our model. Secondly, $P(O|\lambda)$ cannot be computed in the alternative model which renders it impossible to use such HMMs for text classification which seems possible, in principle, in our model. The Baum-Welch algorithm of [12] requires all tokens to be labeled with exactly one state. However, the modifications proposed here could be applied to the alternative model as well. See [14] for a more detailed comparison of the two models.

A restriction of the model discussed here is that it is not possible to enclose a token sequence in nested tags. Only when all tokens are labeled with a tag that corresponds to exactly one state, then cascaded Markov models [2] solve this problem. For the more general case of partially labeled documents, the hierarchical hidden Markov model [6] has to be adapted to partially labeled token sequences, analogously to our adaptation of the standard HMM.

References

1. T. Berners-Lee. Semantic web road map. Internal note, World Wide Web Consortium, 1998.
2. T. Brants. Cascaded markov models. In *Proceedings of the Ninth Conference of the European Chapter of the Association for Computational Linguistics*, 1999.
3. D. Cohn, Z. Ghahramani, and M. Jordan. Active learning with statistical models. *Journal of Artificial Intelligence Research*, 4:129–145, 1996.
4. Mark Craven, Dan DiPasquo, Dayne Freitag, Andrew K. McCallum, Tom M. Mitchell, Kamal Nigam, and Seán Slattery. Learning to construct knowledge bases from the World Wide Web. *Artificial Intelligence*, 118(1-2):69–113, 2000.
5. L. Eikvil. Information extraction from the world wide web: a survey. Technical Report 945, Norwegian Computing Center, 1999.
6. S. Fine, Y. Singer, and N. Tishby. The hierarchical hidden markov model: Analysis and applications. *Machine Learning*, 32:41–64, 1998.
7. Ralph Grishman and Beth Sundheim. Message understanding conference - 6: A brief history. In *Proceedings of the International Conference on Computational Linguistics*, 1996.
8. Thomas Hofmann and Joachim M. Buhmann. Active data clustering. In *Advances in Neural Information Processing Systems*, volume 10, 1998.
9. N. Hsu and M. Dung. Generating finite-state transducers for semistructured data extraction from the web. *Journal of Information Systems, Special Issue on Semistructured Data*, 23(8), 1998.
10. Anders Krogh and Jesper Vedelsby. Neural network ensembles, cross validation, and active learning. In *Advances in Neural Information Processing Systems*, volume 7, pages 231–238, 1995.
11. N. Kushmerick. Wrapper induction: efficiency and expressiveness. *Artificial Intelligence*, 118:15–68, 2000.
12. Andrew McCallum, Dayne Freitag, and Fernando Pereira. Maximum entropy Markov models for information extraction and segmentation. In *Proceedings of the Seventeenth International Conference on Machine Learning*, 2000.
13. L. Rabiner. A tutorial on hidden markov models and selected applications in speech recognition. *Proceedings of the IEEE*, 77(2):257–285, 1989.
14. T. Scheffer, S. Hoche, and S. Wrobel. Learning hidden markov models for information extraction actively from partially labeled text. Technical report, University of Magdeburg, 2001.
15. Kristie Seymore, Andrew McCallum, and Roni Rosenfeld. Learning hidden markov model structure for information extraction. In *AAAI'99 Workshop on Machine Learning for Information Extraction*, 1999.
16. V. Vapnik. *Statistical Learning Theory*. Wiley, 1998.

Adaptive Lightweight Text Filtering

Gabriel L. Somlo and Adele E. Howe

Colorado State University
Computer Science Dept.
Fort Collins, CO 80523, U.S.A.
{howe, somlo}@cs.colostate.edu

Abstract. We present a lightweight text filtering algorithm intended for use with personal Web information agents. Fast response and low resource usage were the key design criteria, in order to allow the algorithm to run on the client side. The algorithm learns adaptive queries and dissemination thresholds for each topic of interest in its user profile. We describe a factorial experiment used to test the robustness of the algorithm under different learning parameters and more importantly, under limited training feedback. The experiment borrows from standard practice in TREC by using TREC-5 data to simulate a user reading and categorizing documents. Results indicate that the algorithm is capable of achieving good filtering performance, even with little user feedback.

1 Introduction

Text filtering makes binary decisions about whether to disseminate documents that arrive from a dynamic incoming stream. *Adaptive* filtering systems [3] start out with little or no information about the user's needs, and the decision of whether to disseminate a document must be made when the document becomes available. The system is given feedback on each disseminated document to update its user profile and improve its filtering performance.

In this paper, we present a lightweight filtering system designed for use in personal Web information agents. We assess how well the algorithm works with little feedback, a requirement for its application as a *personal* information gathering agent. Also, we assess the effect of algorithm parameters on robustness.

We make two key contributions in our research. First, our algorithm adapts standard filtering techniques to the needs of personalized web applications: lightweight, privacy protecting and responsive to user provided examples. The algorithm learns a profile of user information interests. Second, we adapt a rigorous evaluation method to web systems by using text filtering benchmarks to simulate user behavior. Traditionally, Web systems have been evaluated with user studies, with the disadvantages of slow data collection, little experimental control and decreased objectivity of conclusions. Relying on simulated user feedback allows us to test many alternative design decisions *before* subjecting the system to a user study, which means we are less likely to waste subjects' time and are more likely to produce a well tuned system.

F. Hoffmann et al. (Eds.): IDA 2001, LNCS 2189, pp. 319–329, 2001.

2   Filtering Algorithm

At its core, our filtering algorithm uses TF-IDF vectors [8] to represent documents and topic queries. As documents arrive from the incoming stream, they are first transformed into TF (term frequency) vectors, by associating each distinct word in the document with its frequency count, and then are weighted by each word's IDF (inverse document frequency). The components of the vectors are computed using the well-known formula: $TFIDF_{d,t} = TF_{d,t} \cdot IDF_t$ where $TF_{d,t}$ is the term frequency of t in document d. For any given term, its IDF is computed based on the fraction of the total documents that contain this term at least once: $IDF_t = \log \frac{D}{DF_t}$ where D is the number of documents in the collection, and DF_t is the number of documents containing t (document frequency).

Traditionally, D and DF values assume the existence of a fixed corpus, in which documents are known a priori. In filtering, however, we do not assume random access to all documents. Thus, D and DF are approximated: every time a new incoming document is processed, we increment D and the entries into DF that correspond to the terms contained in the new document.

The similarity between two TF-IDF vectors (e.g., between a query and a document vector) is computed using the cosine metric [8], i.e., the normalized dot product between the two vectors. This measures the cosine of the "angle" between the two vectors, and ranges from 1 (when the vectors have the same direction) to 0 (when the vectors have nothing in common).

For filtering, topic queries generalize subjects of interest to a user. In addition to a TF-IDF vector, each topic query in the profile maintains a dissemination threshold, the minimum similarity between the vector of an incoming document and the topic's query vector required for disseminating the document. Dissemination thresholds range between $[0, 1]$. When a new document arrives, its vector is compared to each topic query vector; if the similarity exceeds the dissemination threshold for at least one topic, the document is shown to the user.

Filtering algorithms start with profile queries created by converting a description of each topic into a TF-IDF vector; the vectors are updated based only on feedback from documents that have been disseminated [3]. Our algorithm starts with an empty profile. Learning occurs from positive examples of relevant document-topic pairs provided by the user. For our application, new topics are created in the user profile when the user submits a relevant document for a nonexistent topic; the document's TF-IDF vector forms the initial query vector for the topic and the dissemination threshold is initialized to 0.5. Earlier, pilot experiments showed that the initial value did not matter; feedback quickly moves the value within a short distance of its apparent ideal.

For existing topics, the query vector is learned using a mechanism inspired by relevance feedback [7]. Both the original query vector and the document vector are normalized to eliminate the influence of document length, after which the document vector is weighted by a document weight w and added to the query vector. The document weight controls how much influence the original query vector and the new document vector will each have on the updated query.

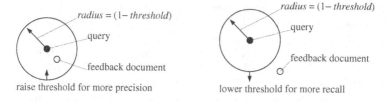

Fig. 1. Learning the dissemination threshold

```
AddDocToTopic(doc, topic)
    s = similarity(doc.vector, topic.vector);
    normalize(doc.vector);
    normalize(topic.vector);
    topic.vector + = w · doc.vector;
    normalize(topic.vector);
    topic.threshold + = α · (s − topic.threshold);
```

Fig. 2. Topic learning mechanism

Like the query vector, the dissemination threshold also needs to be adapted. Callan [1] proposes a dissemination threshold that starts low and grows in time to improve precision. The system presented in [9,10] features dissemination thresholds that can both increase and decrease, but the algorithm heavily relies on numerous cached documents to optimize the dissemination threshold.

To keep our filtering system lightweight, we prohibit the algorithm from caching past document vectors. Instead, the threshold tracks the similarity between the original query vector and the latest feedback document, with a learning rate α. We can imagine the original query vector to be at the center of a "bubble", with a radius given by the dissemination threshold. If the new feedback document is inside this bubble, we can raise the threshold (shrink the bubble), to improve precision. For feedback documents outside the bubble, we must lower the threshold to improve recall. This process is illustrated graphically in Figure 1. Unlike Callan's method, ours permits non-monotonic changes to the threshold vector. The learning algorithm, for both the query vector and the dissemination threshold, is presented in pseudocode in Figure 2. The pseudocode includes two critical parameters: w (document weight) and α (learning rate).

3   Personal Web Information Agents

Personal Web information agents, such as Letizia [6], WebWatcher [4] and Web-Mate [2], can be tailored, over time, to help their users find information that matches their own interests. The helper agent intercedes between the user and their Web access, learning a model of user requests to be used in modifying requests and/or filtering results. Letizia and WebWatcher recommend links that are likely to lead to interesting documents by pre-fetching pages (Letizia) or comparing the profile to a fixed collection (WebWatcher). WebMate uses greedy incremental clustering to build its user profile.

Web information agents are similar to text filtering systems. Both types of systems examine an incoming document stream; agents often generate the stream for themselves by crawling the Web or by extracting links from a Web page during browsing. Both types of systems maintain user profiles. Both systems make binary decisions about the documents in the stream: whether to recommend them to the user based on their similarity to the user profile. Thus, we propose that filtering algorithms such as the one presented in Section 2 be used with Web helper agents, and that performance evaluation of these agents be done with rigorous methods borrowed from IR, and specifically from text filtering.

We simulate the behavior of the user with data from a text filtering benchmark test. The agent starts with an empty profile. The user can create new topics by showing the agent a relevant document and providing a topic name. The agent will compare all new documents from its incoming stream against each topic in the profile, and decide whether to disseminate them to the user. How the incoming stream is generated is beyond the scope of this paper, but any method used by the Web agents mentioned earlier in this section will do.

Usually, in text filtering, only disseminated documents are used for feedback [3]. Our scenario relaxes this requirement by allowing users to present documents they encountered independently when surfing the Web. At any time, a user can create a new topic or reinforce an existing one. It is likely though that users will stop providing feedback on a given topic after a certain number of documents, and expect the agent to continue its filtering task without further instruction.

4   Performance Evaluation

The purpose of our study is to determine the robustness of our filtering system, in particular its sensitivity to the amount of feedback and to parameter settings. The experiment proceeds as follows: For each new incoming document, the system predicts relevance based on the current user profile. The new document is compared to each internal profile topic vector, and a record is made when any of the dissemination thresholds have been passed. Then, the "user" provides feedback for relevant documents, which is simulated with the relevance ratings in the datasets. The new document is added to all topics for which it is known to be relevant. Our independent variables are:

w document weight that controls how much influence the new document vector exerts on the updated query. Values below one favor the original query vector. Our experiments used: 0.1, 0.15, 0.2, 0.25, 0.5, 1.0, 2.0, and 4.0.

α the learning rate for the dissemination threshold, with values of 0.1, 0.3, 0.5, 0.7, and 0.9.

U_{max} the maximum number of updates (documents provided as feedback) for each topic, with values of 5, 10, 20, 30, 40, 50, 60, 70, 100, 200, and ∞ (as many as possible).

We use two datasets: Foreign Broadcasting Information Service (FBIS) collection, and Los Angeles Times (LATIMES) collection from TREC[1] disk #5. The FBIS and the LATIMES collections consist of 48527 and 36330 news articles, classified into 194 topics. Because the first document that is relevant to a topic is always used to create an entry for the topic in the user profile, we only use topics with more than one relevant document, which reduces the total number of topics to 186.

For each combination of independent variables, we record:

RF number of relevant documents found correctly as relevant;

RM number of relevant documents missed (incorrectly identified as non-relevant);

NF number of non-relevant documents found as relevant;

These values are used to compute the following performance metrics:

Recall	number of relevant documents correctly disseminated	$recall = \frac{RF}{RF+RM}$
Precision	number of relevant disseminated documents	$precision = \frac{RF}{RF+NF}$
LGF	performance metric proposed by Lewis and Gale [5], assume equal weights for recall and precision.	$LGF = \frac{2 \cdot precision \cdot recall}{precision+recall}$
LF1, LF2	linear utility functions from TREC-8 evaluation [3]	$LF1 = 3RF - 2NF$ $LF2 = 3RF - NF$

5   Results

First, we test the influence of algorithm parameters under ideal conditions, when all possible learning information is used ($U_{max} = \infty$). Then, we evaluate robustness by reducing the amount of feedback to the algorithm, i.e., smaller values of U_{max}. A single user building their own web profile (our application) cannot be expected to provide thousands of relevance judgments for every topic.

5.1   Impact of Parameters α and w when $U_{max} = \infty$

A two-way Analysis of Variance indicates that both α and w, individually and jointly, strongly influence the value of recall ($p < 0.00001$). Recall is consistently good for values of $\alpha \geq 0.7$ and $w \leq 0.5$, but strongly decreases outside this region. A surface plot as well as an ANOVA table for each dataset are shown in Figure 3. Performance depends primarily on α and is close to optimal as long as w is within a range of values. The best parameter values are in Table 1.

Precision depends strongly on α ($p < 0.00001$). As one would expect, precision is highest for the smallest tested value of $\alpha = 0.1$, the most conservative

---

[1] TREC (Text REtrieval Conference) document collections can be ordered from http://trec.nist.gov

	FBIS					LATIMES				
Metric:	Rec.	Prec.	*LGF*	*LF*1	*LF*2	Rec.	Prec.	*LGF*	*LF*1	*LF*2
α	0.9	0.1	0.1	0.1	0.1	0.9	0.1	0.3	0.1	0.1
w	0.5	0.5	0.2	0.5	0.25	0.5	1.0	0.5	1.0	0.25
Value	0.511	0.516	0.453	4533	8441	0.493	0.510	0.386	1012	2097

Table 1. Best parameter values for each metric

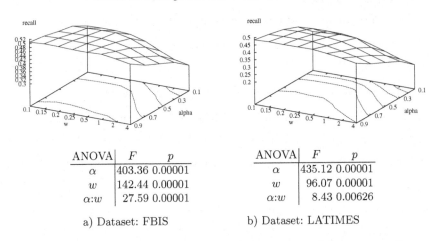

ANOVA	F	p
α	403.36	0.00001
w	142.44	0.00001
$\alpha{:}w$	27.59	0.00001

a) Dataset: FBIS

ANOVA	F	p
α	435.12	0.00001
w	96.07	0.00001
$\alpha{:}w$	8.43	0.00626

b) Dataset: LATIMES

Fig. 3. Effects of learning rate α and document weight w on recall

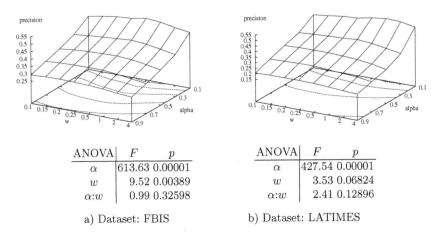

ANOVA	F	p
α	613.63	0.00001
w	9.52	0.00389
$\alpha{:}w$	0.99	0.32598

a) Dataset: FBIS

ANOVA	F	p
α	427.54	0.00001
w	3.53	0.06824
$\alpha{:}w$	2.41	0.12896

b) Dataset: LATIMES

Fig. 4. Effects of learning rate α and document weight w on precision

update. The document weight w also has a significant effect on precision, albeit much smaller. The best values for precision seem to occur for $w \in [0.25, 1.0]$. The interaction effects are negligible. 0.5104 at $\alpha = 0.1$ and $w = 1.0$ for LATIMES.

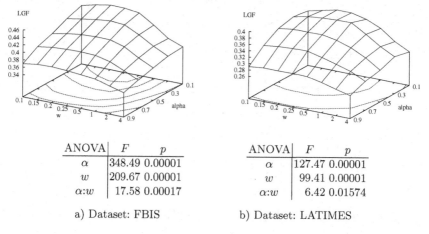

ANOVA	F	p
α	348.49	0.00001
w	209.67	0.00001
$\alpha{:}w$	17.58	0.00017

a) Dataset: FBIS

ANOVA	F	p
α	127.47	0.00001
w	99.41	0.00001
$\alpha{:}w$	6.42	0.01574

b) Dataset: LATIMES

Fig. 5. Effects of learning rate α and document weight w on LGF

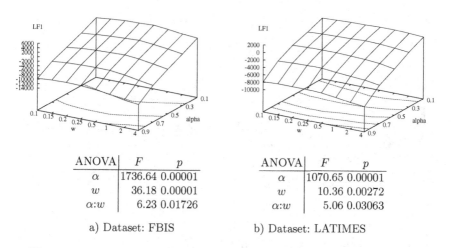

ANOVA	F	p
α	1736.64	0.00001
w	36.18	0.00001
$\alpha{:}w$	6.23	0.01726

a) Dataset: FBIS

ANOVA	F	p
α	1070.65	0.00001
w	10.36	0.00272
$\alpha{:}w$	5.06	0.03063

b) Dataset: LATIMES

Fig. 6. Effects of learning rate α and document weight w on TREC8 $LF1$

The previous analysis considers recall and precision in isolation, which is clearly unrealistic for our application. The measures that encompass both recall and precision (LGF, $LF1$, $LF2$) show strong effects of w and α (see Figures 5,6,7). LGF and $LF1$ show interaction effects. LGF shows almost best performance within the interval, $\alpha \le 0.3$ and $w \le 0.5$.

In conclusion, with unlimited training, the best learning parameters for our algorithm are $w \in [0.2, 1.0]$, and $\alpha \le 0.3$ for everything except recall, where values of $\alpha \ge 0.9$ give the best results. Fortunately for our application, performance is always pretty good within the same interval on *both* datasets; thus, we would expect our best parameter settings to generalize to other data sets.

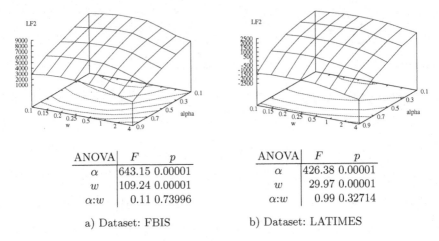

ANOVA	F	p
α	643.15	0.00001
w	109.24	0.00001
$\alpha{:}w$	0.11	0.73996

a) Dataset: FBIS

ANOVA	F	p
α	426.38	0.00001
w	29.97	0.00001
$\alpha{:}w$	0.99	0.32714

b) Dataset: LATIMES

Fig. 7. Effects of learning rate α and document weight w on TREC8 $LF2$

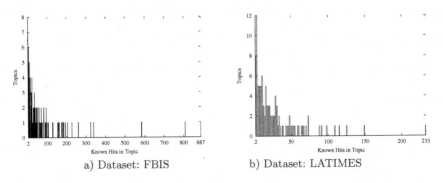

a) Dataset: FBIS   b) Dataset: LATIMES

Fig. 8. Distribution of known relevant documents across topics

5.2   Robustness with Less Feedback (U_{max})

The two datasets vary wildly in the number of relevant documents per topic. Histograms of the distribution of topics according to the number of known relevant documents they contain are given in Figure 8. To determine the effects of limiting training, we had to focus on topics for which many documents were available. We used the topics in the fourth quartile of the above distributions: this includes 36 topics from the FBIS dataset, with 75 or more known relevant documents, and 36 topics from the LATIMES dataset, with 31 or more known relevant documents. These topics include the majority of known relevant documents in our datasets: 7,678 for FBIS, and 2,368 for LATIMES.

We plotted recall, precision, and LGF against U_{max} for each combination of α and w in Figure 9. These plots show that the interaction effects between U_{max} and the two parameters are small for both datasets. Additionally, the effect of

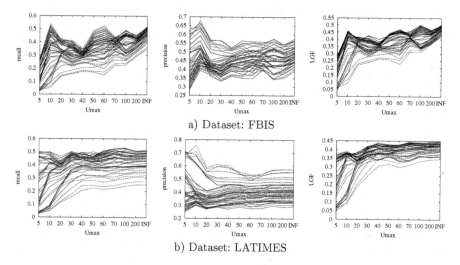

a) Dataset: FBIS

b) Dataset: LATIMES

Fig. 9. Influence of U_{max} on recall, precision, and LGF for combinations of α and w

U_{max} is non-monotonic, and that while no parameter combination dominates all values of U_{max}, some are consistently among the top few.

ANOVA tests indicate that U_{max} strongly influences recall: $F = 128.34, p < 0.00001$ for FBIS, and $F = 36.30, p < 0.00001$ for LATIMES. In comparison, the influence on precision is significant only on the FBIS dataset ($F = 47.36, p < 0.00001$ for FBIS, $F = 1.02, p < 0.31$ for LATIMES). The U_{max} on LGF is significant for both datasets: $F = 100.59, p < 0.00001$ for FBIS, and $F = 31.07, p < 0.00001$ for LATIMES. In contrast to the unlimited updates, limited updates favor high values of α, while w remains in the same range. This result follows from the need to make the most out of the available data.

Interestingly, our plots show that very good results can be obtained after as few as 10 updates per topic. After 10 updates per topic for FBIS, we reached 0.4489 recall and 0.4501 precision, resulting in 0.4495 LGF at $\alpha = 0.9$ and $w = 0.5$; LATIMES showed 0.4055 recall, 0.3818 precision, and LGF of 0.3933, at $\alpha = 0.7$ and $w = 1.0$. As an example, Figure 10 displays the surface plots of LGF at $U_{max} = 10$ and $U_{max} = \infty$.

6   Conclusions

We presented a lightweight filtering algorithm intended for use with Web information agents, and an evaluation method for such algorithms based on TREC benchmarks. We ran a factorial experiment to test the algorithm's robustness. We found that performance is robust within a relatively wide interval of parameter values. Importantly for our target application, we found that the algorithm

328   Gabriel L. Somlo and Adele E. Howe

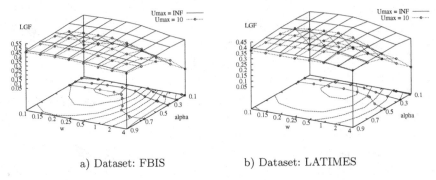

a) Dataset: FBIS   b) Dataset: LATIMES

Fig. 10. Influence of U_{max} on LGF for the FBIS dataset

is robust against limited training information; we observed a slow degradation in performance as the amount of feedback was reduced down to 10.

When used in a Web information agent, our algorithm will need to train with up to 10 or 20 positive examples per topic. In consequence, the algorithm must start out with a large learning rate for the dissemination threshold. An idea we intend to test in the future is to lower α as more documents are used to train a particular topic.

Using benchmarks instead of user studies has allowed us to collect large amounts of information in a short period of time. First, we intend to test our algorithm on more datasets, starting with the Financial Times document collection used in the filtering track at recent TREC conferences. Ultimately, when these algorithms have been incorporated into the web agent, we will test the efficacy of the design decisions derived here in a user study.

References

1. Callan, J. Learning While Filtering Documents *Proceedings of the 21st International ACM-SIGIR Conference on Research and Development in Information Retrieval*, Melbourne, Australia, 1998.
2. Chen, L., Sycara, K. WebMate: A Personal Agent for Browsing and Searching. *Proceedings of the Second International Conference on Autonomous Agents*, Minneapolis, MN, 1998.
3. Hull, D.A., Robertson, S. The TREC-8 Filtering Track Final Report *Proceedings of the Eighth Text REtrieval Conference (TREC-8)*, Gaithersburg, MD, 1999.
4. Joachims, T., Freitag, D., Mitchell, T. WebWatcher: A Tour Guide for the World Wide Web. *Proceedings of the 15th International Joint Conference on Artificial Intelligence (IJCAI-97)*, Nagoya, Japan, 1997.
5. Lewis, D.D., Gale, W.A. A Sequential Algorithm for Training Text Classifiers *Proceedings of the 17th International ACM-SIGIR Conference on Research and Development in Information Retrieval*, Dublin, Ireland, 1994.
6. Lieberman, H. Letizia: An Agent That Assists Web Browsing. *Proceedings of the 14th International Joint Conference on Artificial Intelligence (IJCAI-95)*, Montreal, Canada, 1995.

7. Rocchio, J.J. Relevance Feedback in Information Retrieval *Salton, G. (ed.), The SMART Retrieval System: Experiments in Automatic Document Processing* Prentice-Hall, 1971.

8. Salton, G. *Automatic Text Processing: The Transformation, Analysis, and Retrieval of Information by Computer* Addison-Wesley, 1988.

9. Zhai, C., Jansen, P., Stoica, E., Grot, N., Evans, D.A. Threshold Calibration in CLARIT Adaptive Filtering *Proceedings of the Seventh Text REtrieval Conference (TREC-7)*, Gaithersburg, MD, 1998.

10. Zhai, C., Jansen, P., Roma, N., Stoica, E., Evans, D.A. Optimization in CLARIT TREC-8 Adaptive Filtering *Proceedings of the Eighth Text REtrieval Conference (TREC-8)*, Gaithersburg, MD, 1999.

A General Algorithm for Approximate Inference in Multiply Sectioned Bayesian Networks*

Zhang Hongwei, Tian Fengzhan, and Lu Yuchang

Department of Computer Science and Technology Tsinghua University
State Key Laboratory of Intelligent Technology and Systems Beijing China 100084
zhw@s1000e.cs.tsinghua.edu.cn, tfz99@mails.tsinghua.edu.cn,
lyc@tsinghua.edu.cn

Abstract. Multiply Sectioned Bayesian Networks(MSBNs) extend the junction tree based inference algorithms into a coherent framework for flexible modelling and effective inference in large domains. However,these junction tree based algorithms are limited by the need to maintain an exact representation of clique potentials. This paper presents a new unified inference framework for MSBNs that combines approximate inference algorithms and junction tree based inference algorithms, thereby circumvents this limitation. As a result our algorithm allow inference in much larger domains given the same computational resources. We believe it is the very first approximate inference algorithm for MSBNs.

Keywords: Multiply sectioned Bayesian network; Bayesian inference

1 Introduction

Bayesian networks (BNs) [6,16] provide a coherent and effective framework for inference with uncertain knowledge.However,when the problem domains become larger and more complex, modelling a domain as a single BN and conducting inference in it become increasingly more difficult and expensive, even intractable. Multiply Sectioned Bayesian Networks (MSBNs) [13] provide an alternative to meet this challenge by relaxing the single BN paradigm. The framework allows a large domain to be modeled modularly and the inference to be performed distributively while maintaining the coherence. It supports objected-oriented modeling and multi-agent paradigm [11]. Many standard exact inference methods[7,8,5] for BNs are extended to MSBNs [4,10,12].However, these junction tree based inference algorithms are limited by the need to maintain an exact representation of clique potentials, which is a Joint Probability Distribution(JPD) representation of some variables and is exponential in the number of the variables of the JPD. For some large BNs, the cliques are too large to allow this exact representation of the factors. The problem can be even more severe in MSBNs whose problem domains are much larger and more complex.

---

* This work is supported by NSF of china grant 79990580 and National 973 Fundamental Research Program grant G1998030414. Thanks the anonymous reviewers for helpful comments.

F. Hoffmann et al. (Eds.): IDA 2001, LNCS 2189, pp. 330–339, 2001.

A completely different approach is taken by Monte Carlo based inference algorithms [14,15]. The basic idea is to estimate the JPD as a set of (possibly weighted) sampling, but ignoring the structure of the BNs. While these algorithms can deal with some tasks that exact inference algorithm can't do, the convergence of these algorithms can be quite slow in high dimension space, where the samples can explore only a small part of the space.

Some hybrid inference methods[1,2,3] for single BNs are developed, which combines the best features of exact and approximate inference algorithm. In this paper we present a general approach to combine exact and approximate technology for inference in MSBNs. Like the standard MSBNs inference algorithms, our approach builds a LJFBU structure and propagates messages between JT-BUs. However, rather than performing exact inference in JTBUs, our algorithm conducts an approximate one. We believe that our algorithm is the very first approximate inference algorithm for MSBNs.

We introduce the basic concept about MSBNs in section 2 and in section 3 we review three types of algorithms for single BNs: exact inference algorithms based on junction tree; approximate inference algorithms based on Monte Carlo sampling and some hybrid algorithms. We extract these algorithms' common characters. In section 4, we discussed the MSBNs inference algorithms, especially the X.Yiang's algorithm that is the basis of our algorithm. In section 5 we present our general inference algorithm. This section is the center of our paper. In section 6 we discuses some improvements to our general algorithm.

2   Preliminaries

In this section, we present briefly the formal theory of the MSBNs framework. We assume that readers are familiar with the basics about the representation of probabilistic knowledge using BNs and the common inference methods for BNs [5,6,7,9,16]. A MSBN M is a collection of Bayesian subnets that together define a BN. M represents probabilistic dependence of a total universe partitioned into multiple sub-domains each of which is represented by a subnet. The partition should satisfy certain conditions to permit coherent distributed inference. One condition requires that nodes shared by two subnets form a *d-sepset*, defined as below.

Definition 1: let $D_i = (N_i, E_i)$ $i = 0, 1$ be two DAGs such that $\mathcal{D} = D_0 \bigcup D_1$ is a DAG. The intersection $I = N_0 \bigcap N_1$ is a *d-sepset* between D_0 and D_1 if for every $x \in I$ with its parents $\pi \subseteq N_0$ or $\pi \subseteq N_1$. Each $x \in I$ is called a *d-sepnode*.

A d-sepset is a sufficient information channel for passing all relevant evidence from one subnet to another. Formally, a pair of subnets is conditionally independent given their d-sepset. Just as the structure of a BN is a DAG, the structure of a MSBN is a Multiply Sectioned DAG (MSDAG)with a hypertree organization:

Definition 2: A hypertree MSDAG $\mathcal{D} = \bigcup_i D_i$ where each D_i is a DAG, is a connected DAG such that (1) there exists a hypertree over $\mathcal{D}$ and (2) each hyperlink d-separates[6,7] the two subtrees that connects.

It is trivial to prove that if each hypernode D_k of a hypertree MSDAG is replaced by the cluster N_k and each hyperlink between D_i and D_j is replaced by their d-sepset I_{ij}, then the resultant is a junction tree. A hyperlink is a sufficient information channel for passing all relevant evidence from one side of hyperlink to the other. Formally, given a hyperlink, the two subtrees connected through the link are conditionally independent. A MSBN is defined as follows:

Definition 3: A MSBN M is a triplet $M = (\mathcal{N}, \mathcal{D}, \mathcal{P})$. $\mathcal{N} = \bigcup N_i$ is the total universe where each N_i is a set of variables. $\mathcal{D} = \bigcup D_i$ (a hypertree MSDAG) is the structure where nodes of each DAG D_i are labelled by elements of N_i. For each $x \in \mathcal{N}$, its occurrence with the most parents (breaking ties arbitrarily) $\pi(x)$ is associated with a probability distribution $P(x|\pi(x))$, and each other occurrence is associated with a constant (trivial) distribution. $\mathcal{P} = \prod_i P_D(N_i)$ is the JPD, where $P_D(N_i) = \prod_{x \in N} P(x|\pi(x))$ is a local distribution over N_i . Each $S_i = (N_i, D_i, P_i)$ triplet is called a subnet of M . S_i and S_j are adjacent if D_i and D_i are adjacent.

Inference in a MSBN can be performed more effectively on a compiled representation, called Linked Junction Forest (LJF) of Belief Universes (LJFBU). A LJFBU has the same hypertree organization as its deriving MSBN. Each hypernode is a Junction Tree Belief Universe (JTBU) relative to a JT converted from its deriving subnet. Each hyperlink includes a linkage tree converted from its deriving d-sepset. Here is a definition of linkage tree given by Y.Xiang in [4].

Definition 4: Let I be the d-sepset between JTs T_a and T_b in a Linked Junction Tree (LJF). A linkage tree L of T_a with respect to T_b is constructed as follows. Initialize L to T_a . Repeat the following on clusters of I until no variable can be removed:

1. Remove a variable $x \notin I$ if x is contained in a single cluster c.
2. If c becomes a subset of an adjacent cluster d after (1), union c into d.

Each cluster l in L is a *linkage*. Define a cluster in T_a that contains l as its *linkage host* and break ties arbitrarily.

It is trivial to proof that a linkage tree is a JT. A linkage tree is an alternative representation of the d-sepset. A triplet $(L, I, B_L(I))$ is called a Linkage Tree of Belief Universes (LTBU) where $B_L(I)$ is a belief table associated with L. When a d-sepset is represented as a LTBU, it allows more compact representation of belief over the d-sepset, smaller clusters of JTs being linked, and more efficient inference. A LJFBU is then defined as follows:

Definition 5: Let M be a MSBN. A LJFBU F derived from M is a triplet (J, L, P) . J is a set of JTBUs each of which is derived from a subnet in M. The JTBUs are organized into a hypertree isomorphic to the hypertree MSDAG of M . L is a set of LTBUs each of which is derived from a pair of adjacent JTBUs in the hyper tree. $P = \dfrac{\prod_i P_J(N_i)}{\prod_j P_L(N_j)}$ is the joint system belief (JSB) where each $P_J(N_i)$ is the belief table of a JTBU and each $P_L(N_j)$ is the belief table of a LTBU.

3   Inference Algorithms for Single BNs Review

In this section we review briefly three kinds of inference algorithms used in Bayesian networks: exact algorithms based on junction tree, approximate algorithms based on Monte Carlo sampling and hybrid algorithms that combine these exact and approximate algorithms.

3.1   Junction Tree Based Algorithms

There have been developed many kinds exact inference algorithms. However the most commonly used exact algorithm is junction tree based algorithms [5,7,8,9] The basic data structure used by these algorithms is called a junction tree, which can be constructed via the process of moralization and triangulation [7,9]. Every node in junction tree has its internal state and can send /receive message from /to its neighbors. Inference in the Bayesian network is then formulate in terms of messages passing in the junction tree. The message exchange follows the constraints:

1. Each node sends one message to each neighbor
2. Each node can send a message to a neighbor after it has received a message from its all other neighbors.

A message sent by a node is prepared on the basis of the messages received and its internal state. Note that the complexity of calculating the exact joint potential,ϕ_U, of a universe, U, is exponential in the number of variables of the universe. In many real world applications, the calculation of ϕ_U is prohibitive due to the size of the configure space of U.

3.2   Monte Carlo Based Algorithms

In addition exact inference algorithms, there have been developed a range of approximate inference algorithms for single BNs [15]. Of these, the most commonly used approach is Monte Carlo sampling. Sampling is a general-purpose technique for inference in probabilistic models; its applicability goes far beyond Bayesian networks. The basic idea is that we can approximate the distribution by generating independent sampling from it. We can then estimate the value of a quantity relative to the original distribution by computing its value relative to our samplings. Let p be a probability distribution over some space Ω,the expectation of any function f is defined as $E_p(f) = \sum_{\omega \in \Omega} f(\omega)p(\omega)$ .Now assume that we can generate a set of random samples from p , $\omega[1], \omega[2], \ldots\ldots, \omega[M]$. We can view these samples as a compact approximation to the distribution p. Therefore, we can approximate $E_p(f) \approx \frac{1}{M} \sum_i^M f(\omega[i])$ .So we need only define some special functions f and then we can approximate the marginal probability we expected (detailed in [15]).Note that the convergence of these approximate algorithms can be quite slow in high dimensional spaces where the samples can explore only a small part of the space.

3.3   Hybrid Algorithms Combining Exact and Approximate Methods

There have been developed several hybrid algorithms, which combine the best features of exact and approximate algorithms [1,2,3]. Dawid provided a preliminary description of a hybrid between Monte-Carlo sampling methods and exact local computations in junction trees [1]. Utilizing the strengths of both methods, such hybrid methods have the potential of expanding the class of problems which can be solved under bounded resources as well as solving problems which otherwise resist exact solutions. Kjaerulff provided a detailed description of a particular instance of such a hybrid scheme called Hugs[2];namely combination of exact inference and Gibbs sampling in discrete Bayesian networks. The extension involves the introduction of a new kind of belief universe, called a Gibbs universe, One of the benefits of which is the computational complexity is not exponential in the number of variables of the universe. However it has a dramatic impact on both the complication process and the various inference steps performed in a junction tree, especially the message-passing scheme of the ordinary junction trees. This calls for an extension of the message-scheduling vocabulary and management different with standard message passing.

In Hugs the algorithm only samples once in a clique and the variable is then restricted to take on one of the sampled values. If the sampled values are poor (due, for example, to an uninformed sampling distribution prior to message propagation), the resulting value space for variable can potentially badly skew the remainder of the computation. Nor does this algorithm contain an iterative phase, which can help address this limitation.

Koller presented a general algorithm for approximate inference [3]. Like the clique tree algorithm, Koller's algorithm also builds a clique tree and propagates messages from one clique to another. However, rather than computing messages that correspond to exact intermediate results, this algorithm uses sampling as a substitute for exact operations within a clique like Hugs, but more importantly it uses iterations to gradually improve the quality of the estimation. We will use Koller's algorithms to perform local inference in our algorithm presented in section 5.

4   Inference in MSBNS

So far inference algorithms in MSBNs mostly are extensions of junction tree based algorithms such as HUGIN, Shafer-Shenoy, lazy propagation methods [10,12,13]. Y.Xiang presented a framework which simple the theory of MSBN [4]. We will discuss this algorithm dentally in this section, based on which a hybrid algorithm for MSBNs will be presented in the next section.

As shown by Jensen[7], the operations *CollectEvidence* and *DistributeEvidence* bring a JTBU internally consistent. We combine the two operations into a single operation *UnifyBelief* as in [4] for simplicity.

Operation 1 *UnifyBelief:* Let T be a JTBU and C be any cluster(also called clique) in T. When *UnifyBelief* is called on T, initiate *CollectEvidence* at C followed by *DistributeEvidence* from C.

Analogous to single BNs that have two operations *CollectEvidence* and *DistributeEvidence*, we define two operation *CollectBelief* and *DistributeBelief* accordingly. After the two operations, MSBNs will get to a global consistence [4]. However when message passing from one subnetwork to another, it must be through a linkage tree not a sepset like in BNs. So we have to introduce two extra operations: *AbsorbThroughLinkage* and *UpdateBelief* for message pass scheduling.

Operation 2 *AbsorbThroughLinkage:* let l be a linkage in linkage tree L between JTBUs T_a and T_b. Let C_a and C_b be the corresponding linkage host of l in T_a and T_b. Let $B_l^*(l)$ be the linkage belief associated with l, and $B_{C_b}^*(l)$ be the linkage belief on l defined in C_b. When *AbsorbThroughLinkage* is called on C_a to absorb from C_b through l perform the following:

1. Updating host belief $B_{C_a}'(C_a) = B_{C_a}(C_a) * \frac{B_{C_b}^*(l)}{B_l^*(l)}$
2. Updating linkage belief $B_l(l) = B_{C_b}^*(l)$

The operation *UpdateBelief* propagates belief from a JTBU to another adjacent JTBU through multiple linkages (a hyperlink) between them. In the HUGIN method, evidence is propagated from a cluster to an adjacent one through a sepset by an operation called *Absorption* [7]. *UpdateBelief* is analogous to *Absorption* but the sender and the receiver are JTBUs and the channel is a d-sepset.

Operation 3 *UpdateBelief:* Let T_a and T_b be adjacent JTBUs and L be the linkage tree between them. When *UpdateBelief* is called on T_a relative to T_b, perform the following:

1. For each linkage l in L, call the host of l in T_a perform *AbsorbThroughLinkage*.
2. Perform *UnifyBelief* at T_a.

After the *UpdateBelief* operation, T_a and T_b are internally consistent. And L is consistent with T_a and T_b.

CollectBelief recursively propagates belief inwards (from leaves towards an initiating JTBU) on the hypertree of a LJFBU. Just as *UpdateBelief* is analogous to *Absorption* at a higher abstraction level, *CollectBelief* is analogous to *CollectEvidence* in the HUGIN method but at the hypertree level.

Operation 4 *CollectBelief:* Let T be a JTBU. Let caller be an adjacent JTBU or the LJFBU. When caller calls T to *CollectBelief*, T perform the following:

1. If T has no neighbor except caller, it performs *UnifyBelief* and return.
2. Otherwise, for each adjacent JTBU Y except caller, call *CollectBelief* in Y. After Y finishes, T performs *UpdateBelief* relative to Y.

DistributeBelief recursively propagates belief outwards (from an initiating JTBU towards leaves) on the hypertree of a LJFBU. *DistributeBelief* is analogous to *DistributeEvidence* in the HUGIN method but at the hypertree level.

Operation 5 *DistributeBelief:* Let T be a JTBU. Let caller be an adjacent JTBU or the LJFBU. When caller calls T to *DistributeBelief*, T performs the following:

1. If caller is a JTBU, performs *UpdateBelief* relative to caller.
2. For each adjacent JTBU Y except caller, call *DistributeBelief* in Y .

CommunicateBelief combines the previous two operations to bring a LJFBU into consistency. *CommunicateBelief* is analogous to *UnifyBelief* (at the JTBU level) but at the LJFBU/hypertree level.

Operation 6 *CommunicateBelief:* When *CommunicateBelief* is initiated at an LJFBU, *CollectBelief* is called at any JTBU T, followed by a call of *Distribute-Belief* at T.

CommunicateBelief brings a LJFBU into global consistency [4]. Given these above operations, the inference in MSBNs can simply be an operation *Communicate-Belief*. What's more, the initialization of MSBN also can be finished with the *CommunicateBelief* operation [4]. So these operation simplify the theory of the MSBNs.

5   A General Approximate Algorithm for MSBN

In the above section, we have discussed how to perform inference in MSBNs. Analogous to inference in BNs where the basic computation is sum out operation in cliques, the basic computation in MSBNs is *UnifyBelief* operation in JTBUs. One inference in MSBNs can be finished after definitive times (relative to the numbers of hyperlinks) *UnifyBelief* operations. So the complexity of inference of MSBNs is mainly determined by the complexity of the local inference in JTBUs. As we have mentioned above, the junction tree based algorithm will become intractable when the clique size (the number of variables in a clique) is too large. Therefore the limitations that exist in BNs inference also restrict the application of MSBNs inference. What's more, because MSBNs' application domain usually is a much larger and more complex one, the subnet can be very complex. As a result the limitation can be more serve in MSBNs.

Intuitively, if we combine exact and approximation inference algorithms in MSBNs like Hugs in BNs, perhaps we can circumvent the limitations mentioned above and can achieve the best of the two worlds. Based on this simple idea we present a general scheme for approximate inference in MSBNs. Analogous to hybrid inference algorithms in BNs [1,2,3], which use sampling as a substitute for exact operations within a clique that is a node of the junction tree, we use an approximate inference as a substitute for exact inference in the subnets which is a hypernode in the hypertree structure of MSBNs. Like standard MSBN inference algorithms, our approach also builds LJFBUs of MSBNs and propagates messages between JTBUs. However, rather than computing messages that corresponds to exact intermediate results, our algorithms uses approximate inference techniques in JTBUs to compute and manipulate the message between JTBUs.

Because of the approximate inference in JTBUs, the message between them is also an approximate one. Now what we eager to know is whether or not we should modify the message-scheduling scheme between subnets and how if need, in order to perform inference properly in MSBNs. Give a glance to this problem, it seems that we certainly need modify the message-scheduling scheme, like Hugs in BNs which extends the message-scheduling scheme of standard junction tree based algorithms. However if we analysis the problem carefully, we can get the conclusion easily that we need not revised the message -scheduling scheme. In fact, after an executing of approximate inference in JTBUs, very approximate universe, U , will have found an approximate potential $\hat{\phi}_U$ for ϕ_U , which can be view as an exact universe in the sense that all subsequent absorptions and generations of messages will be performed in the usual exact manner using $\hat{\phi}_U$ (or modifications of $\hat{\phi}_U$). And as we have known, the messages passing between JTBUs can be implemented by the *AbsorbThroughLinkage* operation. Before this operation, the JTBUs that send message to its neighbors must have been internally consistent because of the performing of *UnifyBelief* operation. In other words, before a JTBU can send a message to its neighbors, every clique potential in it must have found an approximate one. So the message between subnets can be viewed as an exact one for the reason we have mentioned above. Based on this analysis, we can extend easily the X.Yiang's algorithm to an approximate inference framework. In fact we need only redefine the *UnifyBelief* operation, which executes inference in JTBUs.

Operation 7 *UnifyBelief:* Let T be a JTBU . When *UnifyBelief* is called on T, execute any approximate inference algorithm based on junction tree structure in T. Here Koller's algorithm is used.

In order to guarantee the message's absorptions and generations to be done properly, perhaps we need slightly modify the *AborbThroughLinkage* operation, because for every linkage in linkage tree, its linkage host's belief is an approximate one. However this modification depends on the representation of the universe belief used in the approximate algorithm. So we don't discuses it in detail here. Let's assume that the *AborbThroughLinkage* operation can work properly in spite of the approximate potentials of JTBUs and LTBUs.

Based on the above discussion, our approximate inference scheme is very simple: the approximate inference in MSBNs is simply an operation *CommunicateBelief* like Y.xiang's algorithm but replacing the *UnifyBelief* with ours (operation 7).

In fact our approximate inference algorithm in MSBNs can also be described as follows:

1. *Initialization:* chose a hypernode (JTBU) T arbitrarily as the root of the hypertree LJFBUs. As a result LJFBUs is directed from T to its leaves. All hypernode except T has one parent.
2. *Collect Belief Phase:* call the recursive operation *CollectBelief* in T . For each node X when *CollectBelief* is called in it , X calls *CollectBelief* in its all children (if any, else do *UnifyBelief* operation 7).After each child finishes *CollectBelief*, X performs *UpdateBelief* relative to this child.

3. *Distribute Belief Phase:* call the recursive operation *DistributeBelief* in T . For very node X in LJFBUs, when *DistributeBelief* is called in it , for each child of X call *UpdateBelief* relative to X and then calls *DistributeBelief* in this child.

6   Improvements

As mentioned above, in our inference scheme we use an approximate inference as a substitute for exact inference in JTBUs. And for description purposes we assumed that all JTBUs would perform approximate inference instead of exact inference. In general, however, this scheme will probably sub optimal, since, as we know, the more exact the message between JTBUs, the more reliable the result. Thus, an optimal scheme will perform exact inference in any JTBUs whenever possible. In other words, we need only perform approximate inference in JTBUs with very large cliques in which standard exact inference algorithms can't work effectively or can't work at all. However, in practice, the strategy to determine whether or not a JTBU should use exact inference algorithm, will heavily depend on the computational resources we have and the approximate inference algorithm's performance. So we do not discuss it in detail here.

For the similar reason, we can modify the Koller's general algorithm slightly to improve its performance. In other words, we need not find an approximate $\hat{\phi}_U$ for every clique potential ϕ_U , if only we have enough memory to store these exact potentials. The extra benefit is that it can also speed the convergence of the iterations when computing the messages. This idea is also used by HUGs.

7   Conclusions

Multiply Sectioned Bayesian Networks (MSBNs) allow a large domain to be modeled modularly and the inference to be performed distributively, coherently. Many algorithms have been developed for inference in MSBNs such as Y.Xiang's algorithm. Usually these algorithms work well. However, if one of these subdomains is still very complex, these algorithms will break downs for the well-known reason. This limitation hider the practical application of MSBNs

In this paper we present a general inference approximate algorithm based on X.Yiang's algorithm and Koller's algorithm. Our algorithm uses approximate inference technology to perform local inference in JTBUs while communicating message between subnets. So it takes the advantages of the two worlds and can circumvent the limitation mentioned above. Our algorithm retains the modelling flexibility and reduces the runtime space complexity, allowing inference in much lager and complex domains given the same computational resources.

In our scheme, we use the LJFBU as our algorithm's data structure like Y.Xiang's algorithm, and use Koller's general algorithm for performing local inference in JTBUs. Because Y.Xiang's algorithm unifies the BNs and MSBNs inference algorithms [4], and Koller's algorithm unifies many exact and approximate inference algorithms for single BNs [3], so our algorithm is more general one, which unify many well known approximate and exact inference algorithms used in single BNs and MSBNs and can have many instantiations varied along the approximation algorithm used in the JTBUs.

Supposedly, this paper represents the very first attempt to perform approximate inference in MSBNs. So, obviously, a large number of theoretical as well as practical issues need to be addressed. On the practical side, a thorough comparison with standard exact MSBNs inference methods will be conducted in the very near future.

References

1. Dawid, A. P. Kjaerulff, U. Lauritzen, S.L. (1994). Hybrid propagation in junction trees, proceedings of the Fifth international conference on information processing and management of uncertainty in Knowledge-based systems (IPMU).
2. Kjaerulff, U. HUGS: Combining Exact Inference and Gibbs Sampling in Junction Trees, In proc. UAI,1995.
3. Daphne Koller, Uri Lerner, Dragomir Angelov. A General Algorithm for Approximate Inference and Its application to Hybrid Bayes Nets. In proc. UAI, 1999.
4. Y.Xiang. Belief updating in multiply sectioned Bayesian networks without repeated local propagations. International Journal of Approximate Reasoning 23 (2000).
5. G. Shafer. Probabilistic Expert Systems. Society for Industrial and Applied Mathematics Philadelphia, 1996.
6. F.V. Jensen. An introduction to Bayesian networks, UCL Press, 1996.
7. F.V. Jensen. S.L. Lauritzen, K.G. Olesen, Bayesian updating in causal probabilistic networks by local computations, Computational Statistics Quarterly 4 (1990) .
8. S.L. Lauritzen, D.J. Spiegelhalter. Local computation with probabilities on graphical structures and their application to expert systems, Journal of Royal Statistical Society B 50 (1988).
9. A.L. Madsen, F.V. Jensen. Lazy propagation in junction trees, in: Proceedings of the 14th Conference on Uncertainty in Artificial Intelligence, 1998.
10. Y. Xiang. Optimization of inter-subnet belief updating in multiply sectioned Bayesian networks, In proc. UAI,1995.
11. Y. Xiang. A probabilistic framework for cooperative multi-agent distributed interpretation and optimization of communication, Artificial Intelligence 87 (1/2) (1996).
12. Y. Xiang and F.V. Jensen. Inference in Multiply Sectioned Bayesian Networks with Extended Shafer-Shenoy and Lazy Propagation, In proc. UAI,1999.
13. Y. Xiang, D. Poole, M.P.Beddoes. Multiply sectioned Bayesian networks and junction forests for large knowledge based systems, Computational Intelligence 9 (2) (1993).
14. Dagum P,Luby M. An optimal Approximation for Bayesian inference. Artificial Intelligence 1993, 60: 141 153.
15. Neal R M. Probabilistic inference using Markov chain Monte Carlo methods: [Technical Report CRGTR-93-1], Department of computer Science, university of Toronto, 1993.
16. J. Pearl. Probabilistic Reasoning in Intelligent Systems: Networks of Plausible Inference, Morgan Kaufmann, Los Altos, CA, 1988.

Investigating Temporal Patterns of Fault Behaviour within Large Telephony Networks

Dave Yearling[1] and David J. Hand[2]

[1] BTexact Technologies, Adastral Park
Ipswich, United Kingdom
david.yearling@bt.com
http://www.btexact.com
[2] Department of Mathematics, Imperial College,
180 Queen's Gate London, United Kingdom
d.j.hand@ic.ac.uk
http://www.ma.ic.ac.uk

Abstract. This paper outlines a first pass technique that can be used to pin point non-maintainable plant in large telephony networks for further investigation. Due to their size, complexity and low fault count, recognising rogue plant in these copper wire networks can be very difficult. By just concentrating on their respective fault time series, we may characterise plant items using a combined approach based on SQL and renewal theory. SQL provides a pre-analysis filter and renewal theory differentiates plant in terms of their respective hazard functions. From this we can form interpretations of plant maintainability.

1 Introduction

This paper describes a data mining approach to investigating plant reliability within a UK based copper access telephony network. The BT copper access network principally consists of all wire and plant from customer premises to the exchanges. This represents a service to approximately 20 million residential and 6 million business customers.

The approach taken here is a *first pass* or *rough cut* approach. It is a technique that can be used as primary characterisation of the network with respect to its performance prior to further study. Here we focus directly on customer-reported faults and in particular the respective inter-fault times. Using data mining and reliability techniques, we can characterise specific nodes in the network by investigating patterns in their respective fault report histories. We shall specifically focus on the elevated rates and tendencies for faults to cluster in time. Elevated rates may be indicative of frail plant, whilst clustering may be characteristic of poor maintainability, bad workmanship or both. Clearly, plant may be both frail and non-maintainable. Identification of such plant can be used as a precursor to audits and development of replacement strategies by allowing effective use of resources.

F. Hoffmann et al. (Eds.): IDA 2001, LNCS 2189, pp. 340–349, 2001.

2   Data Description

The fault data used for this study is based on all reactive repair work undertaken over a year long period, reported by customers. For the whole UK this represents a very large database; consequently a small warehouse was constructed for data gathering and cleaning. Although the approach is aimed at large datasets, for confidentiality and clarity reasons we shall confine this paper to a small typical subset.

The relationships between faults and their respective causes are not straightforward. Certainly, geographical location is an important indication of network type and health. Consequently we confine our subset to a single geographic area to avoid confounding plant effects with network heterogeneity.

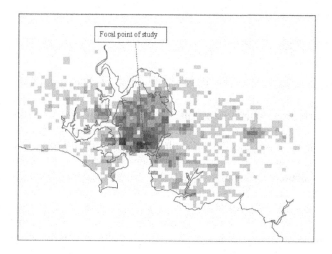

Fig. 1. Density of faults for chosen study area

The focal point of the study was chosen to incorporate the highest number of faults (Fig. 1). The exchange serving the majority of customers within a 1 mile radius of this focal point was determined, and the fault reports were aggregated with respect to the Distribution Points (DP's) which connect customers to the Principal Connection Points (PCP's), (Fig. 2). In this study, we define a 'fault' as any maintenance activity that involves accessing the network to rectify a customer reported problem.

The resulting data consist of daily fault counts by DP over the period with no missing values. This area performs well in terms of fault rate; some 95% of the 1108 DP's experienced less than 4 reported faults for the period. Therefore, the first filtering process removed these DP's allowing us to focus on the remaining 54. The time series for each DP was plotted using a simple visualisation colour grid (Fig. 3). Here DP's are arranged across the top of the plot with time run-

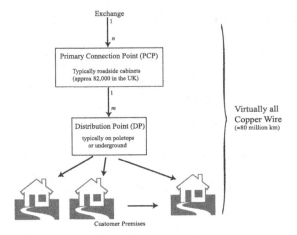

Fig. 2. Schematic description of the network serving the typical customer

ning down the page, as marked. Each cell is coloured by the number of faults for a given day, gradually darkening for counts up to 10. There are clear patterns. Firstly, we note the low counts in each daily cell, those few which are non zero have only 1 or 2 faults (with very few exceptions). Secondly, the DP's substantially differ in the total number of faults associated with them, and thus have different predisposition to faults. Lastly, there is a degree of temporal clustering shown by some of the DP's. Typically, they appear to be very small rarely exceeding a few days. This is indicative of intervention faults; faults caused by engineers entering the network to effect repair but inadvertently causing further defects in the process. We know the vast majority of plant within this area is sound by virtue of an overall low fault count and that all engineers are available to repair all plant. This implies that plant condition rather than poor workmanship may be the principal factor. It is this particular fault pattern that is of direct interest in this paper.

3   Analysis Using Reliability Theory

There are many ways in which we may tackle this problem; arguably the most pertinent lie in epidemiology and reliability applications. Amongst these, two spring readily to mind: an application of an appropriate scan test (for temporal clusters), and a reliability theory approach. The scan test approach [1] appears to be very useful in such applications. However, this technique does not attempt to identify an intuitive explanatory process model, which helps when down streaming solutions to engineers. Therefore, in this paper we concentrate solely on a reliability theory approach.

Reliability theory focuses on the failure times of individuals. Very often we are primarily interested in the first failure, which is usually a catastrophic event (e.g

Year Start

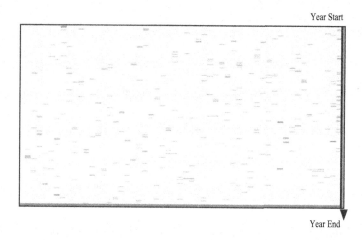

Year End

Fig. 3. Faults by Study DP

death, destruction or machine stoppage). This preoccupation on machine/person morbidity in reliability problems is detailed in [3]. In general we tend to experience situations in which failure maybe inconvenient but certainly not catastrophic. Reliability theory refers to such systems as repairable. That is they can be restored (by either repair or part replacement) to *working order* [3]. Here we simply define *working order* as restoration to a steady state appropriate to the DP. That is, the DP is returned to an operational (but not brand new) state.

Considering our set of DP's as a collection of mutually distinct repairable systems, we now reformulate the data in terms of *fault days* (i.e. a dichotomous variable). Furthermore, we also assume that repairs are instantaneous and as we have daily slots this is reasonable. Indeed, for this period approximately 90% of faults reported are repaired within 24 hours. In this instance, we are principally interested in the pattern of fault rates. That is the change of fault rate given the fact a fault has recently occurred (i.e. a heightened state of risk) in a DP with its own propensity for experiencing faults.

Upon examination of Fig. 3, we are particularly interested in clusters of two or three individual fault days. In the main, reliability techniques applied to repairable systems are concerned with looking for steady deterioration or improvement in reliability. To do this, we refer to the intuitive bathtub curve hazard function operating over the life of the system, and attempt to determine which part of that curve the system is on at the point of study. With respect to this problem, we believe DP systems have a very long lifetime, with a distorted bathtub hazard. A conceptual illustration of the baseline hazard function for a typical DP is shown in Fig. 4, in relation to a year long observation study period. Here, the initial high hazard operates over a period of days or at most weeks, with a gradual decrease in reliability after many years. This position on the bathtub curve may be checked under a Poisson process by testing for trend,

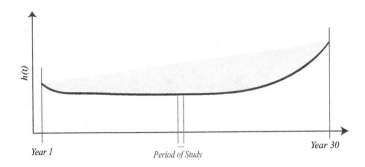

Fig. 4. A conceptual baseline hazard function depicting plant ageing

allowing the rate λ to vary smoothly as a function of time: $\lambda(t)$ [2]. Applying such tests to a selection of the most likely DP's, indicated there was no evidence against the null hypothesis of zero trend.

As we are focusing on small periods of elevated risk immediately after engineer's work on frail plant, we are consequently interested in the transient changes in hazard. A simple approach is to model each DP's fault day history as a renewal process. Here, we have a hazard function that begins at a post-repair 'maximum' swiftly decaying after each repair to the respective DP baseline 'bathtub' hazard. Again, conceptually this is shown in Fig. 5.

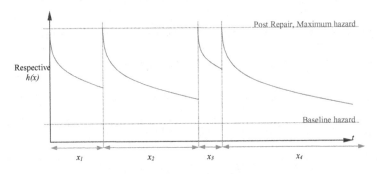

Fig. 5. Proposed renewal process with decaying post repair hazard.

Instead of a Poisson renewal process (with exponential inter-fault times and constant baseline hazard) we need a renewal process with i.i.d inter-fault times which support a view similar to Fig. 5. Considering the most widely used distributions in reliability theory, we can discount several on inspection. The Weibull hazard either decays to zero (for shape parameter less than one) or increases unboundedly (with shape parameters above unity). The lognormal distribution again has an inappropriate hazard rapidly increasing to a maximum followed by a slow decay. Alternatively, the gamma distribution has the desirable property of sharply decaying hazards for shape parameters less then 1, and sharply increasing hazards for shape parameters greater than 1. In both cases, they converge to a constant hazard, the hazard for the unit shape gamma distribution (as with the Weibull, the general exponential distributed case), as illustrated in Fig. 6.

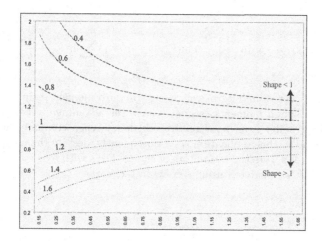

Fig. 6. : Gamma hazard functions with unit scale and various shape values.

4   Gamma Renewal Process as an Exploration Tool

The inter-fault day times for each of the 114 test DP's were taken for the period, omitting the time from study start date to first failure in each case. By fitting gamma distributions to each set of inter-fault times, we can examine the shape parameters and make decisions on the plant stability. To do this we need a computationally economic technique, as the intention is to examine sections of network much larger than the illustrative study set. There are many ways in which we can approach this. We may calculate maximum likelihood estimates (MLE's) for the shape parameters, and then either perform tests or perform parametric bootstraps on the sampling distributions for the shape parameter. Unfortunately there is difficulty with both of these approaches; the inflation of the type I error due to the large number of tests being made, and the large number of maximum likelihood calculations for the bootstrap made especially difficult with small sample sizes (many of the DP's have very small sample sizes). Irrespective of any statistical inference we may wish to make, it is worthwhile to calculate the shapes purely for inspection purposes.

The shape parameters (c, in the gamma density function in equation (1)) were found by maximum likelihood; the log likelihood function is given in (2).

$$f(x) = \left[\frac{x}{b}\right]^{c-1} \frac{e^{-[x/b]}}{b\Gamma(c)} = \frac{x^{c-1}e^{-[x/b]}}{b^c\Gamma(c)} \qquad (c > 0, b > 0) \tag{1}$$

$$\ln L(b,c) = -nc\ln\left[\frac{1}{b}\right] - n\ln[\Gamma(c)] + (c-1)\sum_{i=1}^{n}\ln X_i - \frac{1}{b}\sum_{i=1}^{n}X_i \tag{2}$$

The MLE's were found iteratively using standard moment estimates as starting values. Although moments estimates have the advantage of being trivial to calculate, they suffer from poor relative efficiency when compared with maximum likelihood in this application [7]. Consequently, despite the computational difficulties, MLE's were the estimators of choice. DP's with 5 or more inter-fault times had MLE's calculated using a Newton technique.

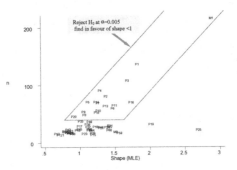

Fig. 7. Gamma shape MLE against sample size, with rejection region.

When examining the shape parameters we are specifically interested in two cases of DP, those with unit shape and those significantly smaller than unity. This will enable us to differentiate between plant experiencing faults with elevated post-repair hazard, from plant which experience faults according to a Poisson process. In order to tackle this, it may be useful to consider possible tests which despite difficulties with multiple testing may provide a useful exploratory aid. One such possibility [4], is based on a one-tailed test for shape parameters which are statistically significant from unity, and employs lookup tables to avoid lengthy maximum likelihood calculations. Under the Null hypothesis of unit shape against the alternative of shape less than one, we calculate the following statistic, [4]:

$T = 2nc \ln[\bar{x}/\tilde{x}]$,   where n is DP sample size

$\bar{x}$ is mean failure time

$\tilde{x}$ is geometric mean failure time

c is calculated from lookup tables, [4]

(3)

Under the null hypothesis T is asymptotically chi-squared, with degrees of freedom (n) taken from lookup tables [4]. We reject the null hypotheses for values of T which exceed this value. Here, the level of the test (α) is largely moot due to the large numbers of tests we may wish to undertake.

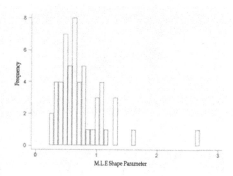

Fig. 8. Distribution of MLE shape parameters over the 54 study DP's.

In order to evaluate this test, a test data set that enjoyed slightly larger sample sizes was used. Several of the units that were non-significant did show noticeable clusters when we examined their inter-fault times in detail. This is put down to the effect of small sample sizes. This is highlighted by the test results obtained ($\alpha = 0.005$ level), where 22 of the 92 units tested have shape parameters significantly less than 1. However, plotting the MLE's against sample size whilst bounding all those which were rejected (Fig. 7), suggests problems applying this test to small sample sizes.

Returning to the MLE's of the study DP's shape parameters, plotting them arrives at the frequency histogram shown in Fig. 8. Apart from the clear skewed nature of Fig. 8, the plot does indicate large differences in the shape parameters with a substantial proportion being less than one. It is these DP's which are of specific interested. In order to identify plant for further study, we need to be able to refer to the sampling distributions of the MLE's in some way. For large samples, the MLE sampling distributions are asymptotically Gaussian, but such distribution theory results are inappropriate for such small sample sizes. What we can do is perform a parametric bootstrap, resampling from gamma distributions with the estimated shapes (remembering that the scale is a nuisance parameter) and recalculating the MLE's for each sample. We choose the parametric bootstrap over a simple data bootstrap primarily for generalisation when this technique is scaled up for larger sets.

Dotted Line
shows unit shape

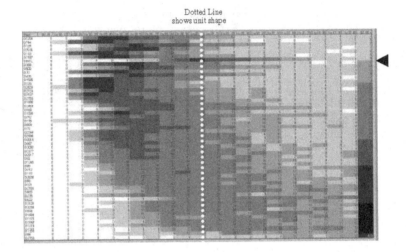

Fig. 9. Shape parameter sampling distributions for each study DP.

Performing the parametric bootstraps, we can construct sampling distributions for each respective DP. These are given in the colour grid shown in Fig. 9, where we have the shape bins ranging across the page (from 0 to 2.0), and each horizontal line on the plot representing one of the 54 DP's. The whole plot was sorted on the final bin size (greater than 2.0) in ascending order. The large values for extreme shape estimates (greater than 2.0) are mainly a feature of the small sample sizes here. Those samples which are small (typically 5) and have one or two larger inter-fault times will have these large modes in their distributions. Despite this, we now have a useful visual aid to decide which DPs to investigate further. Immediately, we can see that DINTL (arrowhead in Fig. 9) has a distribution centred on 1, and with the largest count associated with it we can reasonably conclude that the fault process follows a Poisson process with high rate. Similarly, the DPs with high densities to the left are identified as those units most worthy of further investigation. Furthermore, checking the sample sizes in line with the distributions, we can see no patterns that may raise suspicion of bias in this particular instance.

What we have provided is a structured and semi-automated way of identifying plant that invites further attention by the investigator. Using a combination of standard SQL and statistics, we can filter and drill down relatively quickly to these points of interest.

5   Conclusion

We have illustrated a data mining approach, combining SQL and statistical renewal theory, useful for characterising network plant in terms of temporal fault patterns. The statistical models are intuitive to engineers, and reflect the process of fault propagation in the plant itself. This approach enables engineers and network auditors to evaluate and characterise large homogeneous sections of the network quickly and relatively easily. Also, because it appears in this small study that DP's tend to be similar in sample sizes and parameter shapes, by using the parametric bootstrap we can calculate suitable lookup tables prior to any application.

Although the overall number of faults is small in this study, it must be remembered that we are dealing with a very low level of granularity in the data (almost to the individual customer), and that the small study area we have chosen performs well in terms of fault count. If this approach is scaled up in terms of larger geographical regions, by identifying actual plant items we can begin to attach cost aspects to programs of improvement and quality reforms. It will also highlight ways in which we can create style patterns for DP's that can then give further insight into fault behaviour. Furthermore, this technique can help differentiate poor workmanship from non-maintainable plant.

Clearly, there is a great deal of information that can be obtained form this approach. However, as we have already stated, this is a rough-cut method requiring more work and refinement before it can be used effectively. Consequently, there is a requirement to trial this approach in the field for evaluation and further improvement. This is currently in progress.

References

1. Kulldorff M, Nagarwalla N: Spatial disease clusters: Detection and Inference. Statistics in Medicine. (1995) 14:799-810
2. Cox D R, Lewis P A W: The Statistical Analysis of Series of Events. Methuen (1966)
3. Ascher H, Feingold H: Repairable Systems Reliability Modeling, Inference, Misconceptions and Their Causes. Marcel Dekker,(1984)
4. Bain L J: Statistical Analysis of Reliability and Life-Testing Models, Theory and Methods. Marcel Dekker (1978)
5. Ansell J I, Phillips M J: Practical Problems in the Statistical Analysis of Reliability Data. Applied Statistics (1989) 38, No2, pp205-247
6. Ascher H: Regression Analysis of Repairable Systems Reliability. NATO ASI Series,(1983) Vol F3.
7. Kendal M, Stuart A: Advanced Theory of Statistics - Volume II Inference and Relationship 4th Edition. Griffin and Co ltd. (1979)
8. Cox D R: Renewal Theory. Methuen (1970)
9. Crowder M J, Kimber A C, Smith R L, Sweeting T J: Statistical Analysis of Reliability Data. Chapman and Hall (1991)
10. Hastings NA J, Peacock J B: Statistical Distributions. Butterworth (1975)
11. Tobias P, Trindade D: Applied Reliability 2nd Edition. Van Nostrand (1995)

Closed Set Based Discovery of Representative Association Rules

Marzena Kryszkiewicz

Institute of Computer Science, Warsaw University of Technology
Nowowiejska 15/19, 00-665 Warsaw, Poland
mkr@ii.pw.edu.pl

Abstract. Discovering association rules among items in a large database is an important database mining problem. However, the number of association rules may be huge. The problem can be alleviated by applying concise lossless representations of association rules. There were proposed a few such representations in the late ninetieths. Representative association rules are such an example representation. The association rules, which are not representative ones, may be derived syntactically from representative rules by means of a cover operator. In the paper we show how to discover all representative rules using only closed itemsets and their generators.

1 Introduction

Discovering *association rules* among items in large databases is an important database mining problem. The problem was introduced in [1] for sales transaction database. The association rules can identify sets of items that are purchased together with other sets of items. For example, an association rule may state that 90% of customers who buy butter and bread buy milk as well. Association rules are meaningful also for relational databases [7].

The number of association rules is usually large. A user should not be presented with all of them, but rather with those which are original, novel, interesting. There were proposed several definitions of what is an interesting association rule (see e.g. [7,11,13]. In particular, pruning out uninteresting rules which exploits the information in taxonomies results in the rule number reduction amounting to 60% [13]. The interestingness of a rule is usually expressed by some quantitative measure. Different approach was proposed in [4], where a concise lossless representation of association rules, called *representative rules*, was introduced. As verified experimentally, the number of representative rules can reach even 1% of all association rules [14]. One can derive all or a part of rules of his/her interest (e.g. classification rules for given decision attribute) from representative rules by simple syntactical transformations. There were proposed in [4,5] methods for computing representative rules from *frequent itemsets*. It was shown in [5] that a rule $X \Rightarrow Y$ can be representative only if $X \cup Y$ possesses properties of a *closed itemset*. This observation was applied in the representative

F. Hoffmann et al. (Eds.): IDA 2001, LNCS 2189, pp. 350–359, 2001.

rules generation algorithms offered in [5,12]. The experiments confirm that the closed itemsets are a very compact representation of frequent itemsets in the case of highly correlated data [3]. For alternative concise lossless representations of association rules obtainable from closed itemsets see e.g. [8,9].

In the paper we investigate and prove properties of representative rules. In particular, we prove that predecessors of representative rules are specific itemsets (*generators*). These properties will enable: 1) efficient extraction of a subset of frequent closed itemsets from which representative rules can be generated certainly, 2) decrease in the number of candidate representative rules. A respective algorithm for computing representative rules is proposed as well.

2   Basic Notions and Properties

2.1   Itemsets, Frequent Itemsets, Closures, and Closed Itemsets

Let $I = \{i_1, i_2, \ldots, i_m\}$, $I \neq \emptyset$, be a set of distinct literals, called *items*. In the case of a transactional database, a notion of an item corresponds to a sold product, while in the case of a relational database an item will be an (*attribute,value*) pair. In general, any non-empty set of items is called an *itemset*. Let D be a set of transactions (or tuples, respectively), where each transaction (tuple) T is a subset of I.[1] *Support* of an itemset X is denoted by $sup(X)$ and defined as the number (or the percentage) of transactions in D that contain X. The itemset X is called *frequent* if its support is greater than some user-defined threshold *minSup*. The set of all frequent itemsets will be denoted by F:

$$F = \{X \subseteq I | sup(X) > minSup\}.$$

Closure of the itemset X is denoted by $\gamma(X)$ and is defined as the greatest (w.r.t. set inclusion) itemset that occurs in all transactions in D in which X occurs. Clearly, $sup(X) = sup(\gamma(X))$.

The itemset X is defined *closed* iff $\gamma(X) = X$.

Let X be a closed itemset. A minimal itemset Y satisfying $\gamma(Y) = X$ is called a *generator* of X. By $G(X)$ we will denote the set of all generators of X.

Example 1. Let $T_1 = \{A, B, C, D, E\}$, $T_2 = \{A, B, C, D, E, F\}$, $T_3 = \{A, B, C, D, E, H, I\}$, $T_4 = \{A, B, E\}$ and $T_5 = \{B, C, D, E, H, I\}$ are the only transactions in the database D. $\{A, B, C, D, E\}$ is closed since $\gamma(\{A, B, C, D, E\}) = \{A, B, C, D, E\}$. The itemsets $\{A, C\}$ and $\{A, B, C\}$ are not closed because $\gamma(\{A, C\}) = \{A, B, C, D, E\} \neq \{A, C\}$ and $\gamma(\{A, B, C\}) = \{A, B, C, D, E\} \neq \{A, B, C\}$, respectively. The support of $\{A, C\}$ and $\{A, B, C\}$ is the same as the support of their closure $\{A, B, C, D, E\}$ and is equal to 3. $\{A, C\}$ is a minimal subset occurring in all transactions in which its closure $\{A, B, C, D, E\}$ occurs. Hence, $\{A, C\}$ is a generator of $\{A, B, C, D, E\}$. □

---

[1] Without any loss of generality, we will restrict further considerations to transactional databases.

Property 1 [9]. Let $X \subseteq I$.
a) X is closed iff $\forall Y \supset X, \; sup(X) \neq sup(Y)$.
b) X is a generator iff $\forall Y \subset X, \; sup(X) \neq sup(Y)$.

The next property states that the closure of an itemset X can be computed:
1) as the intersection of the transactions in D that are supersets of X, or
2) as the intersection of the closed itemsets that are supersets of X.

Property 2 [9]. Let $X \subseteq I$.
a) $\gamma(X) = \bigcap \{T \in D | T \supseteq X\}$.
b) $\gamma(X) = \bigcap \{Y \subseteq I | Y$ is *closed* and $Y \supseteq X\}$.

An itemset X is defined *frequent closed* iff X is closed and frequent. In the sequel, the set of frequent closed itemsets will be denoted by *FC*, i.e.:

$$FC = \{X \subseteq I | sup(X) > minSup \text{ and } \gamma(X) = X\}.$$

The itemset X is defined *frequent generator* iff X is a generator and is a frequent itemset. The set of frequent generators will be denoted by *FG*, i.e.:

$$FG = \bigcup \{G(X) | X \in FC\}.$$

The knowledge on *FC* is sufficient to derive all frequent itemsets ($F = \{X \subseteq I | \exists Y \in FC, \; X \subseteq Y\}$). The support of $X \in F$ is equal to the support of its closure. Thus, frequent closed itemsets constitute a concise lossless representation of frequent itemsets. The algorithms for computing frequent closed itemsets require the discovery of frequent generators first. In [8,9,10], generators are treated as seeds of closed itemsets that are determined by intersecting database transactions (according to Property 2a).

In the sequel of the paper, we will apply the following notation:
Let $X \in F$. By $maxSuperSup(X)$ we will denote the maximum from the supports of all frequent closed itemsets that are proper supersets of X, i.e.:

$$maxSuperSup(X) = max\left(\{sup(Y) | Y \in FC \text{ and } Y \supset X\} \cup \{0\}\right).^2$$

Similarly, $minSubSup(X)$ will denote the minimum from the supports of all frequent generators that are proper subsets of X, i.e.:

$$minSubSup(X) = min\left(\{sup(Y) | Y \in FG \text{ and } Y \subset X\} \cup \{\infty\}\right).^3$$

Property 3. Let $X \in F$.
a) $X \in FC$ iff $sup(X) \neq maxSuperSup(X)$.
b) $X \in FG$ iff $sup(X) \neq minSubSup(X)$.
Proof: Follows from Property 1.   □

---

[2] We add $\{0\}$ (for convenience) to handle the case when X is a maximal frequent itemset. Then no information on its proper supersets is available in *FC*. On the other hand, it is known that supports of proper supersets of X are less than $minSup$.

[3] We add $\{\infty\}$ (for convenience) to handle the case when no proper subset of X is in *FG*. Clearly, the support of an empty set is not less than the support of X.

2.2   Association Rules

An association *rule* is an implication $X \Rightarrow Y$, where $\emptyset \neq X, Y \subset I$ and $X \cap Y = \emptyset$. *Support* of the rule $X \Rightarrow Y$ is denoted by $sup(X \Rightarrow Y)$ and defined as $sup(X \cup Y)$. Confidence of $X \Rightarrow Y$ is denoted by $conf(X \Rightarrow Y)$ and defined as $sup(X \Rightarrow Y)/sup(X)$. The problem of mining association rules is to generate all rules that have sufficient support and confidence. In the sequel, the set of all association rules whose support is greater than $minSup$ and confidence is greater than $minConf$ will be denoted by AR.

Association rules are usually generated in two steps (see e.g. [2]):

1. Generate all frequent itemsets F;
2. From each frequent itemset generate association rules the confidence of which is greater than $minConf$. Let $Z \in F$ and $\emptyset \neq X \subset Z$. Then any rule $X \Rightarrow Z \setminus X$ is association one if $sup(Z)/sup(X) > minConf$.

2.3   Rule Cover and Representative Rules

A notion of a *cover operator* was introduced in [4] for deriving a set of association rules from a given association rule without accessing a database. The *cover* C of the rule $X \Rightarrow Y$, $Y \neq \emptyset$, was defined as follows:

$$C(X \Rightarrow Y) = \{X \cup Z \Rightarrow V | Z, V \subseteq Y \wedge Z \cap V = \emptyset \text{ and } V \neq \emptyset\}.$$

Each rule in $C(X \Rightarrow Y)$ consists of a subset of items occurring in the rule $X \Rightarrow Y$. The antecedent of any rule r covered by $X \Rightarrow Y$ contains X and perhaps some items from Y, whereas $r's$ consequent is a non-empty subset of the remaining items in Y. The following properties of the cover operator will be used further in the paper:

Property 4 [4]. Let $r : (X \Rightarrow Y)$ and $r' : (X \Rightarrow Y)$ be association rules.

$$r' \in C(r) \text{ iff } X' \cup Y' \subseteq X \cup Y \text{ and } X' \supseteq X.$$

Next property states that every rule in the cover of another rule has support and confidence not less than the support and confidence of the covering rule.

Property 5 [4]. Let r and r' be association rules.

$$\text{If } r' \in C(r), \text{ then } sup(r') \geq sup(r) \text{ and } conf(r') \geq conf(r).$$

It follows from Property 5 that if r belongs to AR, then every rule r' in $C(r)$ also belongs to AR. The number of different rules in the cover of the association rule $X \Rightarrow Y$ is equal to $3^m - 2^m$, where $m = |Y|$ (see [4]).

Example 2. Let us consider the database from Example 1. Let $r : (B \Rightarrow DE)$. Then, $C(r) = \{B \Rightarrow DE, B \Rightarrow D, B \Rightarrow E, BD \Rightarrow E, BE \Rightarrow D\}$. The support of r is equal to 4 and its confidence is equal to 80%. The support and confidence of all other rules in $C(r)$ are not less than the support and confidence of r. □

Knowing that an association rule covers several other association rules tempts to restrict generation of association rules to those that cannot be derived by applying the cover operator. This type of rules was introduced in [4] and was called *representative association rules*. Formally, a set of *representative association rules* is denoted by RR and defined as follows:

$$RR = \{r \in AR | \neg \exists r' \in AR, \ r' \neq r \text{ and } r \in C\,(r')\}.$$

Representative rules constitute a concise lossless representation of all association rules since the union of covers of rules in RR equals to AR. As verified experimentally, the ratio RR/AR depends strongly on the number of long frequent itemsets, which in turn depends on the $minSup$ value [14]. The more long frequent itemsets in data (smaller $minSup$ value), the smaller the ratio. However, even more important factor than the value of $minSup$ is the character of the mined data. The number of AR is much higher for highly correlated data than for weakly correlated data of the same size. The RR representation compensates this difference since the ratio RR/AR is much smaller in the case of highly correlated data.

Example 3. Given $minSup = 2$ and $minConf = 77\%$, the following representative rules would be found for the database D from Example 1: $RR = \{AC \Rightarrow BDE, AD \Rightarrow BCE, B \Rightarrow CDE, C \Rightarrow BDE, D \Rightarrow BCE, E \Rightarrow BCD, A \Rightarrow BE, B \Rightarrow AE, E \Rightarrow AB\}$. There are 9 rules in RR, whereas the number of all rules in AR is 93. Hence, RR constitutes 9.68% of AR. $\quad\square$

In the literature, there were proposed the following algorithms for computing representative rules: *GenAllRepresentatives* [4], *FastGenAllRepresentatives* [5], and *Generate-RAR* [12].

GenAllRepresentatives. The algorithm generates representative rules from frequent itemsets based on the following property:

Property 6 [4]

$$RR = \{(X \Rightarrow Y) \in AR | \neg \exists (X' \Rightarrow Y') \in AR, \ (X = X' \wedge X \cup Y \subset X' \cup Y') \text{ or }$$
$$(X \supset X' \wedge X \cup Y = X' \cup Y')\}.$$

Property 6 implies that an association rule $X \Rightarrow Z \setminus X$ is representative one if:

- there is no longer rule $X \Rightarrow Z' \setminus X$ in AR such that $Z \subset Z'$, and
- there is no rule with shorter antecedent $X' \Rightarrow Z \setminus X'$ in AR such that $X \supset X'$.

FastGenAllRepresentatives. The algorithm generates representative rules from frequent itemsets based on the two following properties:

Property 7 [5]. Let $X \Rightarrow Z \setminus X \in AR$.

If $sup(Z) = maxSuperSup(Z)$, then $X \Rightarrow Z \setminus X \notin RR$.

Property 7 states that RR cannot be built from frequent itemsets which have proper supersets with the same support as themselves. In other words, no representative rule can be obtained from an itemset that is not closed.

Property 8 [5]. Let $X \Rightarrow Z \setminus X \in AR$.

If $maxSuperSup(Z) / sup(X) > minConf$, then $X \Rightarrow Z \setminus X \notin RR$.

Property 8 states that an association rule $X \Rightarrow Z \setminus X$ is not representative if there is a rule $X \Rightarrow Z' \setminus X$ in AR such that $Z \subset Z'$.

Generate-RAR. The algorithm generates representative rules directly from frequent closed itemsets (according to Property 7). Nevertheless, the antecedents of representative rules may not be closed and thus their supports must be calculated. The supports of the antecedents are determined as the supports of their respective closures. A closure of an itemset X is found as the intersection of all closed itemsets being supersets of X (according to Property 2b), which is time-consuming operation. In this paper, we propose a new algorithm that discovers RR from FC and FG without any itemset support computations.

3 More on Representative Rules in Terms of Closed Itemsets

In this section we investigate properties of RR in order to devise an efficient algorithm for $RR's$ discovery. We already know that RR can be built only from FC itemsets. Nevertheless, we still do not know if each frequent closed itemset will provide at least one representative rule. In fact, it is not guaranteed unless an additional condition is satisfied.

Further on, by *representative itemsets RI* we denote all frequent itemsets from which at least one representative rule can be generated certainly. Property 9 specifies which frequent closed itemsets belong to RI.

Property 9

$$RI = \{X \in FC | sup(X) / minSubSup(X) > minConf \text{ and}$$
$$maxSuperSup(X) / minSubSup(X) \le minConf\}.$$

Proof. Let $X \in FC, Y \in FG$ and $sup(Y) = minSubSup(X)$, i.e. Y is a minimal subset of X with the smallest possible support.

a) The condition $sup(X) / minSubSup(X) > minConf$ states that it is possible to create a rule from X, e.g. $Y \Rightarrow X \setminus Y$ which will belong to AR and will not be covered by any other rule $Y' \Rightarrow X \setminus Y'$ in AR such that $Y' \subset Y$.

b) The next condition $maxSuperSup(X)/minSubSup(X) \leq minConf$ expresses the fact that every longer rule $Y \Rightarrow Z \setminus Y$, $Z \supset X$, which covers $Y \Rightarrow X \setminus Y$, does not belong to AR, even if the confidence of this rule has the highest possible value (i.e. in the case when $sup(Z) = maxSuperSup(X)$ and $sup(Y) = minSubSup(X)$). This means that $Y \Rightarrow X \setminus Y$ is not covered by any longer rule in AR either.

Hence, by Property 6, the rule $Y \Rightarrow X \setminus Y$ belongs to RR. Thus, it is certain that at least one representative rule can be generated from X.   □

Let us note that the representative itemsets can be found in linear time w.r.t. the number of FC provided the information on $maxSuperSup$ and $minSubSup$ is kept for each itemset in FC. The next property specifies sufficient conditions for a rule to be representative.

Property 10. Let $Z \in F$ and $\emptyset \neq X \subset Z$.

$$(X \Rightarrow Z \setminus X) \in RR \text{ iff } Z \in RI, X \in FG, X \subset Z, \text{ and}$$

a) $sup(Z)/sup(X) > minConf$, and
b) $maxSuperSup(Z)/sup(X) \leq minConf$, and
c) $sup(Z)/minSubSup(X) \leq minConf$.

Outline of Proof.

a) $Z \in F$ and $sup(Z)/sup(X) > minConf$ guarantees that $X \Rightarrow Z \setminus X$ belongs to AR;
b) $maxSuperSup(Z)/sup(X) \leq minConf$ guarantees that there is no longer rule $X \Rightarrow Z' \setminus X$, $Z' \supset Z$, in AR that would cover $X \Rightarrow Z \setminus X$;
c) $sup(Z)/minSubSup(X) \leq minConf$ guarantees that there is no rule $X' \Rightarrow Z \setminus X'$, $X' \subset X$, in AR that would cover $X \Rightarrow Z \setminus X$.

Thus, by Properties 6 and 9, $X \Rightarrow Z \setminus X \in RR$.   □

4   The Algorithm

The properties proved in the previous section state that each representative rule $X \Rightarrow Z \setminus X$ must be built from a frequent closed itemset and its antecedent must be a frequent generator. A new algorithm ($GenRR$) we offer computes RR based solely on FC and FG itemsets.

Further on, it is assumed that with every generator it is associated a field storing $minSubSup$ value, while with every closed itemset there are associated fields: $maxSuperSup$, $minSubSup$ and $Generators$.

The $GenRR$ algorithm applies Properties 9 and 10. Property 9 determines frequent closed itemsets from which at least one representative rule can be generated (RI). For each itemset Z in RI, a set G of specific antecedents Y of the candidate rules $Y \Rightarrow Z \setminus Y$ is found. In fact, the antecedents G are not generators, but their closures, which belong to FC. Each $Y \Rightarrow Z \setminus Y$, can be treated as a generic rule that can be instantiated by replacing Y with any of its generators if required. The antecedents of candidate rules should satisfy the conditions a, b, and c of Property 10 as these conditions must be met by RR. In the algorithm, the frequent itemsets are included in G only if they satisfy the conditions a, and b of Property 10. Clearly, each instantiation $X \Rightarrow Z \setminus X$ of the generic rule $Y \Rightarrow Z \setminus Y$, where $Y \in G$, will also satisfy these conditions. On the other hand, the satisfaction of the condition c of Property 10 by a generic rule does not imply the satisfaction of the condition by an instantiated rule. Therefore, the final set of *Antecedents* of candidate rules is created from the generators X of itemsets Y in G that passed the condition c of Property 10. Each rule $X \Rightarrow Z \setminus X$, where $X \in Antecedents$ and $Z \neq X$, is found representative.

Algorithm *GenRR*(frequent closed itemsets FC);
    $RI = \{Z \in FC \mid card\,(Z) \geq 2$ and $Z.sup/Z.minSubSup > minConf$ **and**
            $Z.maxSuperSup/Z.minSubSup \leq minConf\}$;
    /* By Property 9, at least one representative rule $X \Rightarrow Z \setminus X$ will be generated */
    /* from each Z in RI.   */
    forall $Z \in RI$ **do begin**
      $G = \{Y \in FC \mid Y \subseteq Z$ and $Z.sup/Y.sup > minConf$ **and**
            $Z.maxSuperSup/Y.sup \leq minConf\}$;
      $Antecedents = \{X \mid \exists Y \in G,\ X \in Y.Generators$ **and**
               $Z.sup/X.minSubSup \leq minConf\}$;
      /* Antecedents of RR satisfy Property 10 */
      if $\{Z\} \neq Antecedents$ **then begin**
      /* Each rule $X \Rightarrow Z \setminus X$, $X \in Antecedents$, has non-empty consequent $Z \setminus X$*/
        forall $X \in Antecedents$ **do begin**
          print($X \Rightarrow Z \setminus X$, with support: $Z.sup$, **and**
             confidence: $Z.sup/\,(X.Closure)\,.sup$);
        endfor;
      endif;
    endfor;
end;

Example 4. Let D be the database from Example 1. Given $minSup = 2$ and $minConf = 77\%$ we will illustrate how the *GenRR* algorithm determines RR from frequent closed itemsets FC and their generators. Table 1 provides information on frequent closed itemsets, while Table 2 shows $minSubSup$ of the generators. Table 3 contains information on intermediate results (RI, G, *Antecedents*) generated by *GenRR* and final results (RR).

One can check that the number of all frequent itemsets is equal to 31, while there are only 4 frequent closed itemsets, 3 out of which are representative itemsets! There were found 9 RR, whereas the number of AR equals to 93.   □

Table 1. Frequent closed itemsets for $minSup = 2$

FC	sup	$maxSuperSup$	$minSubSup$	Generators
$\{A,B,C,D,E\}$	3	0	3	$\{A,C\}, \{A,D\}$
$\{B,C,D,E\}$	4	3	4	$\{C\}, \{D\}$
$\{A,B,E\}$	4	3	4	$\{A\}$
$\{B,E\}$	5	4	5	$\{B\}, \{E\}$

Table 2. Generators and their $minSubSup$

Generator	$\{A,C\}$	$\{A,D\}$	$\{A\}$	$\{B\}$	$\{C\}$	$\{D\}$	$\{E\}$
$minSubSup$	4	4	∞	∞	∞	∞	∞

Table 3. Intermediate and final results of executing $GenRR$

RI	G	$Antecedents$	RR
$\{A,B,C,D,E\}$	$\{A,B,C,D,E\}$	$\{A,C\}, \{A,D\}$	$AC \Rightarrow BDE\,[3,3/3],$ $AD \Rightarrow BCE\,[3,3/3]$
$\{B,C,D,E\}$	$\{B,C,D,E\},$ $\{B,E\}$	$\{C\}, \{D\}, \{B\}, \{E\}$	$B \Rightarrow CDE\,[4,4/5],$ $C \Rightarrow BDE\,[4,4/4],$ $D \Rightarrow BCE\,[4,4/4],$ $E \Rightarrow BCD\,[4,4/5]$
$\{A,B,E\}$	$\{A,B,E\},$ $\{B,E\}$	$\{A\}, \{B\}, \{E\}$	$A \Rightarrow BE\,[4,4/4],$ $B \Rightarrow AE\,[4,4/5],$ $E \Rightarrow AB\,[4,4/5]$

5   Conclusions

In the paper, we proposed an efficient method for computing representative rules by means of frequent closed itemsets and their generators. We specified simple selective conditions for closed itemsets that restrict the number of itemsets only to those from which representative rules can be generated certainly. We proved that antecedents of representative rules are generators. The algorithm for discovery of RR was proposed. In the algorithm, it was applied a brief representation of antecedents of candidate representative rules. Not only candidate rules themselves, but also their antecedents are initially built only from frequent closed itemsets. Selective conditions quickly remove inappropriate generic candidate rules. Then, the remaining candidates are transformed into the final form by replacing the rule antecedents with the corresponding generators. This approach is different from those presented in [4,5,12]. First, we avoid the problem (present in [4,5]) of creating antecedents of candidate rules. Second, the time-consuming operation (required in [12]) of computing support of antecedents is not performed.

References

1. Agrawal, R., Imielinski, T., Swami, A.: Mining associations rules between sets of items in large databases. In: Proc. of the ACM SIGMOD Conference on Management of Data. Washington (1993) 207–216
2. Agrawal, R., Mannila, H., Srikant, R., Toivonen, H., Verkamo, A.I.: Fast discovery of association rules. In: Advances in Knowledge Discovery and Data Mining. AAAI, Menlo Park (1996) 307–328
3. Boulicaut, J.F., Bykowski, A.: Frequent closures as a concise representation for binary data mining. In: Proc. of PAKDD'00, Kyoto. LNAI **1805**. Springer-Verlag (2000) 62–73
4. Kryszkiewicz, M.: Representative association rules. In: Proc. of PAKDD '98. Melbourne. LNAI **1394**. Springer-Verlag (1998) 198–209
5. Kryszkiewicz, M.: Fast Discovery of representative association rules. In: Proc. of RSCTC'98. Warsaw. LNAI **1424**. Springer-Verlag (1998) 214–221
6. Kryszkiewicz M.: Representative association rules and minimum condition maximum consequence association rules. In: Proc. of PKDD'98. Nantes. LNAI **1510**. Springer-Verlag (1998) 361–369
7. Meo, R., Psaila, G., Ceri, S.: A New SQL-like operator for mining association rules. In: Proc. of the 22nd VLDB Conference. Mumbai (1996) 122–133
8. Pasquier N., Bastide Y., Taouil R., Lakhal L.: Efficient mining of association rules using closed itemset lattices. Information Systems **24** (1999) 25–46
9. Pasquier, N.: Algorithmes d'extraction et de réduction des régles d'association dans les bases de données. These de Doctorat. Univ. Pascal-Clermont-Ferrand II (2000)
10. Pei, J., Han, J., Mao, R.: CLOSET: An efficient algorithm for mining frequent closed itemsets. In: Proc. of the ACM SIGMOD DMKD'00. Dallas (2000) 21–30
11. Piatetsky-Shapiro, G.: Discovery, analysis and presentation of strong rules. In: Piatetsky-Shapiro, G., Frawley, W. (eds.): Knowledge Discovery in Databases. AAAI/MIT Press (1991) 229–248
12. Saquer J., Deogun, J.S.: Using Closed itemsets for discovering representative association rules. In: Proc. of ISMIS'00. Charlotte. LNAI **1932**. Springer-Verlag (2000) 495–504
13. Srikant, R., Agrawal, R.: Mining generalized association rules. In: Proc. of the 21st VLDB Conference. Zurich (1995) 407–419
14. Walczak, Z.: Selected problems and algorithms of data Mining. M.Sc. Thesis. Warsaw University of Technology (1998)

Intelligent Sensor Analysis and Actuator Control

Matthew Easley[1] and Elizabeth Bradley[2*]

[1] Rockwell Scientific
444 High Street, Suite 400
Palo Alto, CA 94301-1671
measley@rwsc.com
[2] University of Colorado
Department of Computer Science
Boulder, CO 80309-0430
lizb@cs.colorado.edu

Abstract. This paper describes a tool called ISAAC (*intelligent sensor analysis and actuator controller*) that autonomously explores the behavior of a dynamical system and uses the resulting knowledge to help build and test mathematical models of that system. ISAAC is a unified knowledge representation and reasoning framework for input/output modeling that can be incorporated into any automated tool that reasons about dynamical models. It is based on two modeling paradigms, *intelligent sensor data analysis* and *qualitative bifurcation analysis*, which capture essential parts of an engineer's reasoning about modeling problems. We demonstrate ISAAC's power and adaptability by incorporating it into the PRET automated system identification tool and showing how input/output modeling expands PRET's repertoire.

1 Introduction

One of the most powerful analysis and design tools in existence—and often one of the most difficult to create—is a good model. Modeling is an essential first step in a variety of engineering problems. Faced with the task of designing a controller for a robot arm, for instance, a mechanical engineer performs a few simple experiments on the system, observes the resulting behavior, makes some informed guesses about what model fragments could account for that behavior, and then combines those terms into a model and checks it against the physical system. One of the hardest and most important steps in this process is the input-output analysis. This procedure requires the expert to decide, based only on partial knowledge about a system, what experiments will augment that knowledge in a manner that is useful to the task at hand. He or she must then figure out which of those are possible in the existing experimental framework, perform them, and interpret the results. The goal of the project described in this paper is to produce a software tool that automates this process.

* Supported by NSF NYI #CCR-9357740, NSF #MIP-9403223, ONR #N00014-96-1-0720, and a Packard Fellowship in Science and Engineering from the David and Lucile Packard Foundation.

F. Hoffmann et al. (Eds.): IDA 2001, LNCS 2189, pp. 360–369, 2001.
© Springer-Verlag Berlin Heidelberg 2001

Automating the input/output reasoning process is a hard and interesting problem, from the standpoints of both artificial intelligence and engineering. Planning, executing, and interpreting experiments requires some fairly difficult reasoning about what experiments are useful and possible, how to manipulate the actuators in order to perform them, and how to interpret the resulting sensor data. It is fairly easy to recognize damped oscillations in sensor data without knowing anything about the system or the sensor, for example, but using an actuator requires a lot of knowledge about both. Different actuators have different characteristics (range, resolution, response time, etc.), and the effects of an actuator on a system depend intimately on how the two are connected. A DC motor, for example, may be viewed as a voltage-to-RPM converter with a linear range and a saturation limit. How that motor affects another device depends not only on those characteristics, but also on the linkage between the two. The first stage of the automatic experiment planning and execution process, then, is equivalent to control theory's *controllability/reachability* problem: given a system, an initial condition, and an actuator configuration, what state-space points are reachable with the available control input? In the course of modeling a driven pendulum, for instance, it might be extremely useful to find out what happens when the device is balanced at the inverted point, but getting the system to that point is a significant real-time control problem in and of itself. Finally, input-output analysis also involves reasoning about *utility*: whether or not anything useful can be learned from a particular experiment. If all of one's observations of a pendulum concerned small-angle motions, for example—where the dynamics looks like a simple harmonic oscillator—it might be useful to investigate initial conditions with larger angles.

The *intelligent sensor analysis and actuator controller* (ISAAC) tool described in this paper, which embodies our solutions to these problems, draws upon techniques and ideas from artificial intelligence, dynamical systems, networks, and control theory. ISAAC is based on two knowledge representation and reasoning paradigms, *intelligent sensor data analysis (ISDA)* and *qualitative bifurcation analysis (QBA)*. The first of these two paradigms captures an engineer's ability to reason about sensor data at an abstraction level that is appropriate to the data and to the problem at hand. Determining something as simple as "the voltage oscillates between 5 and 10 volts," for example, can be difficult if one attempts to scan an ASCII text file, but it is trivial to see when the data is presented on an oscilloscope. Distilling available sensor information into this type of qualitative form is reasonably straightforward, as described in our IDA-97 paper[3], but reasoning about the information so derived is subtle and challenging. The second paradigm, qualitative bifurcation analysis, is based on a new construct called the *qualitative state/parameter space*, an abstraction of regular state space with an added parameter axis, which is a useful way to capture information about actuator signals, sensor data, and different behavioral regimes in a single compact representation. The QBA paradigm combines this representation with a set of reasoning tools that emulate a classic reasoning technique from nonlinear dynamics, in which a person varies a control parameter—the actuator input—

and looks for changes ("bifurcations") in the system behavior (the sensor data). The QBA paradigm was described in our IDA-99 paper[11]. ISAAC uses these two paradigms in tandem, interleaving the actuator- and sensor-related reasoning processes in order to determine what experiments are useful and possible, execute them, and reason about the results.

ISAAC is a unified framework that can add I/O capabilities to any automated tool that reasons about models. In order to demonstrate its adaptability and power, we incorporated it into the computer program PRET, an automated system identification tool that takes a set of observations of the outputs of a nonlinear black-box system and produces an ordinary differential equation (ODE) model of the internal dynamics of that system. Section 3 describes an interesting real-world modeling problem that PRET was able to solve using ISAAC's facilities. To provide context for this example, the following section gives a brief overview of PRET.

2   PRET

System identification (SID) is the process of inferring an internal ordinary differential equation model from external observations of a system. SID has two phases: *structural identification*, in which the number and types of terms forming the equation that govern the unknown dynamics are determined, and *parameter estimation*, in which values for the undetermined coefficients in that equation are derived via comparisons with the data. The computer program PRET[4,6] automates the SID process. Its inputs are a set of observations of the outputs of a lumped-parameter continuous-time nonlinear dynamic system, together with a modeling specification that includes resolutions for important variables and an optional list of hypotheses about the physics involved. Its output is an ordinary differential equation (ODE) model of the internal dynamics of that system. See Fig. 1 for a block diagram. PRET's architecture wraps a layer of artificial

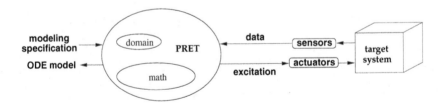

Fig. 1. PRET builds ODE models of lumped-parameter continuous-time nonlinear dynamic systems. The input/output modeling framework described in this paper allows PRET to interact with target systems directly and autonomously.

intelligence (AI) techniques[21] around a set of traditional formal engineering methods. Unlike other AI modeling tools—most of which use libraries to model small, well-posed problems in limited domains and rely on their users to supply detailed descriptions of the target system—PRET works with nonlinear systems in multiple domains. It takes a generate-and-test approach, using a small, powerful domain theory to build models and a larger set of more-general mathematical rules to test those models against the known behavior of the system. The complex interplay of heterogeneous reasoning modes that is involved in this process is orchestrated by a special first-order logic system that uses static abstraction levels, dynamic declarative meta control, and a simple form of truth maintenance in order to test models quickly and cheaply. This approach has met with success in a variety of simulated and real problems, ranging from textbook exercises to practical engineering applications, but it is fundamentally limited by its passive nature. The types of real-world systems that are PRET's ultimate target—high-dimensional, nonlinear, black-box systems, drawn from any domain where system dynamics can be described by ODEs—demand a fully interactive, *input-output* approach, wherein PRET observes systems *actively*, manipulating actuators and reading sensors to perform experiments whose results will augment its knowledge in a useful manner. The ISAAC framework described in this paper provides exactly that functionality. The next section describes one of PRET's more recent successes—a thermistor modeling problem—which was enabled by the knowledge representation and reasoning framework described in this paper.

3   Modeling a Thermistor

Thermistors—temperature sensitive resistors that rely on the large temperature dependence of the resistivity of certain transition-metal oxides[1]—were first created in the 1930s. These simple devices, one of the most fundamental kinds of sensors, are used in a wide range of industrial and consumer applications. As materials scientists find new thermally dependent compounds and engineers continue creating new temperature dependent applications, accurate models of these devices will continue to be an area of active research (see e.g.,[17,20]).

Thermistors are also an interesting application for our work, serving as a proof of concept for the input-output facilities described in the introduction. Fig. 2 shows how a user sets PRET up to model a thermistor. The first step is to specify a modeling domain, which causes PRET to instantiate a small but powerful set of rules that combine model fragments into ODE models. In this case, PRET's `electrical-xmission-line`, which uses networks of linear capacitors and resistors to build models, is the best choice, since engineers routinely model thermistors using these kinds of networks. (Currently, PRET has four other *modeling domains*: `mechanics`, `viscoelastics`, `linear-rotational`, and `linear-mechanics`. Issues regarding the design and use of these domains are discussed in [9,10].) The next step is to specify the state variables of the system to be modeled—in this case, resistance and current—using the `state-variables` line. PRET allows users to suggest hypotheses; in this case, we want to test its ability to build models from scratch, so we leave the `hypotheses` line blank. It also allows users to specify any known observations, either quantitatively or using qualitative keywords like `chaotic`, `constant`, etc. In this case, the system under investigation is connected to an automatic data-acquisition device, so the user need not specify any observations directly. Rather, she or he uses an incantation that instructs the data-acquisition equipment to gather data directly from the system, using an actuator that varies the control voltage `alpha` and a sensor that gathers resistance versus time information. The range of this control parameter and the required resolution of the sensor are prescribed in the `specifications` line of the call. In this case, the thermistor has a precision of $\pm 10\Omega$ and the range of the heat source control is 2 to 5 volts.

```
(find-model
  (domain electrical-xmission-line)
  (state-variables (<resistance> <current>))
  (hypotheses '())
  (drive (<voltage> alpha))
  (observations
    (numeric (<time> <resistance>)
             (data-acquisition (eval *acq-handle*) alpha)))
  (specifications
    (<resistance> absolute-resolution 10.0)
    (control-parameter (alpha absolute-resolution 1.0 (2.0 5.0))) ))
```

Fig. 2. Using PRET to model a thermistor. `alpha` (the applied voltage) is the control parameter. `*acq-handle*` is a short SCHEME procedure which the ISAAC tool uses to gather data from the target system, as described in the text.

This information alone, however, is not enough to let the ISAAC tool execute experiments automatically. Every sensor and actuator is different—whether it is digital or analog, voltage- or current-activated, involves frequency or phase, etc. Because this kind of information is all but impossible to deduce automatically, our implementation requires the user to give a minimal description of the operative sensors and actuators, in the form of a short SCHEME procedure (in this case called `*acq-handle*`) that the user defines in the environment from which he or she calls PRET. This function, which is used to gather data from the target system for specific settings of the parameter `alpha`, is passed to the `find-model` call via the keyword `eval`, which causes the variable `*acq-handle*` to be evaluated in the calling environment. In this case, `*acq-handle*` has one argument—the control parameter `alpha`—and it returns a time series, in this case `<resistance>` vs. `<time>`:

```
(define *acq-handle*
  (let ((length 100)           ; 100 seconds for each alpha
        (time-step 0.5)        ; sample at 2 Hz
        (type '4-wire-resistance)) ; sensor type
    (lambda (alpha)
      (let ((v-control alpha))
        (run-DAQ length time-step type v-control)))))
```

The actual experimental setup for this example is shown in Fig. 3. The ISAAC tool controls the DAQ via an EISA interface;[1] the DAQ interfaces directly with the thermistor via digital and analog input and output channels. The function run-DAQ, which is part of the ISAAC tool's knowledge base, manages this data-acquisition equipment. Its first three arguments specify the length and sample rate of the sensor trace, together with the sensor type. (There are currently four types of sensors—AC-voltage, DC-voltage, 4-wire-resistance, and 4-wire-temperature—and four types of actuators: DC and AC voltage and current sources.) run-DAQ first sends a control voltage (v-control) to the voltage-controlled heat source via the DAQ system's digital-to-analog converter. run-DAQ then uses the DAQ's digital multimeter board to gather a 100-second long time series of the resistance from the thermistor, measured every half second. Under the current implementation, the sample time-step may be set between 1 millisecond and 16 seconds. This driver function operates via system calls to the Standard Instrument Control Library (SICL)[15], which contains high-level procedures that control the data-acquisition hardware.

The ISAAC tool then applies an array of intelligent sensor data analysis tools, described in detail in [3], to identify important qualitative properties of the time series. In the case of the data gathered from the thermistor, shown

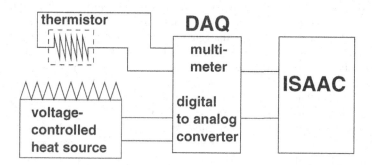

Fig. 3. Experimental setup for automated input-output modeling of a thermistor. The data-acquisition (DAQ) device varies the heat applied to the thermistor via a control voltage and measures the change in resistance via a digital multimeter. ISAAC interfaces to the DAQ via a standard EISA interface.

in the left-most trace of Fig. 4, ISAAC determines that the value of the state variable <resistance> is converging to a fixed point. This kind of information enables PRET to reason about candidate models at a highly abstract level, which allows it to immediately discard the first few models proposed by the electrical-xmission-line domain model generator. The third model—a resistor and capacitor in series—passes the initial testing phase and is passed to

---

[1] The interface, the DAQ, and the host workstation are all Hewlett-Packard equipment, which minimizes interfacing issues.

PRET's parameter estimator for comparison with the time series acquired from the data-acquisition hardware. The parameter estimator succeeds, and the model tester accepts the model, as the result matches the time series to within the desired specifications. The behavior of the model is shown overlaid on the time series data in Fig. 4.

The model-building process is not complete, however, as ISAAC has only explored a single control parameter value, so it next increases the control voltage and gathers another time series. As in the case of the previous time series, ISAAC is able to identify this time series as a transient to a fixed point; because this is the second round of exploration, however, this information gives PRET even more leverage. In particular, similar behavior suggests that the model may be similar, and so this knowledge lets PRET's model generator propose the same (successful) model found in the previous iteration for this new regime. After roughly the same process of qualitative and quantitative tests, the model testing phase quickly accepts the same model, but with slightly different coefficients that account for the faster heating at a higher control voltage.[2] The time series for the last two control parameter values also end at fixed points, so the model generator proposes the identical model for these, and the testing phase succeeds there as well. The series of models is shown in Fig. 4. Note how well the models fit—and do *not* overfit—the data, across the entire control parameter range.

This example just begins to show the power of coupling intelligent sensor analysis tools with autonomous actuator manipulation. Not only was the sensor data transformed into a highly abstract and thus broadly useful form, aiding the model testing phase, but the model *generation* phase was streamlined as well. Once the ISAAC tool identified all of the time series as being qualitatively similar, the rest of PRET (including the model generation and testing phases) had almost no work to do. That is, after the first round of modeling, the search space for the rest of the control parameters was only $O(1)$. This is exactly the kind of knowledge an engineer uses when modeling a system—apply known models as widely as possible, based upon reasoning about observed system behavior. Only when a model is applied to a regime in which it fails should the model search process start again.

4   Related Work

Most of the work in the AI/QR (qualitative reasoning) modeling community builds qualitative models by combining a set of descriptions of state into higher-level abstractions or *qualitative states*[8,14]. Many of these tools reason about equations at varying levels of abstraction, from *qualitative* differential equations (QDEs) in QSIM[18] to ODEs in PRET. ISAAC's approach differs from many of these tools in that it works with noisy, incomplete sensor data from real-world systems, and attempts not to "discover" the underlying physics, but rather to find the simplest ODE that can account for the given observation. In the QR research that is most closely related to PRET, ODE models are built by evaluating

---

[2] that is, the time constant of the system decreases with increased heat

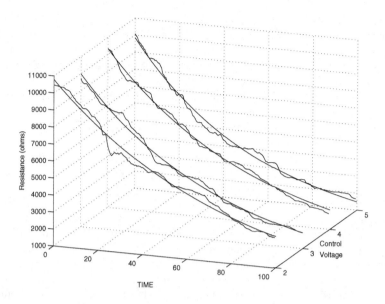

Fig. 4. Responses of a heated thermistor at four different control voltages. The noisy time series are the responses measured using the data-acquisition hardware. The smooth lines are PRET's final models.

time series using qualitative reasoning techniques and then using a parameter estimator to match the resulting model with a given observed system[7]. This modeling tool selects models from a set of pre-enumerated solutions in a very specific domain (linear visco-elastics). PRET is much more general; it works on linear *and nonlinear* lumped-parameter continuous-time ODEs in a variety of domains, uses *dynamic* model generation to handle arbitrary devices and connection topologies, and—thanks to ISAAC—interacts directly and autonomously with the target system using sensors and actuators.

PRET and ISAAC share goals and techniques with several other fields. The former solves the same problems as traditional system identification[16], but in an automated fashion, and it relies upon many of the standard methods and ideas found in basic control theory texts such as controllability and reachability[19]. Both include many of the same concepts that appear in the intelligent data analysis literature[2,13], but they add a layer of AI techniques, such as symbolic data representation and logical inference, on top of these.

5   Conclusion

The approach described in this paper lets automated modeling tools shift fluidly between domain-specific reasoning and general mathematics to process sensor data, manipulate actuators to perform experiments, and use the results to navigate efficiently through an exponential search space of models of nonlinear dynamic systems. Using ISAAC's input/output modeling facilities, PRET has successfully constructed models of real systems in multiple domains, ranging from textbook problems (Rössler, Lorenz, simple pendulum, pendulum on a spring, etc.) to interesting and difficult real-world examples like shock absorbers, well/aquifer systems, and a commercial radio-controlled car[5,6,9,12]. The thermistor example treated in this paper is significant for three reasons. First, it involves direct, autonomous interaction with a physical system. Second, it is a particularly good demonstration of the issues and difficulties involved in the actual physical level implementation of input/output modeling facilities in an automated tool. Lastly, the resulting model duplicates the standard one found in journal-level publications in the associated field[17].

Automating the input-output modeling process is worthwhile for a variety of reasons. It makes expert knowledge useful to novices, allows the corroboration and verification of an expert's results, and it represents an important step towards the type of fully autonomous operation that is required for many real-world AI applications. Planning, executing, and interpreting experiments, however, requires some fairly difficult reasoning about what experiments are useful and possible. Research on these general classes of reasoning problems is ongoing in the control theory, operations research, and AI communities,[3] and many of the specific techniques used in the design and implementation of the ISAAC tool have appeared in one or both of these contexts. The new idea here is the notion of working out a systematic scheme for truly automatic experiment planning, execution, and interpretation *for the purposes of modeling nonlinear systems*.

Acknowledgments: Apollo Hogan, Joe Iwanski, Brian LaMacchia, and Reinhard Stolle also contributed code and/or ideas to this project.

References

1. J. A. Becker, C. B. Green, and G. L. Pearson. Properties and uses of thermistors—thermally sensistive resistors. *Transactions of the American Institute of Electrical Engineers*, pages 711–725, 1946.
2. M. Berthold and D. Hand, editors. *Intelligent Data Analysis: An Introduction*. Springer-Verlag, 2000.
3. E. Bradley and M. Easley. Reasoning about sensor data for automated system identification. *Intelligent Data Analysis*, 2(2):123–138, 1998.
4. E. Bradley, M. Easley, and R. Stolle. Reasoning about nonlinear system identification. *Artificial Intelligence*. To appear; also available as technical report CU-CS-894-99. See www.cs.colorado.edu/~lizb/publications.html.
5. E. Bradley, A. O'Gallagher, and J. Rogers. Global solutions for nonlinear systems using qualitative reasoning. *Annals of Mathematics and Artificial Intelligence*, 23:211–228, 1998.

---

[3] under the rubrics of "experimental design" and "reasoning about action," respectively

6. E. Bradley and R. Stolle. Automatic construction of accurate models of physical systems. *Annals of Mathematics and Artificial Intelligence*, 17:1–28, 1996.

7. A. C. Capelo, L. Ironi, and S. Tentoni. Automated mathematical modeling from experimental data: An application to material science. *IEEE Transactions on Systems, Man and Cybernetics - Part C*, 28(3):356–370, 1998.

8. J. de Kleer and J. S. Brown. A qualitative physics based on confluences. *Artificial Intelligence*, 24:7–83, 1984.

9. M. Easley. *Automating Input-Output Modeling of Dynamic Physical Systems*. PhD thesis, University of Colorado at Boulder, 2000.

10. M. Easley and E. Bradley. Generalized physical networks for automated model building. In *International Joint Conference on Artificial Intelligence (IJCAI-99)*, pages 1047–1052, 1999. Stockholm, Sweden.

11. M. Easley and E. Bradley. Reasoning about input-output modeling of dynamical systems. In *Proceedings of the Third International Symposium on Intelligent Data Analysis (IDA-99)*, 1999.

12. M. Easley and E. Bradley. Meta-domains for automated system identification. In C. Dagli and et al, editors, *Proceedings of Smart Engineering System Design, (ANNIE 00)*, pages 165–170, 2000. ASME Press.

13. A. Famili, W.-M. Shen, R. Weber, and E. Simoudis. Data preprocessing and intelligent data analysis. *Intelligent Data Analysis*, 1(1), 1997.

14. K. D. Forbus. Interpreting observations of physical systems. *IEEE Transactions on Systems, Man, and Cybernetics*, 17(3):350–359, 1987.

15. Hewlett-Packard. *Standard Instrument Control Library Reference Manual*, 1996.

16. J.-N. Juang. *Applied System Identification*. Prentice Hall, Englewood Cliffs, 1994.

17. T. Kineri, T. Kogiso, and Y. Kawaguchi. The characteristics of high temperature thermistor with new materials. *Sensors and Actuators*, pages 57–61, 1989. SAE, SP-771.

18. B. J. Kuipers. Qualitative simulation. *Artificial Intelligence*, 29(3):289–338, 1986.

19. B. C. Kuo. *Automatic Control Systems*. Prentice Hall, seventh edition, 1995.

20. A. Tamaoki, T. Shibata, and H. Sirai. Temperature sensor for vehicle. *Sensors and Actuators*, pages 113–118, 1991. SAE, P-242.

21. P. H. Winston. *Artificial Intelligence*. Addison Wesley, Redwood City CA, 1992. Third Edition.

Sampling of Highly Correlated Data for Polynomial Regression and Model Discovery

Grace W. Rumantir and Chris S. Wallace

School of Computer Science and Software Engineering
Monash University – Clayton Vic 3168 Australia
{gwr,csw}@csse.monash.edu.au

Abstract. The usual way of conducting empirical comparisons among competing polynomial model selection criteria is by generating artificial data from created true models with specified link weights. The robustness of each model selection criterion is then judged by its ability to recover the true model from its sample data sets with varying sizes and degrees of noise.

If we have a set of multivariate real data and have empirically found a polynomial regression model that is so far seen as the right model represented by the data, we would like to be able to replicate the multivariate data artificially to enable us to run multiple experiments to achieve two objectives. First, to see if the model selection criteria can recover the model that is seen to be the right model. Second, to find out the minimum sample size required to recover the right model.

This paper proposes a methodology to replicate real multivariate data using its covariance matrix and a polynomial regression model seen as the right model represented by the data. The sample data sets generated are then used for model discovery experiments.

Keywords: sampling from multivariate normal distribution, polynomial model discovery, tropical cyclone intensity forecasting modelling

1 Motivation and Relationships with Previous Works

Polynomial regression concerns with the task of estimating the value of a target variable from a number of regressors/independent variables. If, on top of the basic independent variables, products of two independent variables are also considered as potential regressors, the estimation models are called second-order polynomials which are represented in the following standardized form:

$$y_n = \sum_{p=1}^{P} \gamma_p u_{np} + \sum_{p=1}^{P}\sum_{q \geq p}^{P} \gamma_{pq} u_{np} u_{nq} + \epsilon_n \Leftrightarrow y_n = \sum_{k=1}^{K} \beta_k x_{nk} + \epsilon_n \quad (1)$$

F. Hoffmann et al. (Eds.): IDA 2001, LNCS 2189, pp. 370–377, 2001.
© Springer-Verlag Berlin Heidelberg 2001

where for each data item n:

y_n	: target variable	x_{nk}	: regressor k;
			$x_{nk} = u_{np}$ or $x_{nk} = u_{np}u_{nq}$; $q \geq p$
u_{np}	: regressor p	β_k	: coeff. for regressor k
γ_p	: coeff. for single regressor	K	$= 2P + P!/2!(P-2)!$
γ_{pq}	: coeff. for compound regressor	ϵ_n	: noise/residual/error term

The values of the error term ϵ is assumed to be uncorrelated, normally and independently distributed $\epsilon_n \sim NID(0, \sigma^2)$.

Model discovery concerns with the task of finding, amongst a large pool of potential regressors, a subset which has the strongest explanatory relationships with the target variable to form a polynomial regression model. The search for the subset of regressors typically follows a stepwise iterative process where, at each step, the degree of influence of a regressor to the target variable is examined based on a certain model selection criterion. In this paper, the polynomial regression model is intended for the task of forecasting. Hence, the regressors measured at one time are to be used to predict the value of the target variable at some future time.

In [4], we have identified five model selection criteria that have been empirically proven to be good candidates for fully automated model selection tasks. This is due to their ability to balance model complexity and goodness of fit resulting in their success in recovering true models from artificially generated data. The five model selection criteria are given in Table 1. In the paper, we also propose a non-backtracking search mechanism which we use in conjunction with the model selection criteria.

Table 1. Summary of the model selection criteria used. i is index for sample item ranging from 1 to n and k is the number of variables in a model

Method		Ref.	Objective Function
Minimum Message Length	MML	[7]	$-\log f(x\|\theta) + \frac{1}{2}\log\|I(\theta)\| - \log h(\theta) - \frac{k+1}{2}\log 2\pi + \frac{1}{2}\log(k+1)\pi - 1 - \log h(\nu, \xi, j, l, J, L)$
Minimum Description Length	MDL	[2]	$-\log f(x\|\theta) + \frac{k}{2}\log n + (\frac{k}{2} + 1) + \log k(k+2)$
Corrected AIC	CAICF	[1]	$-\log f(x\|\theta) + \frac{1}{2}\log\|I(\theta)\| + k + \frac{1}{k}\log n$
Structured Risk Minimisation	SRM	[6]	$\frac{1}{n}\sum_{i=1}^{n} e_i^2 / (1 - \sqrt{\frac{(k+1)(\log \frac{n}{k+1} + 1) - \log \eta}{n}})$
Stochastic Complexity	SC	[3]	$\frac{n}{2}\log\sum_{i=1}^{n} e_i^2 + \frac{1}{2}\log\|X'X\|$

In [5], we use the five model selection criteria given in Table 1 for the task of building tropical cyclone forecasting models from real meteorological data. The target variable is intensity change 72 hours into the future. The pool of potential regressors consists of 36 meteorological variables together with the product of two varibles giving in total $2 * 36 + \frac{36!}{2!(36-2)!} = 702$ potential regressors.

In the experiments in [5], each model selection method converged to a different model with some regressors chosen by most or all of the methods. The methodology for building forecasting models outlined in the paper suggests that the final model is to be built by incrementally including the most frequently chosen regressors one at a time. In effect, this means the five model selection criteria given in Table 1 are all used in an integrated manner to select the final forecasting model.

The performance of each model with increasing complexity is calculated using the model selection criteria given in Table 1 on the training data and the performance criteria on test data given in Section 2.3. Parsimony is reached when adding complexity to an existing model does not improve its performance. The final tropical cyclone forecasting model selected in the paper is given is Table 2 with the explanation of the regressors given in Table 3.

Table 2. The model found in the experiments in [5] as the right model represented by the real multivariate meteorological data sets. Compound variables (Variable1, Variable2) represent the product of two variables. The meanings of the variables are given in Table 3

Chosen Regressor	Normalized Coefficient
(Del12V,Vmax)	-0.099445
POT	0.649806
Del12V	0.165545
(SOI,SOI)	-0.123373
(200Uend,Vmax)	-0.119977
Del200U	0.109351
(SLPA,RainG)	0.108001
(ElNino,Del200U)	0.075820
RainG	0.067429

Table 3. The basic regressors of the Atlantic tropical cyclone intensity change forecasting model shown in Table 2. The target variable is the change of intensity (wind speed) 72 hours into the future

Regressor	Explanation
Del12V	the change of cyclone intensity in the past 12 hours (in knots)
Vmax	the initial cyclone intensity (in knots)
POT	cyclone initial potential intensity
SOI	Surface pressure gradient between Darwin and Tahiti
00Uend	the forecast of eastward wind motion at 200mb at the end of 72 hours
Del200U	the forecast of the change of eastward wind motion at 200mb
SLPA	April-May Caribbean basin Sea Surface Pressure Anomaly
RainG	African Gulf of Guinea rainfall index (0W-10W, 5N-10N)
ElNino	Sea surface temperature anomaly in the eastern equatorial pacific

We are now in the situation where we have a set of multivariate real data and have empirically found a polynomial regression model that is so far seen as the right model represented by the data. We would now like to be able to replicate the multivariate data artificially to enable us to run multiple experiments to achieve two objectives. First, to see if the model selection criteria can recover the model that is seen to be the right model. Second, to find out the minimum sample size required to recover the right model.

This paper proposes a methodology to replicate real multivariate data using its covariance matrix and a polynomial regression model seen as the right model represented by the data. The sample data sets generated are then used for model discovery experiments.

2 Methodology

2.1 Generating Regressor Data from Covariance Matrix

Problem Definition:
Given the covariance matrix A of a set of real data of K variables, we would like to generate data from a set of variables $X_1 \ldots X_k$ such that its covariance matrix is the same as the given covariance matrix A

$$E(X_i X_j) = A_{ij} \tag{2}$$

where A_{ij} is the given covariance of real variables i and j.

Proposition:

Since the real sample data sets come from K variables, we first generate K unit normals $N_1 \ldots N_k = \tilde{N}$. The values of the new variables $X_1 \ldots X_k$ is defined as

$$\tilde{X} = B\tilde{N} \tag{3}$$

We then form some linear combinations of the variables such that Equation 2 holds

$$X^T X = B^T N^T B N \tag{4}$$
$$\Rightarrow E(X^T X) = B^T E(N^T N)B = A \tag{5}$$
$$\Leftrightarrow B^T I B = A$$
$$\Leftrightarrow B^T B = A \tag{6}$$

Now, from Linear Algebra we know that covariance matrix A can be represented by its eigenvalue Λ and eigenvector Q

$$A = Q^T \Lambda Q \tag{7}$$
$$\Leftrightarrow A = Q^T \sqrt{\Lambda}\sqrt{\Lambda}Q \tag{8}$$

Hence, from Equations 6 and 8 we get

$$B = \sqrt{\Lambda}Q \tag{9}$$

Finally, we can generate data for the new variables $X_1 \ldots X_k$ using the unit normals $N_1 \ldots N_k = \tilde{N}$ based on Equations 3 and 9

$$\tilde{X} = \sqrt{\Lambda}Q\tilde{N} \tag{10}$$

That is, for each variable x_i we can generate data using the formula

$$x_i = \sum_j \sqrt{\lambda_i} q_{ji} n_i \tag{11}$$

where:

n_i : a unit normal data point generated from $N(0,1)$
q_{ij} : j^{th} element of eigenvector i
λ_i : eigenvalue i

2.2   Model Discovery Process

The following is the procedure to generate artificial data for the pool of potential regressors and their associated target variable for model discovery:

Step 1: Calculate the covariance matrix of the set of real multivariate data.

Step 2: Generate artificial data for the pool of potential regressors using the formula given in Equation 11.

Step 3: Calculate the values for the target variable using the true model. This is done by multiplying the value of each regressor with its link weight connected to the target variable plus a unit normal noise value.

Step 4: With each model selection criterion given in Table 1 as the cost function, run a search mechanism to rediscover the true model from the data generated in the previous steps.

Step 5: Compare the models found by each criterion using the performance criteria given in Section 2.3.

2.3   Performance Criteria

Like in [4], the performance criteria for the model discovery task is whether or not a model selection method manages to select a model with the same set of variables and corresponding coefficients as those of the true model, as reflected in the following measures:

1. The number of variables selected in the discovered model
2. How close the coefficients of the discovered model are with those of the true model (model error): $1/K * \sum_{k=1}^{K} (\beta_k - \hat{\beta}_k)^2$

3. Model predictive performance (on test data), quantified by
   (a) Root of the mean of the sum of squared deviations: $RMSE = \sqrt{\frac{1}{n} \sum_{i=1}^{n} e_i^2}$
   (b) Coefficient of determination: $R^2 = 1 - (\sum_{i=1}^{n} e_i^2) / \sum_{i=1}^{n} (y_i - \bar{y})^2$

3   Experiments and Results

We use the model given in Table 2 as the true model that we would like to recover from the newly generated data. We generate five categories of data sets with increasing levels of noise, denoted by $DataSet1, \ldots, DataSet5$. For each category, we generate test data sets comprising 100, 500, 1000, 2000, 4000, 6000 and 10000 data points. We then follow the procedure outlined in Section 2.2 to compare the performance of each of the model selection criteria given in Table 1 for the task of recovering the true model from the data sets.

Tables 4 and 5 show the result of the experiments. It is shown that the results of all of the model selection criteria are quite uniform in that they all managed to recover the true model with similar degree of accuracy. Also we can observe that all of the criteria managed to recover the true model of 9 regressors with a data set as small as 100 data points. Increasing the size of the data set does not significantly increase the accuracy of the results. This finding indicates that most of the potential regressors do not have direct influence to the target variable.

This finding is valuable in that it confirms that the integrated model discovery procedure outline in [5] indeed has the ability to select a parsimonious subset of variables from a pool of potential regressors in the absence of prior knowledge of the problem domain.

4   Conclusion

We start with:

1. A set of real data
2. A set of model selection criteria that has previously (in [4]) been empirically proven to be robust for the task of polynomial model discovery
3. A model that has previously (in [5]) been empirically found to be the right model represented by the data using an integrated approach using all of the model selection criteria mentioned above
4. A non-backtracking search engine (outlined extensively in [4])

We would like to be able to replicate the real data so as to be able to conduct multiple experiments to further test the robustness of the integrated model discovery procedure outlined in [5] on two counts. First, to see if the model selection criteria can individually recover the model that is seen to be the right model. Second, to find out the minimum sample size required to recover the right model.

In this paper, we outline the procedure to generate artificial data from the covariance matrix of a set of real data and its true model. The findings from the experiments confirm that the model found by the integrated model discovery

Table 4. Model discovered in data sets DataSet1, DataSet2 and DataSet3

Sample Size	Method	DataSet 1				DataSet 2				DataSet 3			
		nvar	ModelErr	RMSE	R^2	nvar	ModelErr	RMSE	R^2	nvar	ModelErr	RMSE	R^2
100	MML	9	0.0023	1.1968	0.9966	9	0.0100	1.2717	0.9963	9	0.0075	1.1785	0.9959
	MDL	9	0.0023	1.1968	0.9966	9	0.0100	1.2717	0.9963	9	0.0075	1.1785	0.9959
	CAICF	9	0.0023	1.1968	0.9966	9	0.0100	1.2717	0.9963	9	0.0075	1.1785	0.9959
	SRM	9	0.0023	1.1968	0.9966	9	0.0100	1.2717	0.9963	9	0.0075	1.1785	0.9959
500	MML	9	0.0009	1.0173	0.9972	9	0.0008	1.0014	0.9970	9	0.0010	1.0050	0.9972
	MDL	9	0.0009	1.0173	0.9972	9	0.0008	1.0014	0.9970	9	0.0010	1.0050	0.9972
	CAICF	9	0.0009	1.0173	0.9972	9	0.0008	1.0014	0.9970	9	0.0010	1.0050	0.9972
	SRM	9	0.0009	1.0173	0.9972	9	0.0008	1.0014	0.9970	9	0.0010	1.0050	0.9972
1000	MML	9	0.0006	1.0076	0.9973	9	0.0006	1.0222	0.9973	9	0.0013	0.9705	0.9976
	MDL	9	0.0006	1.0076	0.9973	9	0.0006	1.0222	0.9973	9	0.0013	0.9705	0.9976
	CAICF	9	0.0006	1.0076	0.9973	9	0.0006	1.0222	0.9973	9	0.0013	0.9705	0.9976
	SRM	9	0.0006	1.0076	0.9973	9	0.0006	1.0222	0.9973	9	0.0013	0.9705	0.9976
2000	MML	9	0.0004	1.0035	0.9973	9	0.0001	1.0047	0.9974	9	0.0005	0.9861	0.9974
	MDL	9	0.0004	1.0035	0.9973	9	0.0001	1.0047	0.9974	9	0.0005	0.9861	0.9974
	CAICF	9	0.0004	1.0035	0.9973	9	0.0001	1.0047	0.9974	9	0.0005	0.9861	0.9974
	SRM	9	0.0004	1.0035	0.9973	9	0.0001	1.0047	0.9974	9	0.0005	0.9861	0.9974
4000	MML	9	0.0005	1.0106	0.9972	9	0.0002	1.0130	0.9973	9	0.0001	1.0116	0.9973
	MDL	9	0.0005	1.0106	0.9972	9	0.0002	1.0130	0.9973	9	0.0001	1.0116	0.9973
	CAICF	9	0.0005	1.0106	0.9972	9	0.0002	1.0130	0.9973	9	0.0001	1.0116	0.9973
	SRM	9	0.0005	1.0106	0.9972	9	0.0002	1.0130	0.9973	9	0.0001	1.0116	0.9973
6000	MML	9	0.0003	1.0024	0.9973	9	0.0001	0.9982	0.9974	9	0.0001	1.0069	0.9973
	MDL	9	0.0003	1.0024	0.9973	9	0.0001	0.9982	0.9974	9	0.0001	1.0069	0.9973
	CAICF	9	0.0003	1.0024	0.9973	9	0.0001	0.9982	0.9974	9	0.0001	1.0069	0.9973
	SRM	9	0.0003	1.0024	0.9973	9	0.0001	0.9982	0.9974	9	0.0001	1.0069	0.9973
10000	MML	9	0.0001	1.0119	0.9973	9	0.0001	0.9944	0.9974	9	0.0000	1.0163	0.9973
	MDL	9	0.0001	1.0119	0.9973	9	0.0001	0.9944	0.9974	9	0.0000	1.0163	0.9973
	CAICF	9	0.0001	1.0119	0.9973	9	0.0001	0.9944	0.9974	9	0.0000	1.0163	0.9973
	SRM	9	0.0001	1.0119	0.9973	9	0.0001	0.9944	0.9974	9	0.0000	1.0163	0.9973

procedure is indeed a good enough model represently by the real data since it is recovered by all of the model selection criteria individually. The fact that the criteria only requires a relatively small size of data set to recover the true model suggests that the non-backtracking search engine used for the experiments covers a search space extensive enough to converge to the best model. It also suggests that the integrated model discovery procedure indeed has the ability to select a parsimonious subset of variables from a pool of potential regressors in the absence of prior knowledge of the problem domain.

5   Future Work

Based on the experiments in [4], [5] and in this paper, we are satisfied that the integrated model discovery procedure that we have built is robust enough for the task of model discovery in the absence of domain knowledge.

The task of building forecasting systems for tropical cyclone intensity is one of the most difficult area in meteorology for precisely the reason that very little is known about the variables influencing it. We are confident that we can use our procedure to take on this challenge.

Table 5. Continued from Table 4: Model discovered in data sets DataSet4 and DataSet5

Sample Size	Method	DataSet 4				DataSet 5			
		nvar	ModelErr	RMSE	R^2	nvar	ModelErr	RMSE	R^2
100	MML	9	0.0049	1.0821	0.9968	9	3.7055	1.9565	0.9910
	MDL	9	0.0049	1.0821	0.9968	9	3.7055	1.9565	0.9910
	CAICF	9	0.0049	1.0821	0.9968	9	2.5592	1.4493	0.9950
	SRM	9	0.0049	1.0821	0.9968	9	2.5592	1.4493	0.9950
500	MML	9	0.0017	1.0816	0.9972	9	0.0028	1.0112	0.9975
	MDL	9	0.0017	1.0816	0.9972	9	0.0028	1.0112	0.9975
	CAICF	9	0.0017	1.0816	0.9972	9	0.0028	1.0112	0.9975
	SRM	9	0.0017	1.0816	0.9972	9	0.0028	1.0112	0.9975
1000	MML	9	0.0015	1.0576	0.9969	9	0.0022	1.0383	0.9972
	MDL	9	0.0015	1.0576	0.9969	9	0.0022	1.0383	0.9972
	CAICF	9	0.0015	1.0576	0.9969	9	0.0022	1.0383	0.9972
	SRM	9	0.0015	1.0576	0.9969	9	0.0022	1.0383	0.9972
2000	MML	9	0.0004	0.9926	0.9974	9	0.0007	1.0095	0.9973
	MDL	9	0.0004	0.9926	0.9974	9	0.0007	1.0095	0.9973
	CAICF	9	0.0004	0.9926	0.9974	9	0.0007	1.0095	0.9973
	SRM	9	0.0004	0.9926	0.9974	9	0.0007	1.0095	0.9973
4000	MML	9	0.0001	1.0013	0.9974	9	0.0001	1.0102	0.9973
	MDL	9	0.0001	1.0013	0.9974	9	0.0001	1.0102	0.9973
	CAICF	9	0.0001	1.0013	0.9974	9	0.0001	1.0102	0.9973
	SRM	9	0.0001	1.0013	0.9974	9	0.0001	1.0102	0.9973
6000	MML	9	0.0002	1.0075	0.9973	9	0.0001	1.0027	0.9973
	MDL	9	0.0002	1.0075	0.9973	10	0.0004	1.0041	0.9973
	CAICF	9	0.0002	1.0075	0.9973	10	0.0004	1.0041	0.9973
	SRM	9	0.0002	1.0075	0.9973	9	0.0001	1.0027	0.9973
10000	MML	9	0.0001	1.0176	0.9973	9	0.0001	1.0067	0.9973
	MDL	9	0.0001	1.0176	0.9973	9	0.0001	1.0067	0.9973
	CAICF	9	0.0001	1.0176	0.9973	9	0.0001	1.0067	0.9973
	SRM	9	0.0001	1.0176	0.9973	9	0.0001	1.0067	0.9973

References

1. H. Bozdogan. Model selection and akaike's information criterion (AIC): the general theory and its analytical extensions. *Psychometrika*, 52(3):345–370, 1987.
2. J. Rissanen. Modeling by shortest data description. *Automatica*, 14:465–471, 1978.
3. J. Rissanen. Stochastic complexity. *Journal of the Royal Statistical Society B*, 49(1):223–239, 1987.
4. G.W. Rumantir. Minimum Message Length criterion for second-order polynomial model discovery. In T. Terano, H. Liu, A.L.P. Chen, editor, *Knowledge Discovery and Data Mining: Current Issues and New Applications, PAKDD 2000, LNAI 1805*, pages 40–48. Springer–Verlag, Berlin Heidelberg, 2000.
5. G.W. Rumantir. Tropical cyclone intensity forecasting model: Balancing complexity and goodness of fit. In R. Mizoguchi and J. Slaney , editor, *PRICAI 2000 Topics in Artificial Intelligence, LNAI 1886*, pages 230–240. Springer–Verlag, Berlin Heidelberg, 2000.
6. V. Vapnik. *The Nature of Statistical Learning Theory*. Springer, New York, 1995.
7. C.S. Wallace and P.R. Freeman. Estimation and inference by compact coding. *Journal of the Royal Statistical Society B*, 49(1):240–252, 1987.

The IDA'01 Robot Data Challenge

Paul Cohen[1], Niall Adams[2], and David J. Hand[2]

[1] Department of Computer Science. University of Massachusetts
cohen@cs.umass.edu
[2] Department of Mathematics. Imperial College, London
n.adams@ic.ac.uk, d.j.hand@ic.ac.uk

The IDA01 conference featured a *Data Analysis Challenge*, to which all conference participants could respond[1]. The challenge was organized around categorical time series data. A series of vectors of binary data was generated by the perceptual system of a mobile robot, and series of characters was taken from George Orwell's book *1984*. In both cases, the boundaries between meaningful units (activities in the robot data, words in the Orwell data) were absent, and part of the challenge involved inducing these boundaries.

More specifically, in each case we suspect a time series contains several patterns (where a pattern is a structure in the data that is observed, completely or partially, more than once) but we do not know the pattern boundaries, the number of patterns, or the structure of patterns. We suspect that at least some patterns are similar, but perhaps no two are identical. Finally, we suspect that patterns have a hierarchical structure in the sense that shorter patterns can be nested inside longer ones. The challenge is to find the patterns and elucidate their structure.

A *supervised* approach to the problem might involve learning to recognize patterns given known examples of patterns, however, this challenge encourages *unsupervised* solutions, those in which algorithms have no information specific to the data, other than the data itself. One reason for this stringent requirement is to see whether domain-general solutions will be developed. This is also the reason for providing two, quite different datasets: One hopes that an unsupervised pattern-finding algorithm that works on both data sets will provide some insights about general characteristics of patterns.

1 The Robot Dataset

The robot dataset is a time series of 22,535 binary vectors of length 9, generated by a mobile robot as it executed 48 replications of a simple approach-and-push plan. In each trial, the robot visually located an object, oriented to it, approached it rapidly for a while, slowed down to make contact, and attempted to push the object. In one block of trials, the robot was unable to push the object, so it stalled and backed up. In another block the robot pushed until the object bumped into the wall, at which point the robot stalled and backed up. In a third block of

---

[1] Details of the challenge are available at http://genet.cs.umass.edu/dac/

F. Hoffmann et al. (Eds.): IDA 2001, LNCS 2189, pp. 378–381, 2001.

trials the robot pushed the object unimpeded for a while. Two trials in 48 were anomalous.

Data from the robot's sensors were sampled at 10Hz and passed through a simple perceptual system that returned values for nine binary variables. These variables indicate the state of the robot and primitive perceptions of objects in its environment. They are: STOP, ROTATE-RIGHT, ROTATE-LEFT, MOVE-FORWARD, NEAR-OBJECT, PUSH, TOUCH, MOVE-BACKWARD, STALL. For example, the binary vector [0 1 0 1 1 0 1 0 0] describes a state in which the robot is rotating right while moving forward, near an object, touching it but not pushing it. Most of the $2^9 = 512$ possible states do not arise, in fact, only 35 unique states are observed. Fifteen of these states account for more than 97% of the time series. Said differently, more than half of the unique states occur very rarely, and five of them occur fewer than five times. Most of the 512 possible states are not semantically valid; for example, the robot cannot simultaneously be moving backward and moving forward. However, the robot's sensors are noisy and its perceptual system makes mistakes; for example, there are 55 instances of states in which the robot is simultaneously stalled and moving backward.

Because the robot collected ten data vectors every second, and its actions and environment did not change quickly, it is common to see long runs of identical states. The mean, median and standard deviation of run-length are 9.6, 4, and 15.58, respectively; while most runs are short, some are quite long.

Four forms of data are available:

1. 22,535 binary vectors of length 9.
2. 2345 binary vectors of length 9, produced by removing runs from dataset 1. The iterative rule for removing runs is: when two consecutive states are equal, keep the first and discard the second.
3. 22,535 numbers between 0 and 34. The original dataset contains only 35 unique vectors, so we can recode the vectors as numbers betwen 0 and 34, reducing the multivariate problem to a univariate one.
4. 2345 numbers between 0 and 34, obtained by recoding vectors as numbers and reducing runs.

The dataset was segmented into episodes by hand. Each of 48 episodes contained some or all of the following sub-episodes:

A: start a new episode with
     orientation and finding the target
B1: forward movement
B2: forward movement with turning or
     intruding periods of turning
C1: B1 + an object is detected by sonars
C2: B2 + an object is detected by sonars
D: robot is in contact with object (touching, pushing)
E: robot stalls, moves backwards or otherwise ends D

By hand, we associated one of these seven sub-episode type labels with each of the 22535 data items in the robot time series, producing an episode-labelled series of the same length.[2] The dataset contains 355 episodes.

A labelled subset of length 3558 of the original and univariate versions of the dataset is provided as part of the challenge, not to train supervised methods, but to test the performance of methods.

2   The Orwell Dataset

The task here is to take the first 5,000 words of George Orwell's *1984*, where spaces, capitalization and punctuation have been removed, and try to restore the word boundaries. We provide the dataset and also the locations of the word boundaries.

3   Results

At this writing, results are unavailable from IDA participants other than the authors of this report. Our qualitative conclusions are summarized here, and discussed in more detail in technical reports and papers in this volume.

What is a pattern? It is easy to write algorithms to look for structures in time series, but which structures should they look for? In a supervised approach, the answer is, "structures like those in the training data," but unsupervised algorithms must carry some bias to look for particular kinds of structures. Said differently, unsupervised pattern-finding algorithms define "pattern," more or less explicitly, as the sort of thing they find in data.

Most patterns are not meaningful. Patterns found by unsupervised pattern-finding algorithms are usually not meaningful in the domain to which they are applied. Said differently, for most conceptions of "pattern," there are many more patterns than meaningful patterns in a domain. To illustrate the point, consider the notion that patterns are the most frequent subsequences in a series. Listed from most to least frequent, here are the top 100 patterns in Orwell's text:

> th in the re an en as ed to ou it er of at ing was or st on ar and es ic el al om
> ad ac is wh le ow ld ly ere he wi ab im ver be for had ent itwas with ir win gh
> po se id ch ot ton ap str his ro li all et fr andthe ould min il ay un ut ur ve
> whic dow which si pl am ul res that were ethe wins not winston sh oo up ack
> ter ough from ce ag pos bl by tel ain

One sees immediately that most patterns (according to the frequency notion of pattern) are not morphemes in English; most are short, and the longer ones cross word boundaries (e.g., itwas, ethe, andthe). Clearly, if the patterns one

---

[2] This cannot be done algorithmically, as some contextual interpretation of subsequences of the series is required. For example, if the sonars temporarily lose touch with an object, only to reacquire it a few seconds later, we label the intervening data C1 or C2, not B1 or B2, even though the data satisfy the criteria for B1 or B2.

seeks are English words or morphemes (or, as it happens, robot episodes) the notion that patterns are high-frequency subsequences is not sufficient. We can load up our algorithms with bias to find domain-specific patterns, or try to develop a domain-general notion of pattern that has a better success rate than the frequency notion. The article by Cohen and Adams in this volume discusses the latter approach applied to the Challenge datasets.

Induction is necessary. Patterns often have variants and some kind of induction is required to generalize over them. The following image shows two episodes from the robot dataset (with runs removed). Each line represents an interval during which the corresponding value in the 9-vector was 1. The patterns are roughly similar in appearance, and, indeed, semantically similar; but they are not identical. They have different durations and somewhat different morphologies. A paper by Cohen in this volume describes a notion of pattern based on the temporal relationships between events that captures the essential structure of these data.

A session on the Challenge was held at the Intelligent Data Analysis symposium. Results from participants are summarized in technical reports and are available at the Challenge web site, `http://genet.cs.umass.edu/dac/`. There you will also find the Robot and Orwell datasets, as well as others. You are invited to try your methods and compare your results with other participants in the Challenge.

Acknowledgments

Prof. Cohen's work was supported by Visiting Fellowship Research Grant number GR/N24193 from the Engineering and Physical Sciences Research Council (UK). Additional support for Prof. Cohen and for Dr. Adams was provided by DARPA under contract(s) No.s DARPA/USASMDCDASG60-99-C-0074 and DARPA/AFRLF30602-00-1-0529.

Author Index

Lecture Notes in Computer Science

For information about Vols. 1–2084
please contact your bookseller or Springer-Verlag

Vol. 2124: W. Skarbek (Ed.), Computer Analysis of Images and Patterns. Proceedings, 2001. XV, 743 pages. 2001.

Vol. 2125: F. Dehne, J.-R. Sack, R. Tamassia (Eds.), Algorithms and Data Structures. Proceedings, 2001. XII, 484 pages. 2001.

Vol. 2126: P. Cousot (Ed.), Static Analysis. Proceedings, 2001. XI, 439 pages. 2001.

Vol. 2127: V. Malyshkin (Ed.), Parallel Computing Technologies. Proceedings, 2001. XII, 516 pages. 2001.

Vol. 2129: M. Goemans, K. Jansen, J.D.P. Rolim, L. Trevisan (Eds.), Approximation, Randomization, and Combinatorial Optimization. Proceedings, 2001. IX, 297 pages. 2001.

Vol. 2130: G. Dorffner, H. Bischof, K. Hornik (Eds.), Artificial Neural Networks – ICANN 2001. Proceedings, 2001. XXII, 1259 pages. 2001.

Vol. 2132: S.-T. Yuan, M. Yokoo (Eds.), Intelligent Agents. Specification. Modeling, and Application. Proceedings, 2001. X, 237 pages. 2001. (Subseries LNAI).

Vol. 2136: J. Sgall, A. Pultr, P. Kolman (Eds.), Mathematical Foundations of Computer Science 2001. Proceedings, 2001. XII, 716 pages. 2001.

Vol. 2138: R. Freivalds (Ed.), Fundamentals of Computation Theory. Proceedings, 2001. XIII, 542 pages. 2001.

Vol. 2139: J. Kilian (Ed.), Advances in Cryptology – CRYPTO 2001. Proceedings, 2001. XI, 599 pages. 2001.

Vol. 2141: G.S. Brodal, D. Frigioni, A. Marchetti-Spaccamela (Eds.), Algorithm Engineering. Proceedings, 2001. X, 199 pages. 2001.

Vol. 2142: L. Fribourg (Ed.), Computer Science Logic. Proceedings, 2001. XII, 615 pages. 2001.

Vol. 2143: S. Benferhat, P. Besnard (Eds.), Symbolic and Quantitative Approaches to Reasoning with Uncertainty. Proceedings, 2001. XIV, 818 pages. 2001. (Subseries LNAI).

Vol. 2146: J.H. Silverman (Eds.), Cryptography and Lattices. Proceedings, 2001. VII, 219 pages. 2001.

Vol. 2147: G. Brebner, R. Woods (Eds.), Field-Programmable Logic and Applications. Proceedings, 2001. XV, 665 pages. 2001.

Vol. 2149: O. Gascuel, B.M.E. Moret (Eds.), Algorithms in Bioinformatics. Proceedings, 2001. X, 307 pages. 2001.

Vol. 2150: R. Sakellariou, J. Keane, J. Gurd, L. Freeman (Eds.), Euro-Par 2001 Parallel Processing. Proceedings, 2001. XXX, 943 pages. 2001.

Vol. 2151: A. Caplinskas, J. Eder (Eds.), Advances in Databases and Information Systems. Proceedings, 2001. XIII, 381 pages. 2001.

Vol. 2152: R.J. Boulton, P.B. Jackson (Eds.), Theorem Proving in Higher Order Logics. Proceedings, 2001. X, 395 pages. 2001.

Vol. 2153: A.L. Buchsbaum, J. Snoeyink (Eds.), Algorithm Engineering and Experimentation. Proceedings, 2001. VIII, 231 pages. 2001.

Vol. 2154: K.G. Larsen, M. Nielsen (Eds.), CONCUR 2001 – Concurrency Theory. Proceedings, 2001. XI, 583 pages. 2001.

Vol. 2157: C. Rouveirol, M. Sebag (Eds.), Inductive Logic Programming. Proceedings, 2001. X, 261 pages. 2001. (Subseries LNAI).

Vol. 2158: D. Shepherd, J. Finney, L. Mathy, N. Race (Eds.), Interactive Distributed Multimedia Systems. Proceedings, 2001. XIII, 258 pages. 2001.

Vol. 2159: J. Kelemen, P. Sosík (Eds.), Advances in Artificial Life. Proceedings, 2001. XIX, 724 pages. 2001. (Subseries LNAI).

Vol. 2161: F. Meyer auf der Heide (Ed.), Algorithms – ESA 2001. Proceedings, 2001. XII, 538 pages. 2001.

Vol. 2162: Ç. K. Koç, D. Naccache, C. Paar (Eds.), Cryptographic Hardware and Embedded Systems – CHES 2001. Proceedings, 2001. XIV, 411 pages. 2001.

Vol. 2164: S. Pierre, R. Glitho (Eds.), Mobile Agents for Telecommunication Applications. Proceedings, 2001. XI, 292 pages. 2001.

Vol. 2165: L. de Alfaro, S. Gilmore (Eds.), Process Algebra and Probabilistic Methods. Proceedings, 2001. XII, 217 pages. 2001.

Vol. 2166: V. Matoušek, P. Mautner, R. Mouček, K. Taušer (Eds.), Text, Speech and Dialogue. Proceedings, 2001. XIII, 452 pages. 2001. (Subseries LNAI).

Vol. 2170: S. Palazzo (Ed.), Evolutionary Trends of the Internet. Proceedings, 2001. XIII, 722 pages. 2001.

Vol. 2172: C. Batini, F. Giunchiglia, P. Giorgini, M. Mecella (Eds.), Cooperative Information Systems. Proceedings, 2001. XI, 450 pages. 2001.

Vol. 2176: K.-D. Althoff, R.L. Feldmann, W. Müller (Eds.), Advances in Learning Software Organizations. Proceedings, 2001. XI, 241 pages. 2001.

Vol. 2177: G. Butler, S. Jarzabek (Eds.), Generative and Component-Based Software Engineering. Proceedings, 2001. X, 203 pages. 2001.

Vol. 2181: C. Y. Westort (Eds.), Digital Earth Moving. Proceedings, 2001. XII, 117 pages. 2001.

Vol. 2184: M. Tucci (Ed.), Multimedia Databases and Image Communication. Proceedings, 2001. X, 225 pages. 2001.

Vol. 2186: J. Bosch (Ed.), Generative and Component-Based Software Engineering. Proceedings, 2001. VIII, 177 pages. 2001.

Vol. 2188: F. Bomarius, S. Komi-Sirviö (Eds.), Product Focused Software Process Improvement. Proceedings, 2001. XI, 382 pages. 2001.

Vol. 2189: F. Hoffmann, D.J. Hand, N. Adams, D. Fisher, G. Guimaraes (Eds.), Advances in Intelligent Data Analysis. Proceedings, 2001. XII, 384 pages. 2001.

Vol. 2190: A. de Antonio, R. Aylett, D. Ballin (Eds.), Intelligent Virtual Agents. Proceedings, 2001. VIII, 245 pages. 2001. (Subseries LNAI).

Vol. 2191: B. Radig, S. Florczyk (Eds.), Pattern Recognition. Proceedings, 2001. XVI, 452 pages. 2001.

Vol. 2193: F. Casati, D. Georgakopoulos, M.-C. Shan (Eds.), Technologies for E-Services. Proceedings, 2001. X, 213 pages. 2001.